SOLID MECHANICS AND ITS APPLICATIONS
Volume 40

Series Editor: G.M.L. GLADWELL
Solid Mechanics Division, Faculty of Engineering
University of Waterloo
Waterloo, Ontario, Canada N2L 3G1

Aims and Scope of the Series

The fundamental questions arising in mechanics are: *Why?*, *How?*, and *How much?*
The aim of this series is to provide lucid accounts written by authoritative researchers giving vision and insight in answering these questions on the subject of
mechanics as it relates to solids.

The scope of the series covers the entire spectrum of solid mechanics. Thus it
includes the foundation of mechanics; variational formulations; computational
mechanics; statics, kinematics and dynamics of rigid and elastic bodies; vibrations
of solids and structures; dynamical systems and chaos; the theories of elasticity,
plasticity and viscoelasticity; composite materials; rods, beams, shells and
membranes; structural control and stability; soils, rocks and geomechanics;
fracture; tribology; experimental mechanics; biomechanics and machine design.

The median level of presentation is the first year graduate student. Some texts are
monographs defining the current state of the field; others are accessible to final
year undergraduates; but essentially the emphasis is on readability and clarity.

Computational Kinematics '95

Proceedings of the Second Workshop on
Computational Kinematics,
held in Sophia Antipolis, France,
September 4–6, 1995

Edited by

J.-P. MERLET

Institut National de Recherche en Informatique et en Automatique,
Sophia Antipolis, France

and

B. RAVANI

Department of Mechanical Engineering,
University of California,
Davis, California, U.S.A.

SPRINGER-SCIENCE+BUSINESS, MEDIA, B.V.

A C.I.P. Catalogue record for this book is available from the Library of Congress.

ISBN 978-0-7923-3673-0   ISBN 978-94-011-0333-6 (eBook)
DOI 10.1007/978-94-011-0333-6

Printed on acid-free paper

Table of Contents

6 Motion and Grasp Planning

Preface

The aim of this book is to provide an account of the state of the art in *Computational Kinematics*. We understand here under this term that branch of kinematics research involving intensive computations not only of the numerical type, but also of symbolic as well as geometric nature.

Research in kinematics over the last decade has been remarkably oriented towards the computational aspects of kinematics problems. In fact, this work has been prompted by the need to answer fundamental questions such as the number of solutions, whether real or complex, that a given problem can admit as well as computational algorithms to support geometric analysis. Problems of the first kind occur frequently in the analysis and synthesis of kinematic chains, when fine displacements are considered. The associated models, that are derived from kinematic relations known as *closure equations*, lead to systems of nonlinear *algebraic* equations in the variables or parameters sought. The algebraic equations at hand can take the form of multivariate polynomials or may involve trigonometric functions of unknown angles.

Purely numerical methods can be used to solve the problem but they turn out to be too restrictive, especially those involving an iterative process whose convergence cannot, in general, be guaranteed. These drawbacks have been overcome with the development of continuations techniques that are meant to produce all solutions to a given problem. While continuation techniques have provided solutions to a number of problems, they are still difficult to implement and are subject to numerical uncertainties. Hence alternative approach have been sought, that rely on recent advances in algebraic geometry and on modern software for symbolic computations. Current research in kinematics involves symbolic manipulations that were impossible to imagine as recently as ten years ago.

Problems of the second type occur in handling the computations associated with studies of kinematic geometry of motion. Geometric analysis has much of its roots in kinematics and has been the basis for many classical methods of kinematic analysis and synthesis. This includes problems associated with evaluation of singularities of mechanisms and manipulators, rigid body guidance and motion synthesis problems, analysis of the workspace and reachability of manipulators, and generation of trajectories of rigid bodies. Current research in these areas include development of computational algorithms to support such geometric analysis methods.

This book reports the trends and progress attained in Computational Kinematics in a broad class of problems as described above. It has been divided into six parts, namely, *i) kinematics algorithms,* whereby general kinematics problems are discussed in light of their solution algorithms; *ii) kinematics of mechanisms,* in which problems related to specific mechanisms are studied; *iii) singularities,* which is self-descriptive; *iv) workspace,* in which the determination of the workspace of given mechanisms is discussed; *v) parallel manipulators,* in which problems related to this kind of closed-loop mechanisms are addressed; and *vi) motion and grasp planning,* touching upon computational geometry.

The reader will find here a representative sample of the most modern techniques available nowadays for the solution of challenging kinematics problems. In light of it contents, the book should be of interest to researchers, graduate students and practicing engineers working in kinematics or related fields. Especially, roboticists, CAD/CAM specialists, biomechanics specialists, machine designers and computer scientists will find here a useful source of information comprising methods, algorithms and applications.

This book contains the Proceedings of the second Workshop on Computational Kinematics, held at *INRIA Sophia-Antipolis* in France, from September 4 to September 6, 1995. INRIA is herewith given due acknowledgment for its financial and logistical support and encouragement. The decisive financial support of IFToMM and of the European HCM network HEROS must be mentioned. Dr. Karel Nederveen, of Kluwer Academic Publishers, is acknowledged for his encouragement and support in editing the book and publishing it in record time. We wish to thank the organizers of the first Workshop, Pr. J. Angeles and Dr. P. Kovács, for their useful advice. The support of Dr. Nadia Maïzi from Ecole des Mines de Paris and doctorate student Luc Tancredi was decisive in organizing this workshop.

Jean-Pierre Merlet and Bahram Ravani, Editors
INRIA Sophia-Antipolis, France

COORDINATE FREE CRITERIA FOR TESTING THE LINEAR DEPENDENCE OF THE SETS OF SCREWS

D.P. CHEVALLIER *Centre d'Enseignement et de Recherches en Mathématiques Appliquées,*
E.N.P.C La courtine 93167 Noisy le Grand, France.
Tel. 49 14 35 72, Fax. 49 14 35 86

1 Introduction

The verification of the linear dependence and the calculation of the rank of a set of screws are very important tasks in kinematics in the search for singular positions of open chains as as well as in the search for movability conditions of closed loop chains. This mathematical problem is generally solved by standard techniques of linear algebra using determinants of coordinates of the screws relative to more or less arbitrary bases (see for example Hunt [5], Sugimoto and Duffy [10] or Sugimoto [11], Wholhart [12]). However, such methods make no use at all of the specific algebraic structure which can be defined on the set of screws and, as for any coordinate method, the geometrical meaning of the result may be unclear.

It is well known that the checking of the linear dependence of a set of ordinary vectors in three dimensional space can be completely performed by use of coordinate free criteria lying on the properties of the vector product and the triple vector product. Nevertheless, to our knowledge, no similar coordinate free criterion has been exposed in screw theory. In this paper we expose a list of mathematical properties leading to an algorithm for testing the linear dependence or computing the rank of any set of screws. In some sense, this list generalizes to the Lie algebra $\mathfrak{D}$ of the displacement group the two classical criteria valid in ordinary vector algebra (the Lie algebra of the rotation group). Due to the higher dimension of the vector space, many particular cases must be studied for the design of a complete algorithm. They are the concern of specific criteria and so the finite sets of screws are divided into three classes including respectively three, four and eight non obvious subcases.

Several remarks seem to be noteworthy. First, the expression of these criteria requires all the algebraic operations defined in Chevallier [2], in particular the Lie bracket and the module structure of $\mathfrak{D}$ derived from operation V, except operation III. In other words the classical form of screw theory using only the vector space structure and the Klein form should not contain all the necessary tools for this. The remark also meet an idea exposed by Hervé [4]: the mathematical properties of the displacement group and its Lie algebra are a key to the understanding of kinematics.

Second, the form of the general criteria seems to be closely related to properties met in kinematics of overconstraint mechanisms such that the existence of transversals (see Wholhart [12] or Baker and Wholhart [1]); the following results

1

J.-P. Merlet and B. Ravani (eds.), Computational Kinematics, 1–10.
© *1995 Kluwer Academic Publishers.*

contain a purely algebraic technique for pointing out such "transversals screws" in various cases. The eight criteria exposed for the third class sets correspond to simpler applications and one can find corresponding mechanisms with finite or infinitely small mobility.

Third, the classification of vector subspaces of $\mathfrak{D}$ (screw systems), has been recently studied by Rico Martinez and Duffy [8], [9] and by Gibson and Hunt [3]. Here we consider a given set of generators, in practice the data defining a linkage in some configuration, and we solve a rather different problem.

2   The Screw Theory as the Study of a Lie algebra

In this section we explain the relationships between the classical screw theory and the algebraic properties of a Lie algebra endowed with additional operations. We denote by $\mathcal{E}$ and $\mathbf{E}$ the three-dimensional affine Euclidean space and the associated vector space (to every pair (a, b) of points in $\mathcal{E}$ is associated a vector $\overrightarrow{ab} \in \mathbf{E}$ and the scalar product denoted by a dot and the vector product denoted by $\times$ are defined in $\mathbf{E}$).

2.1   Screws and skew-symmetric vector fields

Definition 2.1 *A skew-symmetric vector field on $\mathcal{E}$ is a map $\mathbf{X}$ from $\mathcal{E}$ to $\mathbf{E}$ such that $\overrightarrow{pq} \cdot \mathbf{X}(p) = \overrightarrow{pq} \cdot \mathbf{X}(q)$ for all points p and q or, what is equivalent, such that there exists a unique vector $\omega_{\mathbf{X}}$ with the following property: for all p and q in $\mathcal{E}$, $\mathbf{X}(q) = \mathbf{X}(p) + \omega_{\mathbf{X}} \times \overrightarrow{pq}$.*

The set of all the skew-symmetric vector fields will be denoted by $\mathfrak{D}$, the subset of the constant vector fields over $\mathcal{E}$ will be denoted by $\mathfrak{T}$ (the vector field $\mathbf{X}$ is constant over $\mathcal{E}$ whenever $\omega_{\mathbf{X}} = 0$).

If $\mathbf{X}$ is in $\mathfrak{D}$ but not in $\mathfrak{T}$ (*i.e.* $\omega_{\mathbf{X}} \neq 0$) the set of the points p such that $\mathbf{X}(p)$ is directed as $\omega_{\mathbf{X}}$ (*i.e.* $\mathbf{X}(p) \times \omega_{\mathbf{X}} = 0$) is a straight line $\Delta_{\mathbf{X}}$ directed as $\omega_{\mathbf{X}}$ in $\mathcal{E}$ (the axis of $\mathbf{X}$). Moreover $\mathbf{X}(p)$ takes the same value $f_{\mathbf{X}}\omega_{\mathbf{X}}$ at every point p on $\Delta_{\mathbf{X}}$, where the number $f_{\mathbf{X}}$ is the *pitch* and $f_{\mathbf{X}} = [\mathbf{X} \mid \mathbf{X}]/2(\mathbf{X} \mid \mathbf{X})$ (for the meaning of the notations see subsection 2.2 (IV) and (V)).

For a member of $\mathfrak{T}$ the axis does not exist (or, otherwise stated, the axis is a line at infinity).

The picture of a line Δ, a director ω and an associated real number f is a screw within the common meaning. As we see, it is equivalent to define a screw and to define a skew-symmetric field (taking account of some particular cases). Another picture of a skew-symmetric vector field on $\mathcal{E}$ meets screw theory: if o is any fixed origin point the relation $\mathbf{X} \mapsto (\omega_{\mathbf{X}}, \mathbf{X}(o))$ is one-one and onto; in fact this relation is an isomorphism between the vector spaces $\mathfrak{D}$ and $\mathbf{E} \times \mathbf{E}$. The vectors $\omega_{\mathbf{X}}$ and $\mathbf{X}(o)$ are the Plüker vectors. The pair $(\omega_{\mathbf{X}}, \mathbf{X}(o))$ is the "ray representation" of a screw (the "axis representation" is equivalent and should be $(\mathbf{X}(o), \omega_{\mathbf{X}})$).

2.2   The algebraic structure of $\mathfrak{D}$

We summarize below the algebraic operations defined in $\mathfrak{D}$ (see reference [2]).

(I) **Vector space structure over the real field.**(Obviously defined).

(II) **Lie bracket.** The Lie bracket of two skew-symmetric vector fields $\mathbf{X}$ and $\mathbf{Y}$ is the skew-symmetric vector field $\mathbf{U} = [\mathbf{X}, \mathbf{Y}]$ such that

$$\mathbf{U}(p) = \omega_{\mathbf{X}} \times \mathbf{Y}(p) - \omega_{\mathbf{Y}} \times \mathbf{X}(p) \text{ for all } p \in \mathcal{E}.$$

Endowed with the Lie bracket, the vector space $\mathfrak{D}$ is a Lie algebra, that is $[\mathbf{X}, \mathbf{Y}]$ is skew-symmetric and the Jacobi identity holds:

$$[\mathbf{X}, [\mathbf{Y}, \mathbf{Z}]] + [\mathbf{Y}, [\mathbf{Z}, \mathbf{X}]] + [\mathbf{Z}, [\mathbf{X}, \mathbf{Y}]] = 0.$$

(III) **Adjoint action of D on $\mathfrak{D}$.** (We do not use it in this article.)

Actually $\mathfrak{D}$ is the Lie algebra of the displacement group $\mathbf{D}$ and an algebraic structure defined by operations like (I), (II), (III), exists on the Lie algebra of each Lie group. The last two operations are specific properties of the displacement group.

(IV) **The Klein form.** This is the non degenerate symmetric bilinear form defined by

$$[\mathbf{X} \mid \mathbf{Y}] = \omega_{\mathbf{X}} \cdot \mathbf{Y}(o) + \omega_{\mathbf{Y}} \cdot \mathbf{X}(o) \text{ (with } o = \text{ arbitrary origin in } \mathcal{E}).$$

(V) **The operator Ω and the Killing form.**
For $\mathbf{X} \in \mathfrak{D}$ define $\Omega\mathbf{X} \in \mathfrak{D}$ as the constant vector field such that

$$\Omega\mathbf{X}(p) = \omega_{\mathbf{X}} \text{ for all } p \in \mathcal{E}.$$

Then Ω is a linear operator in $\mathfrak{D}$ and its range and its kernel are equal to $\mathfrak{T}$. Hence $\Omega \circ \Omega = 0$ and, in the dual number setting, the operator Ω is the multiplication by ε. Moreover, the Killing form, a positive degenerate symmetric bilinear form on $\mathfrak{D}$, may be defined from the operator Ω and the Klein form by

$$(\mathbf{X} \mid \mathbf{Y}) = [\mathbf{X} \mid \Omega\mathbf{Y}] \equiv [\Omega\mathbf{X} \mid \mathbf{Y}].$$

The Lie algebra $\mathfrak{D}$ contains remarkable subsets: $\mathfrak{T}$ defined above is a commutative ideal corresponding to the translation group, and for each fixed point p: $\mathfrak{Z}_p = \{\mathbf{X} \mid \mathbf{X}(p) = 0\}$, is the Lie subalgebra corresponding to the rotation group about p. The members of $\mathfrak{Z}_p$ correspond to line vectors through p of screw theory.

Let us note other relations making a link with familiar ones in classical screw theory and allowing geometrical interpretations. Let $\mathbf{X}$ and $\mathbf{Y}$ be in $\mathfrak{D} - \mathfrak{T}$, then the axis of $[\mathbf{X}, \mathbf{Y}]$ is the common perpendicular δ to the axes $\Delta_{\mathbf{X}}$ and $\Delta_{\mathbf{Y}}$. Let $\mathbf{w}$

be a is a normed vector along δ, $\theta = \angle(\Delta_{\mathbf{X}}, \Delta_{\mathbf{Y}})$. If $p = \Delta_{\mathbf{X}} \cap \delta$, $q = \Delta_{\mathbf{Y}} \cap \delta$ and $\overrightarrow{pq} = d\mathbf{w}$ with $d \in \mathbf{R}$, then the value of $[\mathbf{X}, \mathbf{Y}]$ on its axis is:

(1)
$$\begin{cases} [\mathbf{X}, \mathbf{Y}](p) &= [\mathbf{X}, \mathbf{Y}](q) = (f_{\mathbf{X}} + f_{\mathbf{Y}})\omega_{\mathbf{X}} \times \omega_{\mathbf{Y}} + (\mathbf{X} \mid \mathbf{Y})\overrightarrow{pq} \\ &= [(f_{\mathbf{X}} + f_{\mathbf{Y}})\sin\theta + d\cos\theta]\|\omega_{\mathbf{X}}\|\|\omega_{\mathbf{Y}}\|\,\mathbf{w}, \\ [\mathbf{X} \mid \mathbf{Y}] &= (f_{\mathbf{X}} + f_{\mathbf{Y}})(\mathbf{X} \mid \mathbf{Y}) - \overrightarrow{pq} \cdot \omega_{\mathbf{X}} \times \omega_{\mathbf{Y}} \\ &= [(f_{\mathbf{X}} + f_{\mathbf{Y}})\cos\theta - d\sin\theta]\|\omega_{\mathbf{X}}\|\|\omega_{\mathbf{Y}}\|. \end{cases}$$

In particular we have the following result about pairs of screws with parallel axes: if $\mathbf{X}$ and $\mathbf{Y} \in \mathfrak{D} - \mathfrak{T}$ and $[\mathbf{X}, \mathbf{Y}] \in \mathfrak{T}$ then, for all m in $\mathcal{E}$:

(2)
$$[\mathbf{X}, \mathbf{Y}](m) = (\mathbf{X} \mid \mathbf{Y})\overrightarrow{pq}.$$

2.3 The module structure on $\mathfrak{D}$

Let Δ be the dual number ring (that is the ring of the "numbers" of the form $z = x + \varepsilon y$ with x and y in $\mathbf{R}$ and $\varepsilon^2 = 0$; the real and dual parts of z are denoted by $\operatorname{Re} z = x$ and $\operatorname{Du} z = y$, the conjugate of z by $\bar{z} = x - \varepsilon y$). Then $\mathfrak{D}$ has a natural module structure over Δ which extends its real vector space structure, the product of $\mathbf{X} \in \mathfrak{D}$ by the scalar $z \in \Delta$ being defined by:

$$z\mathbf{X} = x\mathbf{X} + y\Omega\mathbf{X}.$$

The *dual inner product* is defined as a dual coefficient combination of the Killing and Klein forms:

$$\{\mathbf{X} \mid \mathbf{Y}\} = (\mathbf{X} \mid \mathbf{Y}) + \varepsilon[\mathbf{X} \mid \mathbf{Y}] \text{ for } \mathbf{X}, \mathbf{Y} \in \mathfrak{D}.$$

With the Lie Bracket we define the *dual triple product* by:

$$\{\mathbf{X}; \mathbf{Y}; \mathbf{Z}\} = \{\mathbf{X} \mid [\mathbf{Y}, \mathbf{Z}]\}.$$

Using this module structure, it makes sense to speak of Δ-linearity or Δ-bilinearity for properties involving scalars belonging to Δ rather than real scalars. For example the Lie bracket in $\mathfrak{D}$ is not only bilinear but is also Δ-bilinear: $[z\mathbf{X}, \mathbf{Y}] = [\mathbf{X}, z\mathbf{Y}] = z[\mathbf{X}, \mathbf{Y}]$ for $z \in \Delta$. The following important properties (see [2]), which look like the classical ones for the dot product and cross product of ordinary vectors sum up rather cumbersome calculations on screws in a very compact form and will play a role in the sequel:

- $\{\cdot \mid \cdot\}$ is a symmetric nondegenerate Δ-bilinear form on $\mathfrak{D}$ (although individually the Killing and Klein forms define nothing particular in the dual number setting).
- $\{\cdot; \cdot; \cdot\}$ is a Δ-trilinear skew-symmetric for on $\mathfrak{D}$.
- The following formula holds: $[\mathbf{X} \mid [\mathbf{Y}, \mathbf{Z}]] = \{\mathbf{X} \mid \mathbf{Z}\}\mathbf{Y} - \{\mathbf{X} \mid \mathbf{Y}\}\mathbf{Z}$.

It is worth noting that any translation of an ordinary vector algebra property into a formally similar statement in the module $\mathfrak{D}$ is not necessarily correct. The results about the linear independence of two or three elements of $\mathfrak{D}$ over Δ and the bases of the module $\mathfrak{D}$ are summarized in the following propositions which play a major role in proving the results of this paper.

Proposition 2.1 *Let* $\mathbf{X}, \mathbf{Y}$ *and* $\mathbf{Z}$ *be in* $\mathfrak{D}$. *The following properties are equivalent*

   i) $\{\mathbf{X}, \mathbf{Y}, \mathbf{Z}\}$ *is a basis of* $\mathfrak{D}$ *over* Δ.

   ii) $\{\mathbf{X}, \mathbf{Y}, \mathbf{Z}\}$ *is linearly independent over* Δ.

   iii) $\{\varepsilon\mathbf{X}, \varepsilon\mathbf{Y}, \varepsilon\mathbf{Z}\}$ *(that is* $\{\omega_\mathbf{X}, \omega_\mathbf{Y}, \omega_\mathbf{Z}\}$*) is linearly independent over* $\mathbf{R}$.

   iv) $\{\mathbf{X}; \mathbf{Y}; \mathbf{Z}\}$ *is invertible in* Δ.

   v) $\{[\mathbf{Y}, \mathbf{Z}], [\mathbf{Z}, \mathbf{X}], [\mathbf{X}, \mathbf{Y}]\}$ *is a basis of* $\mathfrak{D}$ *over* Δ.

Whenever these conditions hold each $\mathbf{U} \in \mathfrak{D}$ *expresses as* $\mathbf{U} = x\mathbf{X} + y\mathbf{Y} + z\mathbf{Z}$ *with:*

$$(3) \quad x = \frac{1}{G}\{\mathbf{U}; \mathbf{Y}; \mathbf{Z}\}, \; y = \frac{1}{G}\{\mathbf{U}; \mathbf{Z}; \mathbf{X}\}, \; z = \frac{1}{G}\{\mathbf{U}; \mathbf{X}; \mathbf{Y}\}, \quad G = \{\mathbf{X}; \mathbf{Y}; \mathbf{Z}\}$$

(Note the nice formula $\{[\mathbf{Y}, \mathbf{Z}]; [\mathbf{Z}, \mathbf{X}]; [\mathbf{X}, \mathbf{Y}]\} = \{\mathbf{X}; \mathbf{Y}; \mathbf{Z}\}^2$ proving the equivalence i) $\Leftrightarrow$ v) as soon as i) $\Leftrightarrow$ iv) is proved.)

Proposition 2.2 *Let be* $\mathbf{X}$ *and* $\mathbf{Y}$ *in* $\mathfrak{D}$, *put* $\mathbf{W} = [\mathbf{X}, \mathbf{Y}]$, *then the following properties are equivalent:*

   i) $\mathbf{X}$ *and* $\mathbf{Y}$ *are linearly independent over* Δ.

   ii) $\varepsilon\mathbf{X}$ *and* $\varepsilon\mathbf{Y}$ *(in other words* $\omega_\mathbf{X}$ *and* $\omega_\mathbf{Y}$*) are linearly independent over* $\mathbf{R}$.

   iii) $[\mathbf{X}, \mathbf{Y}] \notin \mathfrak{T}$.

   iv) $\{[\mathbf{X}, \mathbf{Y}] \mid [\mathbf{X}, \mathbf{Y}]\} = G$ *is invertible in* Δ.

   v) $\{\mathbf{X}, \mathbf{Y}, \mathbf{W}\}$ *is a basis of the* Δ-*module* $\mathfrak{D}$.

Whenever these conditions hold each $\mathbf{U} \in \mathfrak{D}$ *expresses as* $\mathbf{U} = x\mathbf{X} + y\mathbf{Y} + w\mathbf{W}$ *with:*

$$(4) \quad x = -\frac{\{\mathbf{U} \mid [\mathbf{Y}, [\mathbf{Y}, \mathbf{X}]]\}}{G}, \; y = -\frac{\{\mathbf{U} \mid [\mathbf{X}, [\mathbf{X}, \mathbf{Y}]]\}}{G}, \; w = \frac{\{\mathbf{U} \mid [\mathbf{X}, \mathbf{Y}]\}}{G}.$$

In another form, the property $[\mathbf{X}, \mathbf{Y}] \in \mathfrak{T}$ (that is $\omega_\mathbf{X} \times \omega_\mathbf{Y} = 0$) means that the set $\{\mathbf{X}, \mathbf{Y}\}$ is linearly dependent over Δ.

In order to point out the real and dual parts of the dual coordinates of $\mathbf{U}$ in Propositions 2.1 or 2.2 let us define a map $\mathcal{M} : \mathfrak{D} \times \mathfrak{D} \times \mathfrak{D} \to \mathfrak{D}$, which will play an important role below, by

$$\mathcal{M}(\mathbf{X}, \mathbf{Y}, \mathbf{Z}) = \overline{\{\mathbf{X}; \mathbf{Y}; \mathbf{Z}\}}[\mathbf{Y}, \mathbf{Z}] = (\mathbf{X} \mid [\mathbf{Y}, \mathbf{Z}])[\mathbf{Y}, \mathbf{Z}] - [\mathbf{X} \mid [\mathbf{Y}, \mathbf{Z}]]\varepsilon[\mathbf{Y}, \mathbf{Z}].$$

Note that $\mathcal{M}(\mathbf{X}, \mathbf{Y}, \mathbf{Z}) = \mathcal{M}(\mathbf{X}, \mathbf{Z}, \mathbf{Y})$ is reciprocal (orthogonal for the Klein form) to $\mathbf{X}$, $\mathbf{Y}$, and $\mathbf{Z}$. Using the Δ-linearity of the inner product $\{\cdot \mid \cdot\}$, it is readily proved that the formulas (3) write (note that $G\tilde{G} = \text{Re}\{\mathbf{X}; \mathbf{Y}; \mathbf{Z}\}^2 \in \mathbf{R}$):

$$(5) \quad x = \frac{\{\mathbf{U} \mid \mathcal{M}(\mathbf{X}, \mathbf{Y}, \mathbf{Z})\}}{G\tilde{G}}, \; y = \frac{\{\mathbf{U} \mid \mathcal{M}(\mathbf{Y}, \mathbf{Z}, \mathbf{X})\}}{G\tilde{G}}, \; z = \frac{\{\mathbf{U} \mid \mathcal{M}(\mathbf{Z}, \mathbf{X}, \mathbf{Y})\}}{G\tilde{G}}.$$

Now we turn to the basic classification of the finite subsets of $\mathfrak{D}$.

Definition 2.2 *Let* k *be an integer equal to* 0, 1, 2 *or* 3. *A subset* $\mathfrak{S}$ *of* $\mathfrak{D}$ *is said to be of order* k *if the maximal number of elements independent over* Δ *in* $\mathfrak{S}$ *is equal to* k *(in other words the rank of* $\mathfrak{S}$ *over* Δ *is equal to* k*).*

For an *order 2* subset $\mathfrak{S}$, all the axes of the elements of $\mathfrak{S}$ which are not in $\mathfrak{T}$ are parallel to a plane and for all $\mathbf{X}, \mathbf{Y}, \mathbf{Z} \in \mathfrak{S}$ the dual number $\{\mathbf{X}; \mathbf{Y}; \mathbf{Z}\}$ is non invertible (in other words: $(\mathbf{Z} \mid [\mathbf{X}, \mathbf{Y}]) = 0$).

For an *order 1* subset, all the axes of the elements of $\mathfrak{S}$ which are not in $\mathfrak{T}$ are parallel straight lines, all the Lie brackets are in $\mathfrak{T}$ and we have $\{\mathbf{X}; \mathbf{Y}; \mathbf{Z}\} = 0$ for all $\mathbf{X}, \mathbf{Y}, \mathbf{Z} \in \mathfrak{S}$. An *order 0* subset is simply a subset of $\mathfrak{T}$.

Up to a permutation, every finite subset $\mathfrak{S}$ of at most six non zero elements of $\mathfrak{D}$ can be expressed in one and only one of the following *normal forms*:

ORDER 3 NORMAL SETS: $\{\mathbf{X}, \mathbf{Y}, \mathbf{Z}\}$, $\{\mathbf{X}, \mathbf{Y}, \mathbf{Z}, \mathbf{U}\}, \ldots, \{\mathbf{X}, \mathbf{Y}, \mathbf{Z}, \mathbf{U}, \mathbf{V}, \mathbf{W}\}$ where $\{\mathbf{X}, \mathbf{Y}, \mathbf{Z}\}$ is a basis of the Δ-module $\mathfrak{D}$ (see proposition 2.1).

ORDER 2 NORMAL SETS: $\{\mathbf{X}, \mathbf{Y}\}$, $\{\mathbf{X}, \mathbf{Y}, \mathbf{Z}\}, \ldots, \{\mathbf{X}, \mathbf{Y}, \mathbf{Z}, \mathbf{U}, \mathbf{V}, \mathbf{W}\}$ where $[\mathbf{X}, \mathbf{Y}] \notin \mathfrak{T}$ (see proposition 2.2).

ORDER 1 NORMAL SETS: They will be classified according to the number of elements of $\mathfrak{S} \cap (\mathfrak{D} - \mathfrak{T})$ and of $\mathfrak{S} \cap \mathfrak{T}$ and their normal forms are expanded writting at the head the elements which are not in $\mathfrak{T}$; the normal form of an order 1 set of the type (m, n) is

$$\mathfrak{S} = \{\mathbf{X}_1, \ldots, \mathbf{X}_m, \mathbf{U}_1, \ldots, \mathbf{U}_n\} \text{ with } \mathbf{X}_1, \ldots, \mathbf{X}_m \notin \mathfrak{T} \text{ and } \mathbf{U}_1, \ldots, \mathbf{U}_n \in \mathfrak{T}.$$

3 Linear dependence over ℝ of two elements of $\mathfrak{D}$

Proposition 3.1 *A necessary and sufficient condition for a set* $\mathfrak{S} = \{\mathbf{X}, \mathbf{Y}\}$ *of non zero elements of* $\mathfrak{D}$ *be linearly dependent over* **ℝ** *is that one among the following properties holds:*

- *$\mathbf{X}$ and $\mathbf{Y}$ are in $\mathfrak{T}$ and are linearly dependent over* **ℝ**,

- *$\mathbf{X}$ and $\mathbf{Y}$ are not in $\mathfrak{T}$ and $([\mathbf{X}, \mathbf{Y}] = 0$ and $f_\mathbf{X} = f_\mathbf{Y})$.*

In the first case $\mathfrak{S}$ is order 0 and in the second case it is order 1 of the type $(2, 0)$. Note that the condition $f_\mathbf{X} = f_\mathbf{Y}$ also writes: $[\mathbf{X} \mid \mathbf{X}](\mathbf{Y} \mid \mathbf{Y}) - [\mathbf{Y} \mid \mathbf{Y}](\mathbf{X} \mid \mathbf{X}) = 0$.

4 Linear dependence over ℝ of order 3 sets

Since an order 3 subset $\mathfrak{S} = \{\mathbf{X}, \mathbf{Y}, \mathbf{Z}\}$ of $\mathfrak{D}$ is linearly independent over Δ it is always linearly independent over **ℝ**. We only have to treat the cases where the subset contains 4, 5 or 6 elements. Our method based on the use of the function $\mathcal{M}$ and the Klein form for constructing a screw reciprocal to a given set of screws differs from the method proposed by Kerr and Sanger in [6], moreover it has an "Euclidean meaning".

Proposition 4.1 *A necessary and sufficient condition for an order 3 normal set* $\mathfrak{S} = \{\mathbf{X}, \mathbf{Y}, \mathbf{Z}, \mathbf{U}\}$ *be linearly dependent over* **ℝ** *is that* $\mathbf{U}$ *be reciprocal (orthogonal within the meaning of the Klein form) to*

$$\mathcal{M}(\mathbf{X}, \mathbf{Y}, \mathbf{Z}), \quad \mathcal{M}(\mathbf{Y}, \mathbf{Z}, \mathbf{X}), \quad \mathcal{M}(\mathbf{Z}, \mathbf{X}, \mathbf{Y}).$$

Proposition 4.2 *A necessary and sufficient condition for an order 3 normal set* $\{X, Y, Z, U, V\}$ *be linearly dependent over* **R** *is that either* $\{X, Y, Z, U\}$ *or* $\{X, Y, Z, V\}$ *is linearly dependent (test from Prop. 4.1), or* **V** *is reciprocal to*

(6)
$$
\begin{aligned}
&[U \mid \mathcal{M}(X, Y, Z)] \mathcal{M}(Y, Z, X) - [U \mid \mathcal{M}(Y, Z, X)] \mathcal{M}(X, Y, Z), \\
&[U \mid \mathcal{M}(Y, Z, X)] \mathcal{M}(Z, X, Y) - [U \mid \mathcal{M}(Z, X, Y)] \mathcal{M}(Y, Z, X), \\
&[U \mid \mathcal{M}(Z, X, Y)] \mathcal{M}(X, Y, Z) - [U \mid \mathcal{M}(X, Y, Z)] \mathcal{M}(Z, X, Y).
\end{aligned}
$$

The three reciprocity conditions for **V** are not independent (only two of them are). Moreover, the elements (6) span the reciprocal subspace to $\{X, Y, Z, U\}$ and the proposition means that **V** lies in $\text{Span}_{\mathbf{R}}\{X, Y, Z, U\}$ if it lies in the biorthogonal subspace (within the meaning of the Klein form).

Proposition 4.3 *A necessary and sufficient condition for an order 3 normal subset* $\{X, Y, Z, U, V, W\}$ *be linearly dependent over* **R** *is that one among the following properties holds:*

- $\{X, Y, Z, U\}$, $\{X, Y, Z, V\}$, $\{X, Y, Z, W\}$, $\{X, Y, Z, U, V\}$, $\{X, Y, Z, V, W\}$ *or* $\{X, Y, Z, U, W\}$ *is linearly dependent (tests from Prop. 4.1 and Prop. 4.2).*
- **W** *is reciprocal to* $P\,\mathcal{M}(X, Y, Z) + Q\,\mathcal{M}(Y, Z, X) + R\,\mathcal{M}(Z, X, Y)$ *where:*

$$
\left\{
\begin{aligned}
P &= [U \mid \mathcal{M}(Y, Z, X)][V \mid \mathcal{M}(Z, X, Y)] - [V \mid \mathcal{M}(Y, Z, X)][U \mid \mathcal{M}(Z, X, Y)], \\
Q &= [U \mid \mathcal{M}(Z, X, Y)][V \mid \mathcal{M}(X, Y, Z)] - [V \mid \mathcal{M}(Z, X, Y)][U \mid \mathcal{M}(X, Y, Z)], \\
R &= [U \mid \mathcal{M}(X, Y, Z)][V \mid \mathcal{M}(Y, Z, X)] - [V \mid \mathcal{M}(X, Y, Z)][U \mid \mathcal{M}(Y, Z, X)].
\end{aligned}
\right.
$$

5 Linear dependence of order 2 sets

Proposition 5.1 *A necessary and sufficient condition for an order 2 normal set* $\{X, Y, Z\}$ *be linearly dependent over* **R** *is that* **Z** *be reciprocal to*

$$
\mathcal{M}(X, Y, [X, Y]), \quad \mathcal{M}(Y, [X, Y], X), \quad \mathcal{M}([X, Y], X, Y).
$$

Proposition 5.2 *A necessary and sufficient condition for an order 2 normal set* $\{X, Y, Z, U\}$ *be linearly dependent over* **R** *is that either* $\{X, Y, Z\}$ *or* $\{X, Y, U\}$ *is linearly dependent (test from Prop. 5.1), or* **U** *is reciprocal to the following elements*

$$
\begin{aligned}
&[Z \mid \mathcal{M}(Y, [X, Y], X)] \mathcal{M}([X, Y], X, Y) - [Z \mid \mathcal{M}([X, Y], X, Y)] \mathcal{M}(Y, [X, Y], X), \\
&[Z \mid \mathcal{M}([X, Y], X, Y)] \mathcal{M}(X, [X, Y], Y) - [Z \mid \mathcal{M}(X, [X, Y], Y)] \mathcal{M}([X, Y], X, Y), \\
&[Z \mid \mathcal{M}(X, [X, Y], Y)] \mathcal{M}(Y, [X, Y], X) - [Z \mid \mathcal{M}(Y, [X, Y], X)] \mathcal{M}(X, [X, Y], Y).
\end{aligned}
$$

The statement of Proposition 5.2 deserves a remark similar to the one following Proposition 4.2: the three reciprocity conditions are not independent.

Proposition 5.3 *A necessary and sufficient condition for an order 2 normal subset* $\{X, Y, Z, U, V\}$ *be linearly dependent over* **R** *is that one among the following* *properties holds:*

- $\{X, Y, Z\}$, $\{X, Y, U\}$, $\{X, Y, V\}$, $\{X, Y, Z, U\}$, $\{X, Y, Z, V\}$ *or* $\{X, Y, U, V\}$ *is linearly dependent (tests from Prop. 5.1 and Prop. 5.2).*
- V *is reciprocal to* $P\mathcal{M}(X, Y, [X, Y]) + Q\mathcal{M}(Y, [X, Y], X) + R\mathcal{M}([X, Y], X, Y)$, *where*

$$
\left\{
\begin{array}{rl}
P = & [Z \mid \mathcal{M}(Y, [X, Y], X)] [U \mid \mathcal{M}([X, Y], X, Y)] \\
    & -[Z \mid \mathcal{M}([X, Y], X, Y)] [U \mid \mathcal{M}(Y, [X, Y], X)] \\
Q = & [Z \mid \mathcal{M}([X, Y], X, Y)] [U \mid \mathcal{M}(X, [X, Y], Y)] \\
    & -[Z \mid \mathcal{M}(X, [X, Y], Y)] [U \mid \mathcal{M}([X, Y], X, Y)] \\
R = & [Z \mid \mathcal{M}(X, [X, Y], Y)] [U \mid \mathcal{M}(Y, [X, Y], X)] \\
    & -[Z \mid \mathcal{M}(Y, [X, Y], X)] [U \mid \mathcal{M}(X, [X, Y], Y)]
\end{array}
\right.
$$

Proposition 5.4 *Every order 2 set of six elements* $\mathfrak{S}$ *in* $\mathfrak{D}$ *is linearly dependent over* **R**.

6 Linear dependence of order 1 sets

The order 1 sets of more than four elements are always linearly dependent. Hence we a priori need ten criteria for testing order 1 sets and they reduce to eight since two cases, the types $(1, 0)$ and $(1, 1)$, are obvious. We only give some examples of these criteria and some geometrical interpretations.

For an order 1 set $\mathfrak{S}$ let $\Pi = \Pi_{\mathfrak{S}}$ be a plane orthogonal to the axes Δ_X ($X \in \mathfrak{S} - \mathfrak{T}$) and a_X be the intersection of Π and Δ_X. Let $[\mathfrak{S}, \mathfrak{S}]$ be the set of the elements $[X, Y]$ with $X, Y \in \mathfrak{S}$. Using formula (2) we see that, for X and Y in $\mathfrak{S} - \mathfrak{T}$, $[X, Y]$, which is in $\mathfrak{T}$, is the constant field equal to $(X \mid Y) \overrightarrow{a_X a_Y}$ (this property is useful for geometrical interpretations).

A first criterion reduces the testing of the sets of the type $(1, n)$ to ordinary calculations in the three dimensional space $\mathfrak{T}$ (isomorphic to **E**):

Proposition 6.1 *A necessary and sufficient condition for an order 1 normal set* $\mathfrak{S} = \{X, U_1, \ldots, U_n\}$ *of the type* $(1, n)$ *be linearly dependent over* **R** *is that* $\{U_1, \ldots, U_n\}$ *be linearly dependent. In particular, if* $n > 3$, $\mathfrak{S}$ *is linearly dependent and, if* $n = 0$ *or* 1, $\mathfrak{S}$ *is linearly independent.*

Proposition 6.2 *A necessary and sufficient condition for an order 1 normal set* $\mathfrak{S} = \{X, Y, U\}$ *of the type* $(2, 1)$ *be linearly dependent over* **R** *is that either* $\{X, Y\}$ *is linearly dependent over* **R** *(second test from Prop. 3.1), or one among the following properties holds:*

- $[X, Y] = [X, U] = 0$, *[and* $f_X \neq f_Y$*]*.

- $[X, Y] \neq 0$ *and* $[X \mid U] = 0$ *and* $\mathrm{Rank}\{[X, Y], [X, U]\} = 1$ *and* $f_X = f_Y$.

- $[X, Y] \neq 0$ *and* $[X \mid U] \neq 0$ *and* $f_X \neq f_Y$ *and*

$$(X \mid Y)(f_X - f_Y)[X, U] = [X \mid U][X, Y].$$

The geometrical form of the last three conditions is the following:

- $\Delta_X = \Delta_Y$ and U is directed as Δ_X [and $f_X \neq f_Y$].

- $\Delta_X \neq \Delta_Y$, U is orthogonal to the plane (Δ_X, Δ_Y) and $f_X = f_Y$.

- $\Delta_X \neq \Delta_Y$, U is not orthogonal to Δ_X, $f_X \neq f_Y$ and

$$(f_X - f_Y)\omega_X \times U = (\omega_X \cdot U)\overrightarrow{a_X a_Y}.$$

Define the quantity $2\mathcal{D}(X, Y, Z)$ as:

$$(X \mid X)(Y \mid Z)f_X[Y, Z] + (Y \mid Y)(Z \mid X)f_Y[Z, X] + (Z \mid Z)(X \mid Y)f_Z[X, Y].$$

Proposition 6.3 *A necessary and sufficient condition for an order 1 subset $\mathfrak{S} = \{X, Y, Z\}$ of the type (3,0) be linearly dependent over $\mathbf{R}$ is that either $\{X, Y\}$, $\{Y, Z\}$ or $\{Z, X\}$ is linearly dependent over $\mathbf{R}$ (second test from Prop. 3.1), or the following condition holds:*

- Rank $[\mathfrak{S}, \mathfrak{S}](= \text{Rank}\{[X, Y], [X, Z]\}) = 1$ *and* $\mathcal{D}(X, Y, Z) = 0$

In particular, when Rank $[\mathfrak{S}, \mathfrak{S}] = 2$ *the set $\mathfrak{S}$ is linearly independent.*

The geometrical form of the condition is the following:

- Δ_X, Δ_Y, Δ_Z are coplanar and $f_X \overrightarrow{a_Y a_Z} + f_Y \overrightarrow{a_Z a_X} + f_Z \overrightarrow{a_X a_Y} = 0$.

In particular when the axes Δ_X, Δ_Y, Δ_Z are not coplanar the set $\mathfrak{S}$ is linearly independent. The condition $\mathcal{D}(X, Y, Z) = 0$ (or its geometrical form) holds in particular when Rank$[\mathfrak{S}, \mathfrak{S}] = 0$ ($\Delta_X = \Delta_Y = \Delta_Y$) or when Rank$[\mathfrak{S}, \mathfrak{S}] \leq 1$ and $f_X = f_Y = f_Z$, due to the relation

(7)
$$(X \mid X)(Y \mid Z)[Y, Z] + (Y \mid Y)(Z \mid X)[Z, X] + (Z \mid Z)(X \mid Y)[X, Y] = 0.$$

Proposition 6.4 *A necessary and sufficient condition for an order 1 subset $\mathfrak{S} = \{X, Y, Z, U\}$ of the type (4,0) be linearly dependent over $\mathbf{R}$ is that either one three element subset of $\mathfrak{S}$ is linearly dependent or* Rank$[\mathfrak{S}, \mathfrak{S}] \leq 1$ *or one among the properties similar to the following holds:*

- Rank$\{[X, Y], [X, Z]\} = 2$ *(for example) and if α and β are the real numbers such that $[X, U] = \alpha[X, Y] + \beta[X, Z]$, then*

$$\alpha(X \mid Y)(f_Y - f_X) + \beta(X \mid Z)(f_Z - f_X) = (X \mid U)(f_U - f_X).$$

Owing to relation (7), in the last case Rank$[\mathfrak{S}, \mathfrak{S}] = 2$. The last properties holds in particular when the four numbers f_X, f_Y, f_Z and f_U are equal. The geometrical form of the condition is either Δ_X, Δ_Y, Δ_Z, Δ_U are coplanar or:

- Δ_X, Δ_Y, Δ_Z (for example) are not coplanar and if α and, β are the real numbers such that $\overrightarrow{a_X a_U} = \alpha \overrightarrow{a_X a_Y} + \beta \overrightarrow{a_X a_Z}$, then

$$\alpha(f_Y - f_X) + \beta(f_Z - f_X) = f_U - f_X.$$

Conclusion. The principles of an algorithm for testing the rank of any set $\mathfrak{S}$ of non zero members of $\mathfrak{D}$ are now clear. It is compound of two stages:

FIRST STAGE: find the order k and a normal form of $\mathfrak{S}$ (then $\text{Rank}\,\mathfrak{S} \geq k$).

SECOND STAGE: apply specific criteria from sections 4, 5, 6 for the calculation of the value of the rank. For example when $k = 3$ and $\mathfrak{S} = \{X, Y, Z, \dots\}$ one test the sets of the form $\{X, Y, Z, U\}$, $(U \in \mathfrak{S})$ (Prop. 4.1) and stop if the rank is 3. Else one has to test selected sets of the form $\{X, Y, Z, U, V\}$ (Prop. 4.2) and, if this is necessary, selected sets of the form $\{X, Y, Z, U, V, W\}$ (Prop. 4.3).

References

[1] J. E. Baker and K. Wohlhart. On the single Screw Reciprocal to the General Line-Symmetric Six Screw Linkage. *Mech. Mach. Theory* **29** No 1, 169-175 (1994).

[2] D. P. Chevallier. Lie Algebras, Modules, Dual Quaternions and Algebraic Methods in Kinematics. *Mech. Mach. Theory* **26** No 6 p. 613-627 (1991).

[3] C. G. Gibson and K. H. Hunt. Geometry of Screw Systems Part 1 and 2: *Mech. Mach. Theory* Vol. 25 No 1 p. 1-10 and p. 11-27 (1990).

[4] J.M. Hervé. Intrinsic Formulation of Problems of Geometry and Kinematics of Mechanisms. *Mech. Mach. Theory* **17**, 179-184 (1982).

[5] K.H. Hunt. Special Configurations of Robot-arms via Screw Theory. *Robotica*, Part I 4, 171-179, (1986), Part II5, 17-22 (1987).

[6] D. R. Kerr and D. J Sanger. The Inner Product in the Evaluation of Reciprocal Screws. *Mech. Mach. Theory* **24** No 2, 87-92, (1989).

[7] J. Loncaric. Geometric Analysis of Compliant Mechanisms. Ph.D Thesis, Harvard University (1985).

[8] J. M. Rico Martinez and J. Duffy. Classification of Screw Systems Part 1 and 2. *Mech. Mach. Theory* **27** No 4, 459-470 and 471-490 (1992).

[9] J. M. Rico Martinez and J. Duffy. Orthogonal Spaces and Screw Systems. *Mech. Mach. Theory* **27** No 4, 451-458 (1992).

[10] K. Sugimoto and J. Duffy. Applications of linear Algebra to Screw Systems. *Mech. Mach. Theory* **17** No 1, 73-83 (1982).

[11] K. Sugimoto. Existence Criteria for Over-Constrainded Mechanisms, An Extension of Motor Algebra. *J. of Mechanical Design* September 1990, **112**, 295-298.

[12] K. Wohlhart. A New 6R Space Mechanism. Proc 7th World Congress IFTOMM Sevilla Spain 17-22 Sept 1987.

A PSEUDO-DUAL NOTATION FOR KINEMATIC CALCULATIONS

J.M. HERVE,
Ecole Centrale Paris
92295 Chatenay Malabry France

Abstract

We present a new type of algebra which can be used for any kinematic calculation.

1. Introduction

Since the original work by Dimentberg [1], dual numbers and dual vector algebra has been applied to kinematics. We note particular applications to mechanism design by Keler [2], Sugimoto, Duffy and Hunt [3] and numerous papers on robotics. Modern algebraic formulation of the dual algebra is given by Chevallier [4].
In this article, we present a new type of algebra with some analogy to the dual algebra . The pseudo-dual notation makes some calculations in kinematics easier.

2. The pseudo-dual algebra

We recall that a dual number d is a complex mathematical entity composed of two real numbers a and b :

$d = a + \varepsilon\, b$ ε is the dual unit endowed with the characteristic property $\varepsilon^2 = O$

The number 1 may be considered as the identity operator i , then d becomes an operator

$d = i\, a + \varepsilon\, b$ able to act on any vector space.

The pseudo-dual algebra is defined with a base of three operators I, J, E.
The characteric properties of these pseudo-dual units are

$$I^2 = I \, , J^2 = J \, , E^2 = 0$$

$$I J = J I = 0$$

$$I E = E J = E$$

$$E I = J E = 0$$

This system of operators is connected with the dual algebra.

J.-P. Merlet and B. Ravani (eds.), Computational Kinematics, 11–18.

Indeed, we can easily verify

$$I + J = i \,,\, i^2 = i \,,\, i\,E = E\,i = E$$

where i is the identity operator and E may be equated to the dual unit ε .
The operators I, J, E may be represented by matrices acting on any vectors.

$$I = \left[\begin{array}{c|c} i & o \\ \hline o & o \end{array}\right] \qquad J = \left[\begin{array}{c|c} o & o \\ \hline o & i \end{array}\right] \qquad E = \left[\begin{array}{c|c} o & i \\ \hline o & o \end{array}\right]$$

where i is the identity and o the zero operator.

3. Pseudo-dual notation of the affine space

Points of the affine Euclidean space of dimension 3 are denoted by capital letters N, M,
...
Vectors obtained from two ordered points may be represented by the difference of two points

NM = M - N

An origin point, O, is needed to introduce the pseudo-dual notation . The pseudo-dual notation for point O is :

$$\underline{O} = E\,O + J$$

The pseudo-dual notation for a generic point M will be :

$$\underline{M} = E\,OM + J$$

Then, we are able to give the pseudo-dual notation for a vector :

$$\underline{OM} = \underline{M} - \underline{O} = E\,OM$$

$$\underline{M} = \underline{O} + \underline{OM}$$

4. Velocity field

If M is the generic point of a mobile affine Euclidean space relative to a fixed Euclidean space, it is well known that the field of velocity vectors is a field of moments or screw (or twist).

$$dM / dt = V (M) = V + R \times OM$$

where

$V (M)$	is the velocity vector of point M
V	is a given vector equal to the velocity of the point which coincides with point O at the given time t
R	is the angular velocity vector
Rx	is the linear skew-symmetric operator obtained from R and the vector product.

The pseudo-dual notation is

$$E \, dM / dt = E \, V(M) = E (V + R \times OM)$$

It is easy to verify that the expression for $V (M)$ can be split in two factors :

$$d (E \, OM + J) / dt = [I \, R \, x + E \, V] \, [E \, OM + J]$$

$$d \underline{M} / dt = \underline{S} \; \underline{M}$$

$\underline{S} = I \, R \, x + E \, V$ is a pseudo-dual screw

$I \, R \, x$ is the rotation part and $E \, V$ the translation part.

We notice :

$$I \, R \, x \, \underline{M} = E \, R \, x \, OM$$

$$E \, V \, \underline{M} \; = E \, V$$

5. The finite displacement

If the pseudo-dual screw is constant (independent of t), we have a differential equation for the function $\underline{M}$ in the variable t, which can be integrated between times o and t.

$$\underline{M} \, (t) = \exp (t \, \underline{S}) \, \underline{M} \, (o)$$

$$E \, OM \, (t) + J = \exp [t \, (E \, V + I \, R \, x)] \, [E \, OM \, (o) + J]$$

$$\exp [t \, (I \, R \, x + E \, V)] = i + t \, (I \, R \, x + E \, V)$$

$$+ t^2 / 2 ! \, (I \, R \, x + E \, V \,)^2 + \ldots$$

$$+ t^n / n ! \, (I \, R \, x \; E \, V \,)^n + \ldots$$

14

We verify :

$$(I \mathbf{R} x + E \mathbf{V})^2 = I (\mathbf{R} x)^2 + E \mathbf{R} x \mathbf{V}$$

. . .

$$(I \mathbf{R} x + E \mathbf{V})^n = I (\mathbf{R} x)^n + E (\mathbf{R} x)^{n-1} \mathbf{V}$$

. . .

Employing the formula $i = I + J$, we obtain

$$\exp [t (I \mathbf{R} x + E \mathbf{V})] = J + I \exp (t \mathbf{R} x)$$
$$+ E [t \mathbf{V} + t^2 / 2 ! \mathbf{R} x \mathbf{V} + ...$$
$$+ t^n / n ! (\mathbf{R} x)^{n-1} \mathbf{V} + ...]$$

To continue the computation, the vector $\mathbf{V}$ can be decomposed into its projection on $\mathbf{R}$ and on a plane perpendicular to $\mathbf{R}$.

The orthogonal projection of $\mathbf{V}$ on $\mathbf{R}$ is easily obtained employing the scalar product (dot product) $(\mathbf{R} / \mathbf{R}^2) (\mathbf{R} . \mathbf{V})$. The complement of $\mathbf{V}$ is the orthogonal projection of $\mathbf{V}$ on a plane perpendicular to $\mathbf{R}$:

$$\mathbf{V} - (\mathbf{R} / \mathbf{R}^2) (\mathbf{R} . \mathbf{V}).$$

Because of the double vector product, this last expression can be proved to be equal to

$$- (1 / \mathbf{R}^2) (\mathbf{R} x)^2 \mathbf{V}$$

Then, we may write :

$$\mathbf{V} = - (\mathbf{R} x)^2 \mathbf{V} / \mathbf{R}^2 + \mathbf{R} (\mathbf{R} . \mathbf{V}) / \mathbf{R}^2$$

and consequently

$$\mathbf{R} x \mathbf{V} = - (\mathbf{R} x)^3 \mathbf{V} / \mathbf{R}^2$$

The coefficient of E becomes

$$t \mathbf{R} (\mathbf{R} . \mathbf{V}) / \mathbf{R}^2 - t (\mathbf{R} x)^2 \mathbf{V} / \mathbf{R}^2$$
$$- t^2 / 2 ! (\mathbf{R} x)^3 \mathbf{V} / \mathbf{R}^2$$
. . . .
$$- t^n / n ! (\mathbf{R} x)^{n + 1} \mathbf{V} / \mathbf{R}^2$$
. . .
$$= t \mathbf{R} (\mathbf{R} . \mathbf{V}) / \mathbf{R}^2 + (\mathbf{R} x \mathbf{V}) / \mathbf{R}^2 - \exp (t \mathbf{R} x) [(\mathbf{R} x \mathbf{V}) / \mathbf{R}^2]$$

These vectors have a geometric sense. The fixed origin O coincides at $t = o$ with a mobile point O. O becomes O' at time t.

H is the foot of the perpendicular drawn from origin point O on the screw axis.

$$OH = (R \times V) / R^2$$

By rotation of angle tR, $(R = R \, R / \| R \|)$,

OH becomes **O' H'**

$$O'H' = \exp(t \, R \, x) \, OH$$

$$HH' = t \, R \, (R \cdot V) / R^2$$

The coefficient of E is **OO' = OH + HH' + H'O'**

6. The Lie algebra of screws

The vector space of screws at a given point may be endowed with an algebraic structure of Lie algebraic structure in several ways [5], [6].

The commutator or Lie bracket of two pseudo-dual screws provides the closed product of the Lie algebra of pseudo-dual screws.

It may be called a"screw product" because of its analogy to the vector product in classical 3 dimensional vector space.
Let us consider two pseudo-dual screws

$$\underline{S}_1 = IR_1 x + EV_1$$

$$\underline{S}_2 = IR_2 x + EV_2$$

The Lie bracket is

$$[\underline{S}_1, \underline{S}_2] = \underline{S}_1 \underline{S}_2 - \underline{S}_2 \, \underline{S}_1$$

Using the known property (Jacobi identity) of the classical vector product

$$R_1 \times R_2 \, x - R_2 \times R_1 \, x = (R_1 \times R_2) x$$

After some calculation we arrive at :

$$[\underline{S}_1, \underline{S}_2] = I(R_1 \times R_2) x + E(R_1 \times V_2 - R_2 \times V_1)$$

7. Composition product of displacements

A mobile rigid body may undergo two successive displacements. The first displacement is characterized by two vectors S_1 and T_1. S_1 is the rotation vector and T_1 is the translation vector of the origin.
The pseudo-dual representation of this first displacement is the operator

$$\underline{D}_1 = I \exp(S_1 \ x) + E \ T_1 + J$$

For the second displacement, it is :

$$\underline{D}_2 = I \exp(S_2 \ x) + E \ T_2 + J$$

Two successive displacements produce a new displacement, the composition product $\underline{D}_2 \ \underline{D}_1$

$$\underline{D}_2 \ \underline{D}_1 = I \exp(S_2 \ x) \exp(S_1 \ x) + E \ [T_2 + \exp(S_2 \ x) \ T_1] + J$$

The absence of displacement or identity leads to a zero vector for rotation and translation vectors.
The inverse displacement is

$$[I \exp(S \ x) + E \ T + J]^{-1} = I \exp(-S \ x) + E \ [-\exp(-S \ x) \ T] + J$$

The pseudo-dual notation of finite displacements is able to express the algebraic group structure for the set of displacements [7].

8. Description of kinematic pairs

A revolute pair, the axis of which is fixed at the origin, O provides rotations of angle θ about the unit direction vector u :

$$OM \longrightarrow OM'$$
$$[E \ OM' + J] = [I \exp(\theta \ u \ x) + J] [E \ OM + J]$$

If the revolute pair axis is determined by a point $N \neq O$, the corresponding operators are obtained by conjugacy from the previous ones :

$$[I + E \ ON + J] [I \exp(\theta \ u \ x) + J] [I - E \ ON + J]$$
$$= I \exp(\theta \ u \ x) + E \ [1 - \exp(\theta \ u \ x)] \ ON + J$$

A prismatic pair is represented by
$$I + E \, a \, t + J$$

where is the translation amplitude with the unit direction vector t.
All the lower kinematic pairs are easily deduced from the combinations of revolute pairs and prismatic pairs.

9. Series of kinematic pairs

Dealing with this problem in its generality would require extremely lengthy calculations. To simplify, we have chosen the typical example of a series of two revolute pairs. For a given initial configuration of a set of three rigid bodies connected by two R pairs, the successive axes are determined by

$$(N_1 , u_1) \text{ and } (N_2 , u_2).$$

The rotation angles are denoted θ_1 and θ_2 . The series is represented by the composition product of the allowed relative displacements in each pair.

The factor order is from the left to the right for pairs which go from the fixed body to the mobile bodies. A generic point M becomes a point M' :

$$[E \, OM' + J] = [I + E \, ON_1 + J] \, [I \exp (\theta_1 \, u_1 \; x) + J] \, [I - E \, ON_1 + J]$$
$$[I + E \, ON_2 + J] \, [I \exp (\theta_2 \, u_2 \; x) + J] \, [I - E \, ON_2 + J] \, [E \, OM + J]$$

Using pseudo-dual calculations, we obtain :

$$[I - E \, ON_1 + J] \, [EOM' + J] = [I \exp (\theta_1 u_1 \; x) + J] \, [I + EN_1N_2 + J]$$
$$[I \exp (\theta_2 \, u_2 \; x) + J] \, [I - E \, ON_2 + J] \, [E \, OM + J]$$

$$[E \, N_1M' + J] = [I \exp (\theta_1 u \; 1 \; x) + J] \, [I \exp (\theta_2 \, u_2 \; x) + E \, N_1N_2 + J]$$
$$[E \, N_2 \, M + J]$$

One can recognize the pseudo-dual formulation of a result obtained directly in reference [8].

18

10. Conclusion

The pseudo-dual notation for geometry and kinematics is not as powerful a tool as the true dual notation. However, it simplifies the calculation of the affine Euclidian displacement and it may be useful in the establishment of properties of kinematic chains.

11. References

[1] Dimentberg P.M. (1965)*The Screw Calculus and its Applications in Mechanics* , Moscou NAUKA .

[2] Keler M. (1959) *Analyse und Synthese des Raumburbelgetriebe mittels Raumliniengeometrie und dualer Grössen,* Forschung auf dem Gebiete des Ingenieurwesens, Vol 25, pp. 26-32, 51-75.

[3] Sugimoto K. , Duffy J., Hunt K.H. (1982) *Special Configurations of Spatial Mechanisms and Robot Arms,* Mech. Mach. Theory, Vol 17, pp.613-627.

[4] Chevallier D.P. (1991) *Lie Algebras, Modules, Dual Quatermions and Algebraic Methods in kinematics,* Mech. Mach. Theory Vol. 26, N°6, pp. 73-81.

[5] Karger A. , Novak J. (1985) *Space kinematics and Lie Groups,* Gordon and Breach

[6] Hervé J.M. (1994) *The Mathematical Group Structure of the Set of Displacements* , Mech. Mach. Theory Vol. 29, N°1, pp. 73-81.

[7] Hervé J.M. (1978) *Analyse Structurale des Mecanismes par Groupe des Déplacements,* Mech. Mach. Theory Vol. 13 pp. 437-450.

[8] Hervé J.M. (1982) *Intrinsic Formulation of Problems of Geometry and Kinematics of Mechanisms,* Mech. Mach. Theory Vol. 17, N°3, pp. 179-184.

MOBILE ROBOT LOCALIZATION BY CABLE-EXTENSION TRANSDUCERS

D. HONG, K. J. MUELLER AND S. A. VELINSKY
Advanced Highway Maintenance & Construction Technology
Research Center
Department of Mechanical & Aeronautical Engineering
University of California, Davis
Davis, CA 95616-5294
email: savelinsky@ucdavis.edu

Abstract. This paper discusses a novel technique for reference localization of a wheeled mobile robot (WMR) through the use of cable-extension transducers. The robot's planar position is determined through triangulation of two transducers. However, restrictions on the orientation of the transducers require the use of pulleys in the overall system design, and these introduce nonlinearity into the equations governing the reference localization. The theoretical process by which Cartesian coordinates for the WMR are obtained is presented as are the computational aspects of the approach. Additionally, the uncertainty inherent to the linear transducer measurements is discussed.

1. Introduction

The precise localization of wheeled mobile robots (WMRs) has been a very important topic for mobile robot navigation and control problems. Computational kinematic equations are the basis for two distinct approaches for ascertaining robot location: the dead reckoning approach and the reference (or landmark) method (Cox and Wilfong, 1990).

The dead reckoning method for WMRs is the process of position identification by measuring wheel rotations through the use of either in-line encoders, passive measuring wheels (Sugimoto *et al.*, 1988), castor sensing wheels (Culley and Buldar, 1988), or several other methods. For the

19

J.-P. Merlet and B. Ravani (eds.), Computational Kinematics, 19–30.
© *1995 Kluwer Academic Publishers.*

reference method, various sensor systems may be used such as vision systems, infrared sensors, sonar systems, and LIDAR (laser radar) (Cox, 1990; Crowley, 1989; Beckerman and Oblow, 1990; Chen *et al.*, 1993; Figueroa and Mahajan, 1994). The triangulation (or trilateration) principle is essential in the latter method since robot position is based on the simultaneous measurement of the range or bearing to two or more known landmarks.

The University of California, Davis in conjunction with the California Department of Transportation, as part of the Advanced Highway Maintenance and Construction Technology Research Center, is developing a high load WMR for highway maintenance and construction tasks (Winters *et al.*, 1994). The WMR is a differentially steered, self-propelled robot which works in close proximity to a support vehicle. The motion of the mobile robot is controlled relative to the support vehicle, so that a relative position tracking system is essential. Based on the intended applications of the robot system and the corresponding tracking system requirements, a new approach for reference localization is being developed.

The sensor approach that is presented herein is novel in that it uses new technology in the form of cable-extension transducers (CETs) to determine robot position. CETs are linear-displacement sensors which produce electrical signals proportional to the travel of their extension cables. CETs achieve a 0.01% accuracy for up to approximately a 40 meter range, and they are relatively robust, inexpensive, and easy to use compared to other approaches. CETs do require a physical connection between the support vehicle and the mobile robot, but the support vehicle must provide power and materials, and thus a physical connection already exists.

The following section outlines the computational kinematics based on the triangulation principle to determine robot position, and we later summarize the sources and types of errors involved in this technique.

2. Kinematic Position Equations of the CET System

Figure 1 shows a schematic diagram of the sensor system. The cables from both cable-extension transducers are passed around the pulleys A and B and are attached to point P on the robot. The pulleys are required to allow the cable to exit from the transducer with the same orientation, which is a necessity, yet locate an arbitrary position P in the workspace. The size of the pulleys in the figure are exaggerated for illustrative purposes.

The coordinate system is assigned as follows: x passes through the centers of both pulleys and its origin is midway between the two pulleys. The cable length is defined as the distance of P from the point C or D on the pulley circumference as shown in Figure 1. This augmented cable length can be easily obtained by appropriately resetting the counter of the CETs

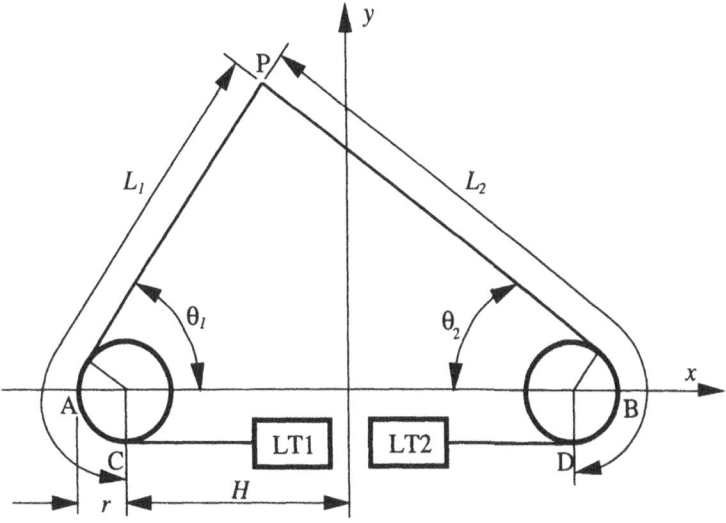

Figure 1. Schematic of the cable-extension transducer system.

or adding an offset value. Due to the pulley effect, it is impossible to find an explicit form for the position equation of point P. Thus, parametric variables, θ_1 and θ_2, which are the inner angles formed by the x-axis and the cables, are used. With these parameters, the following relationships are obtained:

$$L_1 = r(\pi - \theta_1) + (y_P - r\cos\theta_1)/\sin\theta_1, \tag{1}$$

$$y_P \cot\theta_1 = H + x_P + r/\sin\theta_1, \tag{2}$$

$$L_2 = r(\pi - \theta_2) + (y_P - r\cos\theta_2)/\sin\theta_2, \tag{3}$$

$$y_P \cot\theta_2 = H + x_P + r/\sin\theta_2. \tag{4}$$

Rearranging these equations, the position of the end point P of both cables is represented with parametric variables θ_1 and θ_2 given the cable lengths L_1 L_2 as

$$x_P = -H - r/\sin\theta_1 + y_P \cot\theta_1, \tag{5}$$

$$y_P = \{L_1 - r(\pi - \theta_1)\}\sin\theta_1 + r\cos\theta_1, \tag{6}$$

$$x_P = H + r/\sin\theta_2 - y_P \cot\theta_2, \tag{7}$$

$$y_P = \{L_2 - r(\pi - \theta_2)\}\sin\theta_2 + r\cos\theta_2. \tag{8}$$

In these equations, x_P and y_P are the dependent variables that need to be determined. The input variables are the lengths of each cable, L_1 and L_2. However, we can not reduce the equations and get explicit equation forms for x_P and y_P in terms of L_1 and L_2 due to the nonlinearities with respect to the parametric variables θ_1 and θ_2. Consequently, the problem at hand is to numerically solve the nonlinear simultaneous equations (5-8) with four unknowns x_P, y_P, θ_1, and θ_2 given L_1 and L_2. These equations can be solved with existing numerical methods, but some caution needs to be taken.

First, more than one set of solutions may exist. For example, from observation there exists two solutions that simultaneously satisfy the equations with the same L_1 and L_2: when one solution is (x_P, y_P) lying on the positive y plane, another solution may result from wrapping the cables further around the pulleys such that a solution lies in the negative y plane (it is not a mirror image due to the pulleys). However, the latter will be avoided in the real sensor system by placing constraints on the robot's motion. These undesired solutions can be avoided by proper selection of initial values for the numerical nonlinear equation solver. A proper guess of initial values near the desired solution may also accelerate numerical convergence time.

The following approximate trigonometric relationships are useful for optimal estimation of these initial values: the angles A, B, and C can be obtained from the equations:

$$\cos A = (b^2 + c^2 - a^2)/2bc, \tag{9}$$

$$\sin B = b \sin A/a, \tag{10}$$

$$\sin C = c \sin A/a. \tag{11}$$

In these equations, A is the maximum angle, and sides a, b, and c are approximated as $2(r + H)$, L_1 and L_2, respectively. These approximations result by ignoring the pulleys, and this is a reasonable approximation for an initial value calculation since the pulley dimensions are small compared to the sensor base and the cable lengths. The inner angles B and C provide the initial guesses to θ_1 and θ_2, respectively, which then allow the calculation of initial values of x_P and y_P using either equations (5) and (6) or (7) and (8). With this relatively good initial estimate, fast convergence to the desired solution is achieved.

The second possible problem arises from the fact that there are certain combinations of L_1 and L_2 for which no solution exists. This is due to the fact that the sum of the lengths of the two shorter sides of a triangle must be greater than the length of the third side. The geometry of our situation is more complex due to the pulleys. We define pairs of L_1 and L_2 for which no solution exists as *singular combinations*. Singular combinations can be ruled out using appropriate condition statements.

Figure 2 shows 3-dimensional plots of the numerical solutions obtained using a Runge-Kutta solution method. This figure depicts all four variables in terms of the cable lengths, L_1 and L_2. The flat regions at the bottom of each plot depicts the singular combinations of L_1 and L_2.

It would be quite difficult to apply the cable-extension transducer system to the tracking control problem of a high speed mobile robot due to the relatively high speed computation of the nonlinear equations necessary for real-time robot operation. Thus, in this work, the numerical solutions to the kinematic equations are obtained in an off-line manner, and look-up tables are employed for real-time robot operation. The pre-calculated data is stored into memory and is utilized as interpolation parameters. As shown in Figure 2, the solutions of the nonlinear equations are in themselves fairly nonlinear, and thus, a quadratic interpolation function has been selected for the required interpolation accuracy.

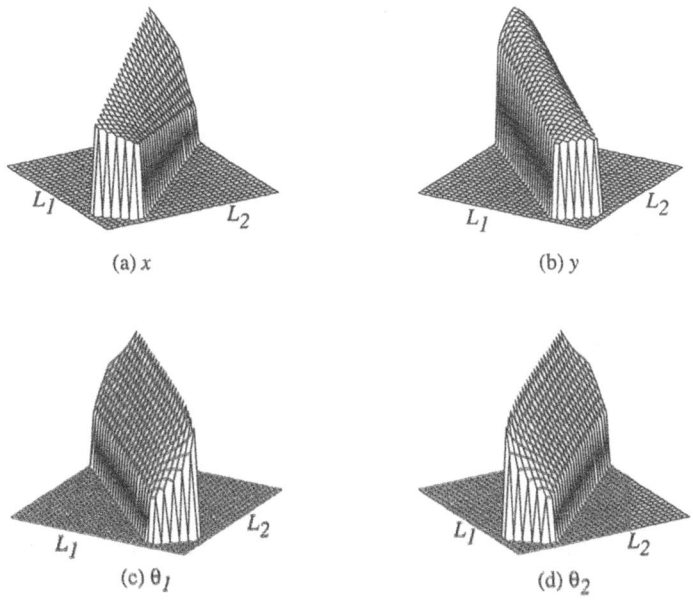

Figure 2. Three dimensional plots of the variables in terms of L_1 and L_2. The flat regions at the bottoms of each plot represents the singular combinations of the L_1, L_2 pair.

3. Quadratic Interpolation

In order to avoid the inversion of the coefficient matrix when using polynomial interpolation, a Lagrangian interpolation function is used. For one-dimensional elements with n nodes, the interpolation function is represented as

$$\mathcal{L}_p(x) \equiv \prod_{\substack{i=1 \\ i \neq p}}^{n} \frac{x - x_i}{x_p - x_i} = \frac{(x - x_1) \cdots (x - x_{p-1})(x - x_{p+1}) \cdots (x - x_n)}{(x_p - x_1) \cdots (x_p - x_{p-1})(x_p - x_{p+1}) \cdots (x_p - x_n)}$$

(12)

where Π indicates the product of the parenthetic binomial expressions. There are $n - 1$ binomial expressions in the numerator and denominator. Note that when $x = x_p$, the pth node, the numerator and denominator becomes identical and the interpolation function becomes unity. However, at any other node where $x \neq x_p$ the function goes to zero. For $n = 3$, representing a quadratic element, the following interpolation functions apply:

$$\mathcal{L}_1(x) = \frac{(x - x_2)(x - x_3)}{(x_1 - x_2)(x_1 - x_3)}, \mathcal{L}_2(x) = \frac{(x - x_1)(x - x_3)}{(x_2 - x_1)(x_2 - x_3)}, \text{ and}$$

$$\mathcal{L}_3(x) = \frac{(x - x_1)(x - x_2)}{(x_3 - x_1)(x_3 - x_2)}.$$

(13)

Note that the C^0 continuity of the field variable will also be preserved with the Lagrange interpolation function.

With this one dimensional Lagrangian interpolation function in mind, consider a two-dimensional normalized rectangular element with coordinates ξ, η as shown in Figure 3. There are nine nodes consisting of one-dimensional quadratic elements, so that quadratic variation of the field variable ϕ exists, and

$$\phi = \sum_{i=1}^{9} N_i \phi_i$$

(14)

where, using Lagrangian polynomials in the ξ and η directions,

$$N_p(\xi, \eta) = [\mathcal{L}_p(\xi)\mathcal{L}_p(\eta)]$$

(15)

at each pth node. Note again that $N_p = 1$ at node p, and that at all other nodes, the interpolation function N_p is zero. The interpolation functions are then as follows:

$$N_1(\xi, \eta) = \frac{(\xi - \xi_2)(\xi - \xi_3)(\eta - \eta_8)(\eta - \eta_7)}{(\xi_1 - \xi_2)(\xi_1 - \xi_3)(\eta_1 - \eta_8)(\eta_1 - \eta_7)},$$

$$N_2(\xi, \eta) = \frac{(\xi - \xi_1)(\xi - \xi_3)(\eta - \eta_9)(\eta - \eta_6)}{(\xi_2 - \xi_1)(\xi_2 - \xi_3)(\eta_2 - \eta_9)(\eta_2 - \eta_6)},$$

$$\vdots$$

$$N_9(\xi, \eta) = \frac{(\xi - \xi_8)(\xi - \xi_4)(\eta - \eta_2)(\eta - \eta_6)}{(\xi_9 - \xi_8)(\xi_9 - \xi_4)(\eta_9 - \eta_2)(\eta_9 - \eta_6)}. \tag{16}$$

4. Error Analysis of the Cable-Extension Transducer System

In order to properly design the linear transducer system for a wheeled mobile robot, one must understand the errors involved in measuring the robot's position. As the robot moves in its workspace, the accuracy of the measured position will not be consistent due to the kinematic nonlinearity.

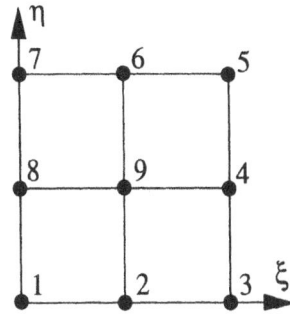

Figure 3. Two-dimensional quadratic element.

Overall, errors with cable-extension transducers may include limitations of sensor resolution, cable sagging, and cable elongation. These errors can be summarized as general uncertainties on the cable lengths, L_1 and L_2. Let us denote these uncertainty bounds on L_1 and L_2 as δL_1 and δL_2, respectively. In the design of our WMR, two assumptions can be employed to ease the error calculation; the uncertainty bounds on both cable lengths are the same, that is $\delta L_1 = \delta L_2 = \delta L$, and the effects of the pulleys can be neglected.

Figure 4 depicts an approximation to the uncertainty bounds based on the following assumption; since the uncertainty bounds are very small compared to the cable lengths, the small region ABCD in which the apex P of the triangle lies can be assumed to be a parallelogram. The maximum possible error due to the uncertainty is then a diagonal line of the parallelogram ABCD, that is $\overline{AP}$ or $\overline{BP}$, such that:

$$e_{max} = \max(\overline{AP}, \overline{BP}). \tag{17}$$

Triangles API and APE are right-angled and they are identical. Also, the angle EPI is the sum of the two inner angles θ_1 and θ_2. Therefore, the diagonal $\overline{AP}$ can be expressed as

$$AP = \frac{\delta L}{\cos\left(\frac{\theta_1+\theta_2}{2}\right)}. \tag{18}$$

Similarly, the diagonal can be expressed as

$$BP = \frac{\delta L}{\sin\left(\frac{\theta_1+\theta_2}{2}\right)}. \tag{19}$$

Let us now denote equations (18) and (19) as $\overline{AP} = f_1(\theta_1, \theta_2)$ and $\overline{BP} = f_2(\theta_1, \theta_2)$ for convenience. Note that the sum of the two inner angles of the triangle, $\theta_1 + \theta_2$, is bounded by π.

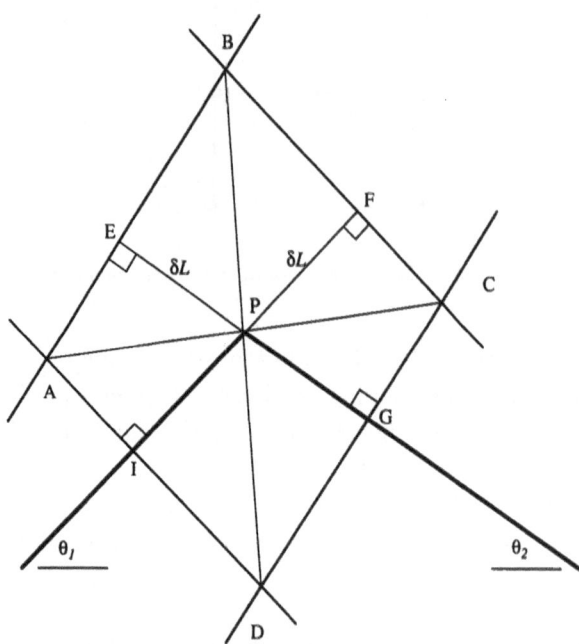

Figure 4. Error lengths included with cable lengths of transducers.

$\overline{AP}$ is unbounded when $\theta_1 + \theta_2$ goes to π, and $\overline{BP}$ is unbounded when $\theta_1 + \theta_2$ goes to 0. Also, the workspace for this sensor system is defined as the finite region that lies in the first two quadrants and is centered along the y axis. It is easily recognized that the finite maximum error occurs on the boundary of the workspace. We can now mathematically define our problem as follows:

<u>Given:</u>   Smooth functions $f_1(\theta_1, \theta_2)$ and $f_1(\theta_1, \theta_2)$:

$\Re^2 \to \Re$ within the finite workspace

<u>Find:</u> $\max[\max f_1(\theta_1, \theta_2), \max f_2(\theta_1, \theta_2)]$

<u>Subject to:</u> $0 < \theta_1 < \pi$

                $0 < \theta_2 < \pi$

                $g(\theta_1, \theta_2) \equiv 0.$

The workspace boundary is imposed by the equality constraint above. The following equations, (20) and (21), relate the Cartesian coordinate system to the inner angles of the triangle, θ_1 and θ_2, as depicted in Figure 5, and these are particularly useful since it is usually far more convenient to use Cartesian coordinates to represent a workspace boundary.

$$x_P = \frac{\sin(\theta_2 - \theta_1)}{\sin(\theta_2 + \theta_1)} H \tag{20}$$

$$y_P = \frac{2 \sin \theta_1 \sin \theta_2}{\sin(\theta_2 + \theta_1)} H \tag{21}$$

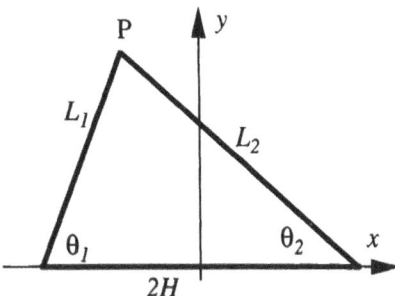

Figure 5. Simplified relation between Cartesian coordinates and the inner angles of the triangle.

Example. Semicircular workspace.

In order to realistically explore the error bound, a numerical example is in order. The example to follow considers a semicircular work space, as shown in Figure 6, which is consistent with the workspace of the wheeled mobile robot system under development. The boundary for the semicircle is split into two parts, the arc MN and the line MN.

i) Along the semi-circle MN: The boundary equations are

$$x_P^2 + (y_P - b)^2 = R^2, \tag{22}$$

$$b \le y_P \le b + R. \tag{23}$$

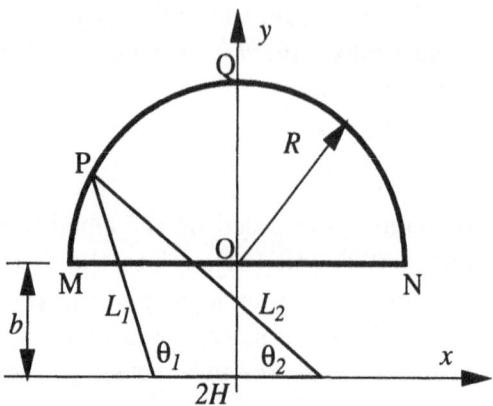

Figure 6. Semicircular workspace.

The Cartesian coordinates of the point P are related to the angles θ_1 and θ_2 from equations (20) and (21), so that equation (22) is modified to the form $g(\theta_1, \theta_2)$ as

$$g(\theta_1, \theta_2) \equiv H^2 \sin^2(\theta_2 - \theta_1) + [2H \sin \theta_1 \sin \theta_2$$
$$- b \sin(\theta_1 + \theta_2)]^2 - R^2 \sin^2(\theta_1 + \theta_2) = 0. \tag{24}$$

In order to find the maximum value for the first equation (18), let us define the following function using the Lagrange multiplier λ:

$$F_1(\theta_1, \theta_2) \equiv f_1(\theta_1, \theta_2) + \lambda g(\theta_1, \theta_2). \tag{25}$$

It is clear that the function f_1 has extreme values when the following equations are satisfied,

$$\frac{\partial F_1}{\partial \theta_1} = 0 \text{ and } \frac{\partial F_1}{\partial \theta_2} = 0. \tag{26}$$

There are now three unknowns, θ_1, θ_2, and λ, and three equations, (24) and (26) which are subject to the non-holonomic constraints (23), so that the maximum values can be algebraically determined. After solving these equations, two solutions are obtained: $\theta_1 = \theta_2$ is at the point Q, and $x_p = \pm R$ are at the points M and N shown in Figure 6. Substituting these points back into equation (18), the error $\overline{AP}$ at Q is determined as

$$\overline{AP}\,|_{atQ} = \delta L \sqrt{1 + [(b + R)/H]^2}, \tag{27}$$

and also at M and N, we have

$$\overline{AP}\,|_{atM} = \overline{AP}\,|_{atN} = \delta L \sqrt{\frac{2L_1 L_2}{L_1 L_2 + H^2 - R^2 - b^2}}, \tag{28}$$

where in each case

$$L_1 = \sqrt{b^2 + (R - H)^2} \text{ and } L_2 = \sqrt{b^2 + (R + H)^2} \tag{29}$$

Similarly the $\overline{BP}$ error can be expressed as

$$\overline{BP}\,|_{atQ} = \delta L \sqrt{1 + [H/(b + R)]^2}, \tag{30}$$

$$\overline{BP}\,|_{atM} = \overline{BP}\,|_{atN} = \delta L \sqrt{\frac{2L_1 L_2}{L_1 L_2 + H^2 - R^2 + b^2}}, \tag{31}$$

where again L_1 and L_2 are represented by equation (29).

ii) Along the straight line MN: The equations for the straight line MN are $y_P = b$ and $-R \leq x_P \leq R$. Following similar procedures as in i) above, it can be concluded that there exists three extreme points M, N, and O along the line MN. From physical intuition, we can see that the extremum at O is the minimum, so that the maximum exists at M or N. Furthermore, the extreme values at M and N are same as those given in equations (28) and (31).

From the equations (28) and (31), we recognize that $\overline{AP}$ at M or N is always greater than $\overline{BP}$ at M or N. Also, $\overline{AP}$ at Q is greater than $\overline{BP}$ at Q as long as H is less than $R + b$. This is reasonable since we do not want the cable to extend beyond the workspace. The maximum error thus occurs on the variable $\overline{AP}$. Let us employ the following parameter values to illustrate numerical results: $H = 33cm, R = 213cm$, and $b = 94cm$. With these parameters, we obtain the maximum values, $\overline{AP}\,|_{atQ} = 9.4\delta L$, and $\overline{AP}\,|_{atM} = AP\,|_{atN} = 17.3\delta L$. Now with a maximum transducer uncertainty of $.03cm$, the maximum error becomes $0.5cm$ at points M and N.

5. Conclusions

This paper has presented a target localization technique for a wheeled mobile robot based on the use of cable-extension transducers and the principle of triangulation. Physical requirements of the cable-extension transducers has resulted in the necessity to solve a system of nonlinear equations to determine robot kinematics. It was determined that solving these equations in real-time would be difficult considering the intended high speed robot operation. However, pre-calculated data points could be implemented in a look-up table and used as quadratic interpolation parameters, and such an approach is applied. A theoretical error analysis was also presented based on uncertainties of the CETs. A procedure for finding maximum errors in a bounded workspace was outlined and example results provided. An actual CET based target localization system has been fabricated, and the next step in our development involves the use of a coordinate measurement system to calibrate the CET based system.

6. Acknowledgment

The authors gratefully acknowledge the Office of New Technology and Research of the California Department of Transportation for the support of this work through the Advanced Highway Maintenance and Construction Technology Research Center at the University of California, Davis.

References

Beckerman, M. and Oblow, E. M. (1990) Treatment of Systematic Errors in the Processing of Wide-Angle Sonar Sensor Data for Robotic Navigation, *IEEE Trans. Robotics and Automation*, **Vol. 6, No. 2**, pp. 137–145.

Chen, Y. D., *et al.* (1993) Dynamic Calibration and Compensation of a 3D Laser Radar Scanning System, *IEEE Trans. Robotics and Automation*, **Vol. 9, No. 3**, pp. 318–323.

Cox, I. J. (1990) Blanche: Position Estimation for an Autonomous Robot Vehicle, *Autonomous Robot Vehicles*, Springer-Verlag, New York, pp. 221–228.

Cox, I.J. and Wilfong, G.T. (1990) *Autonomous Robot Vehicle*. Springer-Verlag, New York.

Crowley, J. L. (1989) World Modeling and Position Estimation for a Mobile Robot Using Ultrasonic Ranging, *Proc. of the IEEE Int. Conf. on Robotics and Automation*, **Vol. 2**, pp. 674–680.

Culley, G. and Buldar, R. (1988) A Free Wheel Approach to AGV Navigation, *Proc. of Int. Computers in Engineering Conference*, **Vol. 2**, pp. 267–275.

Figueroa, F. and Mahajan A. (1994) A Robust Method to Determine the Coordinates of a Wave Source for 3-D Position Sensing, *ASME Journal of Dynamic Systems, Measurement, and Control*, **Vol. 116**, pp. 505–511.

Sugimoto, G., *et al.* (1988) Practical Course Follow Performance of an AGV Without Fixed Guideways, *Proc. of the USA-Japan Sym. on Flexible Automation*, pp. 651–655.

Winters, S.E., Hong, D., Velinsky, S.A., and Yamazaki, K. (1994) A New Robotic System Concept for Automating Highway Maintenance Operations, *Proc. of SPACE '94 - ASCE Conference on Robotics for Challenging Environments*, pp. 374–382.

AN EXTENSION THEOREM FOR
KINEMATIC SYSTEMS OF EQUATIONS

P. KOVÁCS, G. HOMMEL
Institut für Technische Informatik, Sekr. FR 2-2,
Technische Universität Berlin, 10587 Berlin, Germany

Abstract

Before mathematical elimination techniques like resultant methods or the Buchberger Algorithm can be applied to the inverse or direct kinematics problem, some initial elimination steps must be carried out in other ways as the original problem requires too much effort. This initial elimination may yield extraneous roots. The correctness of the simplified system is particularly important when its symbolic solution for several single effector poses is used to identify and predict properties of best "unspecialized" solutions, thus leading the way to optimal symbolic kinematic transformations.

The article investigates the equivalence between the system of kinematic equations of single-chain manipulators and a family of significantly simpler systems of five "prominent" equations in four variables. Precise conditions for their equivalence are proven. The theorem is a prerequisite for the practical application of the Buchberger Algorithm in inverse kinematics. In addition it provides several interesting, original theoretical insights.

1 Introduction

This article investigates symbolic solutions of the inverse kinematics problem. Elementary results and the basic nomenclature are taken from [1].

Powerful universal techniques are available for the symbolic solution of the familiar inverse kinematics problem (IKP). The fundamental *Raghavan-Roth Algorithm* [2] uses - in its final stage - resultant methods to obtain a symbolic *triangulation* of the kinematic system of equations $\mathcal{K}$, i. e. a complete sequence of *characteristic equations* in echelon form whose consecutive solution yields all joint configurations for arbitrary effector poses. As an alternative approach one can use the universal symbolic solution technique for nonlinear systems of equations, the *Buchberger Algorithm* [3]. Because of the complexity of the IKP, the kinematic system of equations $\mathcal{K}$ must be simplified before it can be processed by the Buchberger Algorithm or by standard resultant methods. The breakthrough in [2] is due to an initial simplification yielding a system of 6 equations in 3 variables which is particularily well suited for an application of resultant methods. Independently, the authors suggested in a preliminary report [4] the use of a whole family of systems $\mathcal{E}_{\theta_\nu, \theta_\mu}$ (see below) of 5 equations in 4 variables as an initial simplification before application of the Buchberger Algorithm. This reduces the processing time of the algorithm for an average kinematic system of equations with numerical pose- and link-parameters from several dozen hours of CPU time to a few minutes on a SUN. A similar reduction in magnitude is obtained with respect to memory requirements.

Obviously it is necessary to investigate the correctness of any such simplified system of equations $\mathcal{E}$, i. e. it must be shown that each solution of the kinematic system of equations $\mathcal{K}$ is a solution of $\mathcal{E}$ and that each solution of $\mathcal{E}$ can be *extended* to a solution of $\mathcal{K}$, i. e. suitable values for the remaining variables can be found for any solution of $\mathcal{E}$. The goal of the present article is to rigorously prove exact conditions for the equivalence of $\mathcal{K}$ and the systems $\mathcal{E}_{\theta_\nu, \theta_\mu}$ below. The theorem

31

J.-P. Merlet and B. Ravani (eds.), Computational Kinematics, 31–40.
© 1995 *Kluwer Academic Publishers.*

yields restrictive conditions on the robot structure if some $\mathcal{E}_{\theta_\nu,\theta_\mu}$ is not fully equivalent, thus permitting an easy selection of large numbers of fully equivalent $\mathcal{E}_{\theta_\nu,\theta_\mu}$ for given manipulators. Absolute correctness of solutions is crucial for the following practical application. The two universal solution methods mentioned above can provide symbolic solutions in practice only if all pose- and link-parameters are numerical; in this case we say that $\mathcal{K}$ or $\mathcal{E}$ or the resulting triangulation respectively are (*numerically*) *specialized*. Optimal *un*specialized triangulations can be identified by investigating a sufficient number of specialized triangulations. In particular, one can predict in this way with arbitrarily high probability whether some *un*specialized characteristic equation permits essential simplifications like factorization or decomposition, thus yielding a relevant improvement of the symbolic solution, see [5; 6]. Recent results show that the latter techniques have frequent, substantial applications in the analysis of multi-loop mechanisms. The analysis of single chains is a prerequisite for this investigation. The above improvements for single chains are identified through symbolic solutions for several *generic* effector poses. For this purpose, all corresponding specialized characteristic equations must be free of rounding errors and extraneous solutions.

The $\mathcal{E}_{\theta_\nu,\theta_\mu}$ defined below are a subset of the initial equations used in [2]. The famous additional equation ("$((\tilde{p}\cdot\tilde{p})\,\tilde{l} - (2\tilde{p}\cdot\tilde{l})\,\tilde{p})_3$") introduced in [2] is not contained in $\mathcal{E}_{\theta_\nu,\theta_\mu}$. The theorem below proves that this important equation is not needed for the correctness of $\mathcal{E}_{\theta_\nu,\theta_\mu}$ usually. In the remaining cases it can be shown that the inclusion of the above equation cannot always guarantee correctness either. A single member $\mathcal{E}_{\theta_\nu,\theta_\mu}$ of the family was already used in [7] and a subset was used in [8]. On the basis of the presented results and with explicit reference to their preliminary publication in [4], Weiss [9] used [2] to reduce each $\mathcal{E}_{\theta_\nu,\theta_\mu}$ to four equations in three variables and proved extensibility criteria for these simpler systems. Their solution with modern implementations of the Buchberger Algorithm ("GB") yields a very fast way to obtain specialized symbolic solutions of the IKP. The proof of part IV) of the theorem can serve as an illustrative and interesting example, demonstrating also a simple application of the Buchberger Algorithm.

2 Notation and Theorem

The investigation is restricted to general manipulators, i.e. open or closed *single* loop kinematic chains with n prismatic or revolute joints. The following definitions are based on [1] and [5].

The relation between the joint coordinates and the effector pose of a manipulator with n joints can be expressed by the homogeneous 4×4 *basic kinematic (matrix-) equation*

$$A_1 * A_2 * ... * A_n = T \tag{1}$$

where T is the effector matrix, which specifies the effector pose and the A_i are the *arm matrices*, containing the joint variables. Each A_i is a product of four elementary *Denavit-Hartenberg transformations*, i. e. rotations or translations with respect to the local z- or x-axis

$$A_i = \text{Rot}(z, \theta_i) * \text{Trans}(z, d_i) * \text{Trans}(x, a_i) * \text{Rot}(x, \alpha_i). \tag{2}$$

The kinematic structure of the manipulator is specified by n *Denavit-Hartenberg quadruples*

$$(\theta_1, d_1, a_1, \alpha_1)\,(\theta_2, d_2, a_2, \alpha_2) ... (\theta_n, d_n, a_n, \alpha_n). \tag{3}$$

A complete set of joint values $(\theta_1, ..., d_i, ...\theta_j, ...\theta_n)$ taking the effector to a pose T is called a (*joint-*)*configuration* of T. A *partial configuration* of T is an arbitrary subset of a configuration.

The 12 non-trivial *element equations* of (1) constitute the *basic kinematic system of equations* $\mathcal{K}$, which determines completely all correct configurations for every given effector pose. The *modified kinematic (matrix-) equations* are obtained by consecutively multiplying (1) with the corresponding inverted A_i on the left or right side, which gives matrix equations of the form

$$A_i * A_{i+1} * \dots * A_j \;=\; A_{i-1}^{-1} * \dots * A_1^{-1} * T * A_n^{-1} * \dots * A_{j+1}^{-1}. \tag{4}$$

For each equation of type (4), the simultaneous inversion of the matrix expression on the left and right side yields another modification, the so called *inverse modified equation*. Each of these matrix equations yields a system of 12 non-trivial scalar element equations as above. The basic system $\mathcal{K}$ and all modified systems of equations are equivalent in the sense that they must have exactly the same solution set. The individual element equations of the different systems, however, can vary considerably.

Theorem

For a manipulator with n joints, let θ_ν and θ_μ be arbitrary revolute variables and let

$$\mathrm{Rot}(z,\theta_\nu) * L \;=\; R * \mathrm{Rot}(z,\theta_\mu)^{-1} \tag{5}$$

be a modified kinematic system of equations. Let l_{ij} and r_{ij} be the elements of L and R respectively. The system

$$l_{33} = r_{33} \tag{6}$$

$$l_{34} = r_{34} \tag{7}$$

$$l_{14}^2 + l_{24}^2 = r_{14}^2 + r_{24}^2 \tag{8}$$

$$l_{13} * l_{14} + l_{23} * l_{24} = r_{13} * r_{14} + r_{23} * r_{24} \tag{9}$$

$$l_{13} * l_{24} - l_{23} * l_{14} = r_{13} * r_{24} - r_{23} * r_{14} \tag{10}$$

(which does not contain θ_ν and θ_μ) is denoted $\mathcal{E}_{\theta_\nu,\theta_\mu}$. Then

I)   Each solution of $\mathcal{K}$ is a solution of $\mathcal{E}_{\theta_\nu,\theta_\mu}$.

II) Each solution of $\mathcal{E}_{\theta_\nu,\theta_\mu}$ can be extended to one solution of $\mathcal{K}$ except when a solution of $\mathcal{E}_{\theta_\nu,\theta_\mu}$ satisfies both of the following equations

$$l_{14}^2 + l_{24}^2 = r_{14}^2 + r_{24}^2 = 0 \tag{11}$$

$$l_{33} = r_{33} = \pm 1. \tag{12}$$

and in addition at least one of the variables of the solution vector has a complex value. In this case $\mathcal{E}_{\theta_\nu,\theta_\mu}$ still provides all correct solutions but may also contain extraneous solutions. The characteristic equations for θ_ν and θ_μ can be calculated from $\mathcal{E}_{\theta_\nu,\theta_\mu}$ alone. They are free of extraneous roots.

III) If ((11) and (12)) hold for a *real* solution of $\mathcal{E}_{\theta_\nu,\theta_\mu}$, one of θ_ν or θ_μ can be selected freely and a suitable value of the other variable exists such that $\mathcal{K}$ is satisfied, i. e. the solution can be extended as well.

IV) None of the five equations is algebraically dependent on the others in general, i.e. the omission of any equation from $\mathcal{E}_{\theta_\nu,\theta_\mu}$ can yield extraneous solutions.

Before the theorem is proven, some of its aspects and consequences are discussed.

The theorem states that every $\mathcal{E}_{\theta_\nu,\theta_\mu}$ contains the complete information of $\mathcal{K}$, except for the lower dimensional set of extraneous complex solutions which occur at poses satisfying ((11) and (12)). If ((11) and (12)) hold for a reachable effector pose T_0 the joint axes of joint ν and μ must be identical, i. e. T_0 is an *indeterminate position*. Note, that for all other systems $\mathcal{E}_{\theta_i,\theta_j}$ with $i \notin \{\nu,\mu\}$ or $j \notin \{\nu,\mu\}$, the condition ((11) and (12)) is not satisfied at T_0 (unless axes i and j coincide as well)! Consequently, parts I) and II) of the theorem guarantee that all these $\mathcal{E}_{\theta_i,\theta_j}$ provide partial configurations free of extraneous roots at T_0. In terms of algebraic geometry, the theorem states that fully equivalent systems $\mathcal{E}_{\theta_\nu,\theta_\mu}$ are minimal bases of the corresponding *second elimination ideals* of $\mathcal{K}$.

Full equivalence between $\mathcal{E}_{\theta_\nu,\theta_\mu}$ and $\mathcal{K}$ often follows directly from the manipulator structure. This is important for practical applications since correctness can be *guaranteed a priori*! Individual solutions do not need to be considered in this case. In particular, it becomes easy to select numerical specializations such that $\mathcal{E}_{\theta_\nu,\theta_\mu}$ is free of extraneous roots! For $\mathcal{E}_{\theta_\nu,\theta_{\nu+i}}$ with $i \leq 3$ simple inverse kinematic considerations show that the generation of extraneous roots is an exception. All systems $\mathcal{E}_{\theta_\nu,\theta_{\nu+1}}$ are always fully equivalent to $\mathcal{K}$ for any non-degenerated robot with $n = 6$ because ((11) and (12)) would imply a coincidence of consecutive axes. Full equivalence between $\mathcal{E}_{\theta_\nu,\theta_{\nu+2}}$ and $\mathcal{K}$ is shown easily whenever $a_\nu \neq a_{\nu+1}$ or (alternatively) $\alpha_\nu \neq \alpha_{\nu+1}$ or ($\alpha_\nu = \alpha_{\nu+1} \notin \{0, \pi\} \wedge d_{\nu+1} \neq 0$). Similar conditions result directly for $\mathcal{E}_{\theta_\nu,\theta_{\nu+3}}$ in the practically relevant case α_ν, $\alpha_{\nu+1}, \alpha_{\nu+2} \in \{0, \pm\pi/2, \pi\}$. E. g., full equivalence follows immediately if one of $\alpha_\nu, \alpha_{\nu+1}$, $\alpha_{\nu+2}$ is in $\{\pm\pi/2\}$ and the other two are in $\{0, \pi\}$. If $\alpha_\nu, \alpha_{\nu+1}, \alpha_{\nu+2} \in \{\pm\pi/2\}$ extraneous roots *may* exist only if $a_\nu = \pm d_{\nu+2}$ and additionally $a_{\nu+2} = \pm d_{\nu+1}$. If $\alpha_\nu, \alpha_{\nu+1} \in \{\pm\pi/2\}$ and $\alpha_{\nu+2} \in \{0, \pi\}$ then $(a_\nu \pm a_{\nu+1})^2 + d_{\nu+1}{}^2 \neq a_{\nu+2}{}^2$ implies full equivalence. Due to limited space these and the remaining simple cases are not proven explicitly. The proof of part IV) gives an example. When (8) and (9) are replaced by

$$l_{14}{}^2 + l_{24}{}^2 + l_{34}{}^2 = r_{14}{}^2 + r_{24}{}^2 + r_{34}{}^2 \quad \text{and} \tag{13}$$

$$l_{13} * l_{14} + l_{23} * l_{24} + l_{33} * l_{34} = r_{13} * r_{14} + r_{23} * r_{24} + r_{33} * r_{34} \tag{14}$$

respectively one obtains an equivalent system which is denoted $\mathcal{E}'_{\theta_\nu,\theta_\mu}$. Seemingly paradox, (13) and (14) are known to be less complex usually than (8) and (9).

When (5) is replaced by the *inverse* modified matrix equation, equations (6) to (10) provide a different system $\mathcal{E}_{\theta_\mu,\theta_\nu}$ instead of $\mathcal{E}_{\theta_\nu,\theta_\mu}$. Corresponding to $\mathcal{E}'_{\theta_\nu,\theta_\mu}$ we get a system $\mathcal{E}'_{\theta_\mu,\theta_\nu}$. Extensive but straightforward formula manipulation shows that all four systems are equivalent.

3 Proof of the Theorem

PART I)

This simple statement holds because the expressions on each side of the equations of $\mathcal{E}_{\theta_\nu,\theta_\mu}$ and their related geometrical quantities are invariant under rotations about the axes of joint ν and μ. It is sufficient to show that each of (6) - (10) can be derived from the original system (5). Let

$$\bar{L} = \text{Rot}(z, \theta_\nu) * L \quad \text{and} \tag{15}$$

$$\bar{R} = R * \text{Rot}(z, \theta_\mu)^{-1}, \tag{16}$$

i.e. (5) can be rewritten $\bar{L} = \bar{R}$. Each equation of $\mathcal{E}_{\theta_\nu,\theta_\mu}$ can be obtained from the following system of equations, which can be derived directly from (5) and which corresponds to (6) to (10)

$$\bar{l}_{33} = \bar{r}_{33}, \tag{(6')}$$

$$\vdots$$

$$\bar{l}_{13}\,\bar{l}_{24} - \bar{l}_{23}\,\bar{l}_{14} = \bar{r}_{13}\,\bar{r}_{24} - \bar{r}_{23}\,\bar{r}_{14} \tag{(10')}$$

The equations $\bar{l}_{33} = l_{33}$, $\bar{r}_{33} = r_{33}$, $\bar{l}_{34} = l_{34}$, $\bar{r}_{34} = r_{34}$ are obtained directly and a simple, straightforward calculation shows

$$\bar{l}_{14}{}^2 + \bar{l}_{24}{}^2 = l_{14}{}^2 + l_{24}{}^2 \tag{17}$$

$$\bar{l}_{13}\,\bar{l}_{14} + \bar{l}_{23}\,\bar{l}_{24} = l_{13}\,l_{14} + l_{23}\,l_{24} \tag{18}$$

$$\bar{l}_{13}\,\bar{l}_{24} - \bar{l}_{23}\,\bar{l}_{14} = l_{13}\,l_{24} - l_{23}\,l_{14}. \tag{19}$$

We prove the most difficult case (19) as an example. With the usual abbrevations c_ν and s_ν for $\cos(\theta_\nu)$ and $\sin(\theta_\nu)$ respectively, we get from (15) for $i \in \{1, 2, 3, 4\}$

$$\bar{l}_{1i} = c_v l_{1i} - s_v l_{2i} \quad \text{and} \quad \bar{l}_{2i} = s_v l_{1i} + c_v l_{2i} \qquad \text{and consequently} \tag{20}$$

$$\bar{l}_{13} \bar{l}_{24} - \bar{l}_{23} \bar{l}_{14} =$$

$$c_v s_v l_{13} l_{14} - s_v^2 l_{23} l_{14} + c_v^2 l_{13} l_{24} - s_v c_v l_{23} l_{24} -$$

$$s_v c_v l_{13} l_{14} - c_v^2 l_{23} l_{14} + s_v^2 l_{13} l_{24} + c_v s_v l_{23} l_{24} \quad = \quad l_{13} l_{24} - l_{23} l_{14} . \tag{21}$$

The three equations for the right sides, corresponding to (17) - (19), are proven as easily. Thus, each solution of $\mathcal{K}$ must be a solution of $\mathcal{E}_{\theta_v, \theta_\mu}$.

PART II)

The proof is divided into two sub-steps. First the existence of a unique θ_v is shown such that the element equations of the third and fourth column of (5) are satisfied. θ_μ does not appear in these columns. In the second sub-step, it is proven that a unique θ_μ exists such that the remaining equations are satisfied without changing the former ones.

Assume that an arbitrary solution of $\mathcal{E}_{\theta_v, \theta_\mu}$ is given, i.e. all joint values of the manipulator with exception of θ_v and θ_μ are determined. The given n-2 joint values determine each of the matrices L and R completely. As both of these matrices can be obtained by multiplication of the elementary transformation matrices that correspond to the respective elementary Denavit-Hartenberg transformations, all inherent laws of kinematic, homogeneous 4×4 matrices must hold for each of them. In particular, the upper left 3×3 sub-matrices will be orthogonal. At the beginning, the only information about the relationship between L and R is provided by the five equations (6) to (10).

Step II.1

θ_μ does not appear in the third and fourth column of $\bar{R}$, i.e. $r_{i3} = \bar{r}_{i3}$ and $r_{i4} = \bar{r}_{i4}$. We show now that a θ_v exists, such that $\bar{l}_{i3} = r_{i3}$ and $\bar{l}_{i4} = r_{i4}$ for $i \in \{1, 2\}$. For $i = 3$, the two equations are obtained directly from (6) and (7), independent of the value of θ_v.

We consider two cases, corresponding to the two exceptional cases (11) and (12).

Case 1 $\quad (l_{14}^2 + l_{24}^2 \neq 0 \neq r_{14}^2 + r_{24}^2)$

The two equations (9) and (10) can be written in matrix form as follows

$$\begin{bmatrix} l_{14} & l_{24} \\ l_{24} & -l_{14} \end{bmatrix} * \begin{bmatrix} l_{13} \\ l_{23} \end{bmatrix} = \begin{bmatrix} r_{14} & r_{24} \\ r_{24} & -r_{14} \end{bmatrix} * \begin{bmatrix} r_{13} \\ r_{23} \end{bmatrix} \tag{22}$$

Let $\qquad \hat{L} = \begin{bmatrix} l_{14} & l_{24} \\ l_{24} & -l_{14} \end{bmatrix} \quad \text{and} \quad \hat{R} = \begin{bmatrix} r_{14} & r_{24} \\ r_{24} & -r_{14} \end{bmatrix} \tag{23}$

The precondition of case 1 yields $\det(\hat{R}) = -r_{14}^2 - r_{24}^2 \neq 0$. Thus, (22) can be transformed into

$$(\hat{R}^{-1} * \hat{L}) * \begin{bmatrix} l_{13} \\ l_{23} \end{bmatrix} = \begin{bmatrix} r_{13} \\ r_{23} \end{bmatrix} \quad \text{and thus} \tag{24}$$

$$\det(\hat{R}^{-1} * \hat{L}) = \det(\hat{R}^{-1}) * \det(\hat{L}) = \frac{\det(\hat{L})}{\det(\hat{R})} = \frac{-l_{14}^2 - l_{24}^2}{-r_{14}^2 - r_{24}^2} = 1. \tag{25}$$

The last identity is obtained from (8). Definition (23) shows, that $\hat{R}$ and $\hat{L}$ are orthogonal and thus, $\hat{R}^{-1} * \hat{L}$ must be orthogonal as well. This and (25) proves that $\hat{R}^{-1} * \hat{L}$ represents a rotation in the plane. Thus, equation (24) directly yields a unique θ_v with $\bar{l}_{i3} = r_{i3} (= \bar{r}_{i3})$ for $i = 1, 2$. It remains to show

$$(\hat{R}^{-1} * \hat{L}) * \begin{bmatrix} l_{14} \\ l_{24} \end{bmatrix} = \begin{bmatrix} r_{14} \\ r_{24} \end{bmatrix}. \qquad \text{Explicit calculation gives} \qquad (26)$$

$$\hat{R}^{-1} = \frac{1}{-r_{14}^2 - r_{24}^2} \begin{bmatrix} -r_{14} & -r_{24} \\ -r_{24} & r_{14} \end{bmatrix} \qquad \text{and because of (8) we get} \qquad (27)$$

$$(\hat{R}^{-1} * \hat{L}) * \begin{bmatrix} l_{14} \\ l_{24} \end{bmatrix} = \hat{R}^{-1} * \begin{bmatrix} l_{14}^2 + l_{24}^2 \\ 0 \end{bmatrix} = \frac{l_{14}^2 + l_{24}^2}{-r_{14}^2 - r_{24}^2} \begin{bmatrix} -r_{14} \\ -r_{24} \end{bmatrix} = \begin{bmatrix} r_{14} \\ r_{24} \end{bmatrix} \qquad (28)$$

This gives $\bar{l}_{i4} = r_{i4} \ (= \bar{r}_{i4})$ for $i \in \{1, 2\}$. Thus, case 1 of step II.1 is proven.

The explicit characteristic equation for θ_v is obtained easily by expanding $\hat{R}^{-1} * \hat{L}$ with the aid of (27). The first column corresponds to $(\cos(\theta_v), \sin(\theta_v))$ and this gives

$$\theta_v = \text{Atan2}\left(\frac{r_{24}\, l_{14} - r_{14}\, l_{24}}{r_{14}^2 + r_{24}^2} , \quad \frac{r_{14}\, l_{14} + r_{24}\, l_{24}}{r_{14}^2 + r_{24}^2} \right). \qquad (29)$$

Whenever all joint values in a solution of $\mathcal{E}_{\theta_v, \theta_\mu}$ are real, θ_v is real as well. Note that the denominators in (29) can be replaced by 1 for real r_{14} and r_{24}, i.e. for reachable joint configurations.

Case 2 $\quad (l_{33} \neq \pm1 \neq r_{33})$

The proof is very similar to case 1. The two equations (9) and (10) can be rewritten in a form which is slightly different from (22)

$$\begin{bmatrix} l_{13} & l_{23} \\ -l_{23} & l_{13} \end{bmatrix} * \begin{bmatrix} l_{14} \\ l_{24} \end{bmatrix} = \begin{bmatrix} r_{13} & r_{23} \\ -r_{23} & r_{13} \end{bmatrix} * \begin{bmatrix} r_{14} \\ r_{24} \end{bmatrix} \qquad (30)$$

We denote the new matrices by

$$\check{L} = \begin{bmatrix} l_{13} & l_{23} \\ -l_{23} & l_{13} \end{bmatrix} \quad \text{and} \quad \check{R} = \begin{bmatrix} r_{13} & r_{23} \\ -r_{23} & r_{13} \end{bmatrix} \qquad (31)$$

The precondition of case 2 yields again $\det(\check{R}) \neq 0$ and thus equation (30) can be transformed into

$$(\check{R}^{-1} * \check{L}) * \begin{bmatrix} l_{14} \\ l_{24} \end{bmatrix} = \begin{bmatrix} r_{14} \\ r_{24} \end{bmatrix}, \quad \text{yielding} \qquad (32)$$

$$\det(\check{R}^{-1} * \check{L}) = \det(\check{R}^{-1}) * \det(\check{L}) = \det(\check{L}) / \det(\check{R}) = \qquad (33)$$

$$(l_{13}^2 + l_{23}^2) / (r_{13}^2 + r_{23}^2) = (1 - l_{33}^2) / (1 - r_{33}^2).$$

The last equality in (33) holds, because the normal (unconjugated) scalar product of the third column vector of L or R with itself must equal 1. Equation (6) gives $\det(\check{R}^{-1} * \check{L}) = 1$. We have

$$\check{R}^{-1} = \frac{1}{r_{13}^2 + r_{23}^2} \begin{bmatrix} r_{13} & -r_{23} \\ r_{23} & r_{13} \end{bmatrix} \qquad (34)$$

and as before we deduce $\check{R}^{-1} * \check{L} = \text{Rot}(z, \theta_v)$ for some real or complex θ_v. Now, (32) yields $\bar{l}_{i4} = r_{i4}$ for $i = 1, 2$. Finally we get

$$(\check{R}^{-1} * \check{L}) * \begin{bmatrix} l_{13} \\ l_{23} \end{bmatrix} = \check{R}^{-1} * \begin{bmatrix} l_{13}^2 + l_{23}^2 \\ 0 \end{bmatrix} = \frac{l_{13}^2 + l_{23}^2}{r_{13}^2 + r_{23}^2} \begin{bmatrix} r_{13} \\ r_{23} \end{bmatrix} = \begin{bmatrix} r_{13} \\ r_{23} \end{bmatrix}. \qquad (35)$$

This yields $\bar{l}_{i3} = r_{i3}$ for $i \in \{1, 2\}$. Obviously, a characteristic equation for θ_ν which is similar to (29) can be derived in this case as well. Thus, everything is proven for step II.1.

Step II.2

The multiplication of R by Rot(z, θ_μ)$^{-1}$ does not affect the third and fourth column of the resulting matrix $\bar{R}$. Consequently, a θ_μ remains to be found which guarantees that the first two columns of $\bar{L}$ and $R * $ Rot(z, θ_μ)$^{-1}$ are identical.

Let $O_{\bar{L}}$, O_L and O_R be the orthogonal upper left 3×3 sub-matrices of $\bar{L}$, L and R respectively and let $a_{\bar{L}}$ and a_R be the third column vectors of $\bar{L}$ and R. Step II.1 yields $a_{\bar{L}} = a_R$. A product of orthogonal matrices $O_{\bar{L}}^{-1} * O_R$ must be orthogonal. We show that it must be a rotation about the local z-axis. For the real case it would be sufficient to prove that matrix element (3,3) equals 1. However, as the entries of the matrices might be complex, it must be excluded that the third row or column is of the form $(z, \pm z \sqrt{-1}, 1)$ for some complex value z.

Let v be the third column vector of $O_{\bar{L}}^{-1} * O_R$. We get

$$v = O_{\bar{L}}^{-1} * a_R = (O_L^{-1} * \text{Rot}(z, \theta_\nu)^{-1}) * a_R =$$
$$(O_L^{-1} * \text{Rot}(z, \theta_\nu)^{-1} * \text{Rot}(z, \theta_\nu)) * a_L = O_L^T * a_L = (0,0,1)^T . \tag{36}$$

The scalar product of v with the other two columns of $O_{\bar{L}}^{-1} * O_R$ shows that the third row of this matrix must be of the form (0,0,1) as well. Thus $O_{\bar{L}}^{-1} * O_R = \text{Rot}(z, \theta_\mu)$ for some real or complex θ_μ. If I is the unity matrix we get

$$I = O_{\bar{L}}^{-1} * O_R * \text{Rot}(z, \theta_\mu)^{-1} , \qquad \text{and thus} \tag{37}$$

$$O_{\bar{L}} = O_R * \text{Rot}(z, \theta_\mu)^{-1}, \tag{38}$$

which completes the proof of part II) of the theorem. An explicit characteristic equation for θ_μ can be derived directly from $O_{\bar{L}}^{-1} * O_R$. It depends on θ_ν.

PART III)

Assume that a solution of $\mathcal{E}_{\theta_\nu, \theta_\mu}$ is given where all joint values are real. As all link parameters and the entries of T are real, the matrices L and R must be real as well.

First, we concentrate on the left side of (5). For real entries, the condition (11) yields $l_{14} = l_{24} = 0$. The third column vector of L must be of length one and thus

$$l_{13}^2 + l_{23}^2 = 1 - l_{33}^2. \tag{39}$$

The precondition $l_{33} = \pm 1$ gives $l_{13} = l_{23} = 0$. Elements of T do not appear on the left side of (5) and thus, $\det(O_L) = +1$ always. O_L must also be orthogonal. In order to emphasize the particular range of values of l_{33}, we introduce a new symbol $\sigma = l_{33} = \pm 1$. Now L must be of the form

$$L = \begin{bmatrix} l_{11} & l_{12} & 0 & 0 \\ l_{21} & l_{22} & 0 & 0 \\ 0 & 0 & \pm 1 & l_{34} \\ 0 & 0 & 0 & 1 \end{bmatrix} = \begin{bmatrix} l_{11} & -\sigma l_{21} & 0 & 0 \\ l_{21} & \sigma l_{11} & 0 & 0 \\ 0 & 0 & \sigma & l_{34} \\ 0 & 0 & 0 & 1 \end{bmatrix}. \tag{40}$$

Let Rot(ϕ) denote an arbitrary 2×2 rotation matrix and let

$$J = \begin{bmatrix} 1 & 0 \\ 0 & -1 \end{bmatrix}. \tag{41}$$

Then, there must be a suitable real ϕ_L such that the upper left 2×2 matrix $\tilde{L}$ of O_L is of the form

$$\tilde{L} = \text{Rot}(\phi_L) * X_\sigma \quad \text{with } X_\sigma = I \text{ if } \sigma = 1 \quad \text{and} \quad X_\sigma = J \text{ if } \sigma = -1 \tag{42}$$

Thus, the corresponding sub-matrix on the left side of (5) will be of the form $\text{Rot}(\theta_\nu) * \text{Rot}(\phi_L) * X_\sigma$. Similar considerations for the right side of (5) yield the matrix equation

$$\text{Rot}(\theta_\nu) * \text{Rot}(\phi_L) * X_\sigma = \text{Rot}(\phi_R) * X_\sigma * \text{Rot}(\theta_\mu)^{-1} \tag{43}$$

with a suitable real ϕ_R. As $\sigma = l_{33} = r_{33}$, X_σ is always the same on both sides. Let us consider the case $\sigma = -1$, i.e. $X_\sigma = J$. Multiply (43) by $J^{-1} = J$ from the left. The following factor on the right side can be simplified straightforward

$$J * \text{Rot}(\theta_\mu)^{-1} * J = J * \text{Rot}(-\theta_\mu) * J = \text{Rot}(\theta_\mu) . \tag{44}$$

Thus, (43) can be written in the form

$$\text{Rot}(\theta_\nu) * \text{Rot}(\phi_L) = \text{Rot}(\phi_R) * \text{Rot}(-\sigma \theta_\mu). \tag{45}$$

As all matrices in (45) commute, the equation can be transformed into

$$\text{Rot}(\theta_\nu + \sigma \theta_\mu) = \text{Rot}(\phi_R - \phi_L). \tag{46}$$

This shows directly the existence of infinitely many pairs (θ_ν, θ_μ) which satisfy (5) and thus, all solutions of $\mathcal{E}_{\theta_\nu, \theta_\mu}$ must be correct partial configurations in this case. As the right side of (46) is determined by the elements of L and R, it is possible to derive a characteristic equation from (46) which provides the unique value of θ_ν for every solution of $\mathcal{E}_{\theta_\nu, \theta_\mu}$ and an arbitrary selected θ_μ.

PART IV)

This part is proven by a single example which shows that the omission of any of the five equations of $\mathcal{E}_{\theta_\nu, \theta_\mu}$ changes the solution set. It can be shown that the occurence of extraneous roots after omission of one of the five equations is the general rule.

An interesting example is the simple manipulator with Denavit-Hartenberg specification

$$(\theta_1, 0, a_1, -\pi/2) (\theta_2, 0, a_2, \pi/2) (\theta_3, 0, a_3, \pi/2) (\theta_4, 0, 0, \pi/2) (\theta_5, 0, 0, \pi/2) (\theta_6, 0, 0, 0).$$

The desired statement can be proven numerically by homotopy continuation methods [8] or symbolically by the Buchberger Algorithm [3]. The latter yields any desired triangulation (the so called *lexicographical Gröbner Bases*) for any polynomial system of equations. Thus, the actual proof merely becomes a demonstration of the algorithm. As it is sufficient to investigate the outcome for a single parameter set, we can chose e. g. $a_1 = 2$, $a_2 = 7$, $a_3 = 5$. To demonstrate the generation of extraneous roots in a second example later, we choose a non-generic effector pose

$$T_0 = \begin{bmatrix} 1 & 0 & 0 & 4 \\ 0 & 1 & 0 & 6 \\ 0 & 0 & 1 & 9 \\ 0 & 0 & 0 & 1 \end{bmatrix} . \tag{47}$$

θ_2 and θ_5 are selected arbitrarily as the joint variables to be eliminated, i.e. in terms of elementary D-H transformations $\mathcal{E}_{\theta_2, \theta_5}$ must be derived from the modified kinematic matrix equation

$$\text{Rot}(z, \theta_2) * \text{Trans}(x, a_2) * ... * \text{Rot}(x, \pi/2) = \text{Rot}(x, -\pi/2)^{-1} * \text{Trans}(x, a_1)^{-1} *$$

$$\text{Rot}(z, \theta_1)^{-1} * T_0 * \text{Rot}(z, \theta_6)^{-1} * \text{Rot}(x, \pi/2)^{-1} * \text{Rot}(z, \theta_5)^{-1}. \tag{48}$$

In this case we have $\alpha_2 = \alpha_3 = \alpha_4 = \pi/2$ and thus the corresponding statements at the end of section 2 can be applied. The inequality $7 = a_2 \neq d_4 = 0$ guarantees a priori full equivalence between $\mathcal{K}$ and $\mathcal{E}_{\theta_2, \theta_5}$, without checking (11) and (12) for all individual solutions of $\mathcal{E}_{\theta_2, \theta_5}$.

We inspect $\mathcal{E'}_{\theta_2,\theta_5}$ instead of $\mathcal{E}_{\theta_2,\theta_5}$ because it consists of simpler equations. In explicit form, after setting all parameters to their numerical values, $\mathcal{E'}_{\theta_2,\theta_5}$ becomes

$$\{ s_3 s_4 = -s_1 s_6 + c_1 c_6, \quad 5 s_3 = -4 s_1 + 6 c_1, \quad 70 c_3 + 74 = -24 s_1 - 16 c_1 + 137,$$

$$5 c_3 c_4 - 7 c_4 = -9 s_1 c_6 - 9 s_6 c_1, \quad -7 s_4 c_3 - 5 s_4 = 2 s_1 c_6 + 2 s_6 c_1 - 4 s_6 - 6 c_6 \}. \tag{49}$$

The system $\mathcal{E'}_{\theta_2,\theta_5}$ can be converted to polynomial form by the trivial substitution $\sin(\theta_i) \to s_i$ and $\cos(\theta_i) \to c_i$ for $i \in \{1, 3, 4, 6\}$, where s_i and c_i are considered as independent new variables. After including the four equations

$$s_i^2 + c_i^2 = 1 \quad \text{with } i \in \{1, 3, 4, 6\} \tag{50}$$

the resulting system of equations S is equivalent to $\mathcal{E'}_{\theta_2,\theta_5}$ (cf. [5]).

The lexicographical Gröbner Basis G of S was calculated within 56 seconds on a Micro-VAX II (via a "total degree basis" and subsequent "basis conversion" with variable sequence (s_4, c_4, s_3, c_3, s_1, c_1, s_6, c_6); cf. [3; 5]. The triangulations for all other variable sequences can be obtained from the same "total degree basis" by repeating the "basis conversion" for the desired variable sequence.)

$$G = \{34441444 \, s_4^4 - 27336700 \, s_4^2 + 5405625 = 0,$$

$$17220722 \, s_4^3 - 6853775 \, s_4 + 325500 \, c_4 = 0, \quad 1718750 \, s_3^2 + 60272527 \, s_4^2 - 23687925 = 0,$$

$$17220722 \, s_4^2 - 962500 \, c_3 - 5890925 = 0,$$

$$550000 \, s_3 + 51662166 \, s_4^2 + 1430000 \, s_1 - 20271525 = 0, \tag{51}$$

$$-412500 \, s_3 - 17220722 \, s_4^2 - 715000 \, c_1 + 6757175 = 0,$$

$$258310830 \, s_3 s_4^3 - 105317625 \, s_3 s_4 + 68882888 \, s_4^3 - 85884200 \, s_4 + 66495000 \, s_6 = 0,$$

$$86103610 \, s_3 s_4^3 - 35105875 \, s_3 s_4 - 51662166 \, s_4^3 + 64413150 \, s_4 - 33247500 \, c_6 = 0\}.$$

(51) is a triangulation of S. It can be used to determine consecutively all joint variables. θ_4 can be determined e. g. from the first two equations of (51). Tangent half angle substitution with $x_4 = \tan(\theta_4/2)$ and calculation of the GCD of the two equations yields the characteristic equation for x_4

$$2325 - 560 \, x_4 - 18798 \, x_4^2 + 560 \, x_4^3 + 2325 \, x_4^4 = 0. \tag{52}$$

After solving (52), θ_3 can be determined in the same way from the third and fourth equation of (51) and subsequently θ_1 and θ_6 can be calculated from the remaining equations of (51).

Now each of the five original equations is omitted one by one from $\mathcal{E'}_{\theta_2,\theta_5}$, and the corresponding lexicographical Gröbner Bases of the five resulting systems, consisting of four equations each, are calculated. For all Gröbner Bases, the same variable ordering as above is used. We list only the resulting characteristic equations for x_4. Equation (53) is obtained when the first equation of (49) is omitted. When the second equation is omitted, the resulting system turns out to have infinitely many solutions! There is no characteristic equation for x_4! Equation (54) results from excluding the third equation of (49). (55) results identically when the fourth or fifth equation is omitted.

$$35100885375 - 108072770400 \, x_4 - 600033247160 \, x_4^2 + 995451426144 \, x_4^3 +$$

$$2773036306738 \, x_4^4 - 995451426144 \, x_4^5 - 600033247160 \, x_4^6 + \tag{53}$$

$$108072770400 \, x_4^7 + 35100885375 \, x_4^8 = 0$$

$$58125 - 14000 \, x_4 - 246750 \, x_4^2 - 39760 \, x_4^3 - 2309133 \, x_4^4 + 189280 \, x_4^5 + 4995516 \, x_4^6 -$$

$$189280 \, x_4^7 - 2309133 \, x_4^8 + 39760 \, x_4^9 - 246750 \, x_4^{10} + 14000 \, x_4^{11} + 58125 \, x_4^{12} = 0 \tag{54}$$

$$5405625 - 87724300 \, x_4^2 + 364803254 \, x_4^4 - 87724300 \, x_4^6 + 5405625 \, x_4^8 = 0 \tag{55}$$

As all polynomial degrees are greater than in (52), none of the corresponding system of equations can be equivalent to G, and thus the theorem is proven completely.

Finally we inspect the modified system $\mathcal{E}_{\theta_4,\theta_6}$ with the same link parameters and for the same pose T_0. Inspection of an additional $\mathcal{E}_{\theta_i,\theta_j}$ after $\mathcal{E}_{\theta_v,\theta_\mu}$ is necessary if the missing *univariate* characteristic equations for θ_v and θ_μ are desired. In our example, the characteristic equations for θ_2 and θ_5 resulting from $\mathcal{E}_{\theta_4,\theta_6}$ are not as interesting as the one for $x_1 = \tan(\theta_1/2)$ which can be factored

$$(485 - 2592\, x_1 - 182\, x_1^2 + 1248x_1^3 + 933\, x_1^4) *$$

$$(1201 + 12576\, x_1 + 29154\, x_1^2 - 34272\, x_1^3 + 8433\, x_1^4) = 0 \tag{56}$$

When the univariate characteristic equation for x_1 is calculated from the system (49), it turns out to be $485 - 2592\, x_1 - 182\, x_1^2 + 1248x_1^3 + 933\, x_1^4 = 0$. Thus, the last factor of (56) must consist of extraneous solutions of the robot. A closer inspection shows, that $\theta_5 = 0$ is a solution of $\mathcal{K}$ for the pose T_0. This causes joint axes 4 and 6 to be identical and consequently the statement of part II) of the theorem does not guarantee equivalence. The example shows that extraneous solutions tmay really be generated if the criterion (11) and (12) is satisfied. The roots of the last factor of (56) are

$$x_1 \approx -0.16575 \pm 0.04098\ \sqrt{-1} \quad \text{and} \quad x_1 \approx 2.19777 \pm 0.23436\ \sqrt{-1}, \tag{57}$$

i.e. all four corresponding extraneous solutions have imaginary components according to part II). $\mathcal{E}_{\theta_4,\theta_6}$ contains extraneous roots for $\theta_5 \in \{0, \pi\}$ in accordance with the statements immediately following the theorem in section 2. The same statements show that all other $\mathcal{E}_{\theta_v,\theta_{v+i}}$ with $i \leq 3$ are completely free of extraneous roots for all specializations with $a_1, a_2, a_3 \neq 0$ and $(a_1 \pm a_2)^2 \neq a_3^2$.

References

[1]　Paul, R.P.: *Robot Manipulators: Mathematics, Programming, and Control*. MIT Press, Cambridge, 1981

[2]　Raghavan, M., Roth, B.: *Kinematic Analysis of the 6R Manipulator of General Geometry*. Proc. 5th Intl. Symp. Robotics Research, MIT Press, Cambridge (1990), pp. 263 - 269

[3]　Buchberger, B.: *Applications of Gröbner Bases in Non-Linear Computational Geometry*. In: Trends in Computer Algebra (Ed. R. Janßen). Springer Lecture Notes 296 (1987) pp. 52 - 80

[4]　Kovács, P., Hommel, G.: *Reduced Equation Systems for the Inverse Kinematics Problem*. Technical Report 90-20, TU Berlin, FB 20, 1990

[5]　Kovács, P.: *Minimum Degree Solutions for the Inverse Kinematics Problem by Application of the Buchberger Algorithm*. Proc. 2nd. Intl. Conf. on Advances in Robot Kinematics, Linz (1990), Springer, Berlin (1991), pp. 326 - 334

[6]　Kovács, P., Hommel, G.: *Fast Functional Decomposition of SC-Polynomials*. Proc. IEEE Conf. Robotics and Automation, Atlanta (1993), pp. 980 - 987

[7]　Manseur, R., Doty, K.: *A Fast Algorithm for the Inverse Kinematic Analysis of Robot Manipulators*. Robotics Research 7, No. 3, (1988) pp. 52 - 63

[8]　Tsai, L.W., Morgan, A.P.: *Solving the Kinematics of the Most General Six and Five-Degree-of-Freedom Manipulators by Continuation Methods*. ASME J. of Mechanics, Transmission and Automation in Design 107 (1985) pp. 189 - 200

[9]　Weiss, J.: *Resultant Methods for the Inverse Kinematics Problem*. In: Computational Kinematics (Eds. Angeles et al.), Kluwer, Dordrecht, 1993, pp. 41 - 52

APPLICATIONS OF CLIFFORD ALGEBRAS IN ROBOTICS

B. MOURRAIN, N. STOLFI

SAFIR[†]

2004 route des Lucioles

06565 Valbonne

FRANCE

mourrain,nstolfi@sophia.inria.fr

The aim of this work is to present a tool for computations in robotics. In the first part, we present a mathematical structure which allows us to manipulate points, lines and spheres in the same environment, called a Clifford Algebra. We show how it is related to quaternions, dual quaternions and displacements and we give examples of symbolic manipulations of these objects. We illustrate this formalism, showing how the usual geometrical objects (linear spaces, spheres and displacements) can be manipulated in a same and coherent way.

1. Clifford Algebra

1.1. DEFINITION

Let $\mathbb{E}$ be a n-dimensional vector space on the field $\mathbb{K}$ ($\mathbb{K} = \mathbb{R}$ or $\mathbb{C}$). The exterior algebra $\wedge\mathbb{E}$ is the quotient of the tensor algebra $\otimes(\mathbb{E})$ by the two-sided ideal generated by ($x \otimes x = 0$, $x \in \mathbb{E}$) (see (Lang, 1980)). A linear subspace of $\mathbb{E}$ is naturally represented by an element of $\wedge\mathbb{E}$. As shown in (Barnabei *et al.*, 1985; Mourrain, 1994), two operators can be defined in this algebra (the "join" $\wedge$ for the sum of subspaces and the "meet $\vee$ for the intersection), and many properties in projective geometry (like incidence relations and intersections) can be described in terms of these two operators. If we replace the ideal ($x \otimes x$) by the ideal generated by ($x \otimes x - Q(x)$), where $Q(x)$ is a quadratic form defined on $\mathbb{E}$, we obtain the Clifford algebra.

[†]SAFIR is a common project to INRIA (Sophia-Antipolis), Univ. De Nice-Sophia-Antipolis, CNRS

J.-P. Merlet and B. Ravani (eds.), Computational Kinematics, 41–50.

© *1995 Kluwer Academic Publishers.*

Definition 1 — *The Clifford algebra of* $\mathbb{E}$ *associated with the quadratic form* Q *is the quotient of* $\otimes(\mathbb{E})$ *by the two-sided ideal generated by the relations* $x \otimes x - Q(x) = 0$ *for* $x \in \mathbb{E}$. *We will denote it by* $\mathcal{C}(\mathbb{E}, Q)$ *or* $\mathcal{C}(\mathbb{E})$.

Clifford algebra is clearly a generalization (or a deformation) of exterior algebra. If we add a parameter t in the relations $x \otimes x - t\,Q(x)$, the terms of lower degree $(t = 0)$ in t will correspond to the exterior algebra, and the rest allows us to "deform" this algebra. The vector space $\mathbb{E}$ is embedded into $\mathcal{C}(\mathbb{E})$, and "squares" of vectors of $\mathbb{E}$ are the values of the quadratic form. Note that the previous relations show that for all $x, y \in \mathbb{E}$, $x\,y + y\,x = 2Q(x, y)$ where $(x, y) \mapsto Q(x, y) = (x|y)$ is the symmetric bilinear form associated with Q.

Let us denote by $C^+(\mathbb{E})$ (resp. $C^-(\mathbb{E})$) the subspace of $\mathcal{C}(\mathbb{E})$ obtained as the quotient of $\otimes^+(\mathbb{E}) = \oplus_p \otimes^{2p}(\mathbb{E})$ (resp. $\oplus^-(\mathbb{E}) = \oplus_p \otimes^{2p+1}(\mathbb{E})$) by the same relations as before. It is easy to see that C^+ is a sub-algebra of $\mathcal{C}(\mathbb{E})$. Here are two properties of this algebra (for more details see (Crumeyrolle, 1990)) :

 - *Let* $(e_1, \ldots, e_n)$ *be an orthogonal basis of* $\mathbb{E}$ *for* Q. *Then* $\mathcal{C}(\mathbb{E})$ *is an algebra of dimension* 2^n, *and* $(1, e_{i_1} \ldots e_{i_k}, 1 \leq i_1 < \cdots < i_k \leq n)$ *is a basis of* $\mathcal{C}(\mathbb{E})$.

 - *If* $(e_1, \ldots, e_n)$ *is an orthogonal basis for* Q

$$\lambda : \wedge(\mathbb{E}) \quad \rightarrow \quad \mathcal{C}(\mathbb{E}) \tag{1}$$
$$e_{i_1} \wedge \ldots \wedge e_{i_k} \quad \mapsto \quad e_{i_1} \ldots e_{i_k}$$

   is an isomorphism independent of the orthogonal chosen basis (e_i) *of* $\mathbb{E}$.

1.2. COMPUTATIONS IN CLIFFORD ALGEBRA

For all elements $x, y \in \mathbb{E}$, the Clifford product is :

$$x\,y = x \wedge y + (x|y)$$

Proof : Let $(e_1, \ldots, e_n)$ be an orthogonal basis of $\mathbb{E}$, $x = \sum_i x_i e_i$ and $y = \sum_j y_j e_j$. We have seen that $xy + yx = 2(x|y)$, and the difference $x\,y - y\,x$ can be expanded as $2 \sum_{i,j} l_{ij} e_i e_j$, where the coefficients l_{ij} are 2×2 determinants. Therefore, we have $x\,y - y\,x = 2\,x \wedge y$ and $x\,y = \frac{1}{2}(x\,y + y\,x + x\,y - y\,x) = x \wedge y + (x|y)$. $\square$

The product in $\mathcal{C}(\mathbb{E})$ is neither commutative, nor anti-commutative. Any element A in $\mathcal{C}(\mathbb{E})$ can be written as :

$$A = \sum_{k=0}^{n} A_k$$

where $A_k = \langle A \rangle_k \in \wedge^k \mathbb{E}$ is an extensor of length k. Here are the classical rules for computations in $\mathcal{C}(\mathbb{E})$:

1. $A_r \wedge B_s = (-1)^{rs} B_s \wedge A_r$
2. $(x|x_1 \wedge \cdots \wedge x_r) = \sum_{k=1}^{r} (-1)^{k+1}(x|x_k)x_1 \wedge \cdots \wedge x_{k-1} \wedge x_{k+1} \wedge \cdots \wedge x_r$
3. $(x_1 \wedge \cdots \wedge x_r|y_1 \wedge \cdots \wedge y_s) = (x_1 \wedge \cdots \wedge x_{r-1}|(x_r|y_1 \wedge \cdots \wedge y_s))$
4. $(A_r|B_s) = (-1)^{r(s+1)}(B_s|A_r)$
5. $x A_r = (x|A_r) + x \wedge A_r$
6. $(A_r|B_r) = \langle A_r B_r \rangle_0$

For more details, see (Hestenes, 1987) or (Dress and Havel, 1991). These rules define by linearity $(A|B)$ for any elements $A, B \in \mathcal{C}(E)$. Note that it is not necessarily a scalar. Applying these rules, we find, for instance, that $\forall x, y, z, t \in \mathbb{E}$:

$$
\begin{aligned}
x\,y\,z &= x\,(y \wedge z + (y|z)) \\
&= (x|y \wedge z) + x \wedge y \wedge z + x(y|z) \\
&= x \wedge y \wedge z + (x|y)z - (x|z)y + x(y|z)
\end{aligned}
$$

$$
\begin{aligned}
x\,y\,z\,t &= x \wedge y \wedge z \wedge t \\
&\quad + x \wedge y(z|t) - x \wedge z(y|t) + x \wedge t(y|z) + y \wedge z(x|t) - y \wedge t(x|z) \\
&\quad + z \wedge t(x|y) \ + (x|y)(z|t) - (x|z)(y|t) + (x|t)(y|z)
\end{aligned}
$$

$$
(x \wedge y|y \wedge x) = (x|(y|(y \wedge x))) = x^2 y^2 - (x|y)^2 = \begin{vmatrix} (y|y) & (y|x) \\ (x|y) & (x|x) \end{vmatrix}
$$

This last determinant corresponds in fact, to the square of the surface of $x \wedge y$. It can be generalized to :

$$
(x_1 \wedge \cdots \wedge x_r|y_1 \wedge \cdots \wedge y_r) = (-1)^{\frac{r(r-1)}{2}} \begin{vmatrix} (x_1|y_1) & \cdots & (x_1|y_r) \\ \vdots & \vdots & \vdots \\ (x_r|y_1) & \cdots & (x_r|y_r) \end{vmatrix}
$$

It is the Gramm-Schmidt determinant.

1.3. CLIFFORD GROUP

Let $\mathcal{C}(\mathbb{E})^*$ be the multiplicative group of invertible elements of $\mathcal{C}(\mathbb{E})$. We construct the following map[1], for $g \in \mathcal{C}(\mathbb{E})^*$:

$$
\begin{aligned}
\psi_g : \mathcal{C}(\mathbb{E}) &\rightarrow \mathcal{C}(\mathbb{E}) \\
x &\rightarrow \overline{(g\tilde{x}g^{-1})}
\end{aligned}
$$

[1]Usually the following map is constructed:

$$
\begin{aligned}
\phi_g : \mathcal{C}(\mathbb{E}) &\rightarrow \mathcal{C}(\mathbb{E}) \\
x &\mapsto s(g)\,x\,g^{-1}
\end{aligned}
$$

where s is an automorphism of $\mathcal{C}(\mathbb{E})$ defined by $s(x) = -x$, $\forall x \in \mathbb{E}$.

where ~ is the Hodge operator defined by :

$$
\begin{array}{rcl}
\tilde{} \ : \wedge(\mathbb{E}) & \longrightarrow & \wedge(\mathbb{E}) \\
1 & \longmapsto & \epsilon = e_1 \wedge \cdots \wedge e_n \\
e_{i_1} \wedge \ldots \wedge e_{i_k} & \longmapsto & (-1)^{i_1+\cdots+i_k-k(k+1)/2} e_{j_1} \wedge \ldots \wedge e_{j_{n-k}}
\end{array}
$$

where $i_1 < \ldots < i_k$, $j_1 < \ldots < j_{n-k}$ and $\{i_1,\ldots,i_k\} \cup \{j_1,\ldots,j_{n-k}\} = \{1,\ldots,n\}$. Therefore we have for instance $U \wedge \tilde{U} = \epsilon \left(\sum_{I=(i_1,\ldots,i_k)} U_I^2 \right)$.
We denote by G the subgroup of elements $g \in \mathcal{C}(\mathbb{E})^*$ such that $\forall x \in \mathbb{E}, \psi_g(x) \in \mathbb{E}$. Let $O(\mathbb{E})$ be the group of isometries of $\mathbb{E}$ and $SO(\mathbb{E})$ the sub-group of isometries of determinant equal to 1. We have the following properties:

- ψ_g *is an isometry* $(g \in G)$.
  - If Q is not a degenerate quadratic form: let $(e_1,\ldots,e_n)$ be an orthonormal basis of $(\mathbb{E}, Q)$. Using property (6), we get :

$$
(g\tilde{x}g^{-1}|g\tilde{x}g^{-1}) = \langle g\tilde{x}g^{-1} g\tilde{x}g^{-1}\rangle_0 = \langle g\tilde{x}\tilde{x}g^{-1}\rangle_0
$$

The vector x can be expressed as $x = \sum_{i=1}^{n} x_i e_i$. Then $\tilde{x}\tilde{x} = \sum_{i,j} x_i x_j \tilde{e}_i \tilde{e}_j$. A simple computation shows that $\tilde{e}_i \tilde{e}_j = e_i \wedge e_j$ if $i \neq j$, and $\tilde{e}_i \tilde{e}_j = (-1)^{\frac{n(n-1)}{2}}$ if $i = j$. So we obtain :

$$
\begin{aligned}
(g\tilde{x}g^{-1}|g\tilde{x}g^{-1}) &= (-1)^{\frac{n(n-1)}{2}}(x|x) + \sum_{i \neq j} x_i x_j \langle g.e_i \wedge e_j.g^{-1}\rangle_0 \\
&= (-1)^{\frac{n(n-1)}{2}}(x|x)
\end{aligned}
$$

Now remark that for an extensor A_{n-1} of length $n-1$, we have the result : $(\tilde{A}_{n-1}|\tilde{A}_{n-1}) = (-1)^{\frac{(n-1)n}{2}}(A_{n-1}|A_{n-1})$ (just compute $(e_{i_1} \ldots e_{i_{n-1}}|e_{i_1} \ldots e_{i_{n-1}}) = (-1)^n (e_{i_1} \ldots e_{i_{n-2}}|e_{i_1} \ldots e_{i_{n-2}}) = \cdots = (-1)^{\frac{n(n-1)}{2}}$, where $(i_1,\ldots,i_{n-1}) \subset \{1,\ldots,n\}$). Then with this property we get $(g\tilde{\tilde{x}}g^{-1}|g\tilde{\tilde{x}}g^{-1}) = (-1)^{\frac{n(n-1)}{2}}(g\tilde{x}g^{-1}|g\tilde{x}g^{-1}) = (x|x)$.
  - if Q is degenerate : suppose that the rank of Q is n_1 and $\mathbb{E} = E_1 \oplus F$, where $dim(E_1) = n_1$, and F is the radical of $(\mathbb{E}, Q)$. Then the Clifford algebra is isomorphic to $\mathcal{C}(E_1, Q_1) \otimes (\wedge F)$. The Hodge operator respects this structure : $\overline{\mathcal{C}(E_1,Q_1) \otimes (\wedge F)} = \overline{\mathcal{C}(E_1)} \otimes \overline{(\wedge F)}$. Thus applying the previous point, we get the result.

- *Let* $g \in \mathbb{E} \cap G$ *such that* $Q(g) \neq 0$. *Then,* ψ_g *is the symmetry relative to* $g^{\perp}$. We first notice that : $g\tilde{x} = (g|\tilde{x}) + g \wedge \tilde{x}$ and $\tilde{x}g = (\tilde{x}|g) + \tilde{x} \wedge g$. In addition, $\tilde{x}g = (-1)^{n-2}(g|\tilde{x}) + (-1)^{n-1}g \wedge \tilde{x}$ (applying (1) and (4)), so $(-1)^{n-1}\tilde{x}g = -(g|\tilde{x}) + g \wedge \tilde{x}$ and, adding the two equalities, we get

$g\tilde{x} = 2g \wedge \tilde{x} + (-1)^n \tilde{x}g$. Let us compute $\psi_g(x)$ to make the symmetry relatively to $g^\perp$ appear.

$$
\begin{aligned}
(-1)^n g\tilde{x}g^{-1} &= (-1)^n g\tilde{x}\frac{g}{Q(g)} = \frac{(-1)^n}{Q(g)}(2g \wedge \tilde{x} + (-1)^n \tilde{x}g)g \\
&= \tilde{x} + (-1)^n \frac{2}{Q(g)}(g \wedge \tilde{x})g
\end{aligned}
$$

For the computation of $g \wedge \tilde{x} = \sum_i g_i e_i \wedge \sum_j x_j \tilde{e}_j$, Note that $e_i \wedge \tilde{e}_j = 0$ if $i \neq j$, and $e_i \wedge \tilde{e}_i = (-1)^{i-1}(-1)^{i-1}e_1 \dots e_n = e_1 \dots e_n$. Thus $e_i \wedge \tilde{e}_j = \delta_{i,j}\epsilon$ and $g \wedge \tilde{x} = \sum_{i=1}^n g_i x_i \epsilon = (g|x)\epsilon$.

$$
\begin{aligned}
\psi_g(x) &= \tilde{x} + (-1)^n \frac{2}{Q(g)}\overline{(g \wedge \tilde{x})g} = x + (-1)^n \frac{2}{Q(g)}(g|x)\overline{\tilde{e_1 \dots e_n}g} \\
&= x + (-1)^n \frac{2}{Q(g)}(g|x)\overline{(-1)^{n-1}(\tilde{g|e_1 \dots e_n})} \\
&= x - \frac{2}{Q(g)}(g|x)\overline{\sum_{k=1}^n (-1)^{k+1}(g|e_k)e_1 \dots e_{k-1}e_{k+1} \dots e_n} \\
&= x - \frac{2}{Q(g)}(g|x)\overline{\sum_{k=1}^n (g|e_k)\tilde{e}_k} = x - \frac{2}{Q(g)}(g|x)\sum_{k=1}^n (g|e_k)e_k \\
&= x - \frac{2}{Q(g)}(g|x)g
\end{aligned}
$$

Thus, ψ_g is the symmetry relative to the hyperplane $g^\perp$. When Q is non-degenerate, any element of $O(\mathbb{E})$ can be obtained by compositions of such symmetries.

When Q is degenerated, if we decompose $\mathbb{E}$ in $E_1 \oplus F$ as in the previous point, we obtain the isometries whose restriction on F is the identity.

2. Examples of Clifford Algebras

2.1. QUATERNIONS

Let us consider the case $\mathbb{E} = \mathbb{E}_3 = \mathbb{K}^3$ and $Q(a,b,c) = a^2 + b^2 + c^2$. Denote by (e_1, e_2, e_3) the orthogonal basis of $\mathbb{E}_3$. The Clifford algebra $\mathcal{C}(\mathbb{E}_3)$ has $(1, e_1, e_2, e_3, e_1e_2, e_1e_3, e_2e_3, e_1e_2e_3)$ for basis. With the previous definition, we find that $G^+ = G \cap \mathcal{C}^+(\mathbb{E}_3) = \langle 1, e_1e_2, e_1e_3, e_2e_3 \rangle$. Let us define $\mathbf{i} = e_2e_3$, $\mathbf{j} = e_3e_1$, $\mathbf{k} = e_1e_2$. These elements are such that :

$$\mathbf{ij} = -\mathbf{ji} = -\mathbf{k}, \ \mathbf{ik} = -\mathbf{ki} = \mathbf{j}, \ \mathbf{jk} = -\mathbf{kj} = -\mathbf{i}, \ \mathbf{i}^2 = \mathbf{j}^2 = \mathbf{k}^2 = -1$$

This is the well known *quaternions algebra* $\mathbb{H}$. Note that this space has a natural automorphism, which is called the conjugation. For any quaternion $\mathbf{q} = q_0 + q_1\mathbf{i} + q_2\mathbf{j} + q_3\mathbf{k}$, we define $\bar{\mathbf{q}} = q_0 - q_1\mathbf{i} - q_2\mathbf{j} - q_3\mathbf{k}$. We have the following properties:

$$\mathbf{q}\bar{\mathbf{q}} = \bar{\mathbf{q}}\mathbf{q} = (\mathbf{q}|\bar{\mathbf{q}}) = \langle \mathbf{q}\bar{\mathbf{q}} \rangle_0$$

We can describe isometries in $\mathbb{E}$ with computations in $\mathbb{H}$. Examples :

– *A rotation of angle α and axis $w = w_1\,e_1 + w_2\,e_2 + w_3\,e_3 \in \mathbb{E}_3$ with $\|w\| = 1$ is defined by the quaternion*

$$\mathbf{q} = \cos(\frac{\alpha}{2}) + w_1 \sin(\frac{\alpha}{2})\mathbf{i} + w_2 \sin(\frac{\alpha}{2})\mathbf{j} + w_3 \sin(\frac{\alpha}{2})\mathbf{k} = q_0 + q_1\mathbf{i} + q_2\mathbf{j} + q_3\mathbf{k}.$$

The image of the vector $X \in \mathbb{E}_3$ by the rotation is $X' = \psi_{\mathbf{q}}(X)$ where $\mathbf{q}^{-1} = \bar{\mathbf{q}} = q_0 - q_1\mathbf{i} - q_2\mathbf{j} - q_3\mathbf{k}$. Expanding the product $\psi_{\mathbf{q}}(X)$, we find the usual matrix (expressed relative to e_1, e_2, e_3) :

$$\begin{pmatrix} q_0^2 + q_1^2 - q_2^2 - q_3^2 & 2q_1q_2 - 2q_0q_3 & 2q_1q_3 + 2q_2q_0 \\ 2q_1q_2 + 2q_0q_3 & q_0^2 - q_1^2 + q_2^2 - q_3^2 & 2q_2q_3 - 2q_0q_1 \\ 2q_1q_3 - 2q_2q_0 & 2q_2q_3 + 2q_0q_1 & q_0^2 - q_1^2 - q_2^2 + q_3^2 \end{pmatrix}$$

– $\forall u, v \in \mathbb{E}$, ψ_{uv} is the rotation axis $u \times v$, and the angle α is defined by $\cos(\frac{\alpha}{2}) = \frac{(u|v)}{\|u\|\|v\|}$. The product $u \times v$ is the vector product used in physics [2]. note that $uv = (u|v) + u \wedge v = (u|v) + (u \times v)_1\mathbf{i} + (u \times v)_2\mathbf{j} + (u \times v)_3\mathbf{k}$. We can interpret this geometrically as follows. Let u, v be two unitary vectors of the space $\mathbb{E}_3$. Let θ be the angle between u and v. Then ψ_{uv} is the rotation of angle $2\,\theta$ around the axis orthogonal to the two vectors.

2.2. DUAL QUATERNIONS

Let us consider the case $\mathbb{E} = \mathbb{E}_4$ and $Q(a, b, c, d) = a^2 + b^2 + c^2$ which is a degenerate quadratic form. Let $(e_0, \ldots, e_3)$ be the canonical basis of $\mathbb{E}$. The sub-algebra $C^+(\mathbb{E}_4, Q)$ is generated by the 8 elements :

$$1,\ e_0e_1,\ e_0e_2,\ e_0e_3,\ e_1e_2,\ e_1e_3,\ e_2e_3,\ e_0e_1e_2e_3.$$

Let us define $\epsilon = e_0e_1e_2e_3$. A simple computation shows that $\epsilon^2 = 0$. As for the quaternions algebra, we define as before : $\mathbf{i} = e_2e_3$, $\mathbf{j} = e_3e_1$, $\mathbf{k} = e_1e_2$ and we get :

$$\epsilon\,\mathbf{i} = \mathbf{i}\epsilon = -e_0e_1, \quad \epsilon\,\mathbf{j} = \mathbf{j}\epsilon = -e_0e_2, \quad \epsilon\,\mathbf{k} = \mathbf{k}\epsilon = -e_0e_3.$$

Let us denote $\mathbb{H} = \langle 1, \mathbf{i}, \mathbf{j}, \mathbf{k} \rangle$ and $\epsilon\,\mathbb{H} = \langle \epsilon, e_0e_1, e_0e_2, e_0e_3 \rangle$, then $C^+(\mathbb{E}_4) = \mathbb{H} + \epsilon\,\mathbb{H}$.

Rotations: The image of $x \in \mathbb{E}$ by the rotation R about the axis (O, w) (where $O = e_0$ is the origin and $w = w_1e_1 + w_2e_2 + w_3e_3$ is a unitary vector) throught the angle α is the point $x' = \psi_{\mathbf{q}}(x)$, where

$$\mathbf{q} = \cos(\frac{\alpha}{2}) + w_1 \sin(\frac{\alpha}{2})\mathbf{i} + w_2 \sin(\frac{\alpha}{2})\mathbf{j} + w_3 \sin(\frac{\alpha}{2})\mathbf{k} = q_0 + q_1\mathbf{i} + q_2\mathbf{j} + q_3\mathbf{k}.$$

---

[2]It corresponds to $u \times v = \widetilde{u \wedge v}$.

It is the same thing as in the previous section (quaternions). We do not use the part $\epsilon\,\mathbb{H}$ of the algebra. here. For a rotation of angle α around the axis (a,b) $(a,b \in \mathbb{E}_4)$, we denote by w the unitary vector along (a,b) and u,v two unitary vectors, orthogonal to w such that the angle between u and v is $\frac{\alpha}{2}$. This rotation R is the conjugate by the translation of vector $\vec{Oa}$ of the rotation R_0 of axis (O,w) and angle α: $R = T_{\vec{Oa}} R_0 T_{-\vec{Oa}}$. As we have $e_0 a = \epsilon(a_1\mathbf{i} + a_2\mathbf{j} + a_3\mathbf{k})$, this rotation corresponds to the quaternion

$$
\begin{aligned}
\mathbf{q} &= (1 - \tfrac{1}{2}e_0 a) u\, v (1 + \tfrac{1}{2}e_0 a) = u\,v - \frac{1}{2}(e_0\,a\,u\,v - u\,v\,e_0\,a) \\
&= u\,v + \sin(\frac{\alpha}{2})\,\epsilon\, a \wedge w
\end{aligned}
$$

(we use the fact that in $\mathbb{E}_3 = \langle e_1, e_2, e_3 \rangle$, we have $\tilde{u}\,\tilde{v} - \tilde{v}\,\tilde{u} = -2\,u \wedge v$).

Translations: The image of x by the translation of vector $T = (t_1, t_2, t_3)$, is given by $\psi_t(x)$ where $\mathbf{t} = 1 + \frac{1}{2}\epsilon\,(t_1\mathbf{i} + t_2\mathbf{j} + t_3\mathbf{k})$. Here we only use the part $\epsilon\,\mathbb{H}$ of $\mathcal{C}^+(\mathbb{E}_4, Q)$.

Rotations and Translations: The image of x by the displacement $[R, T]$ is the composition of the rotation R by the translation T. We use the application : $\psi_{t\,q}(x)$ where $\mathbf{t\,q} = (1 + \frac{1}{2}\epsilon\,(t_1\mathbf{i} + t_2\mathbf{j} + t_3\mathbf{k}))(q_0 + q_1\mathbf{i} + q_2\mathbf{j} + q_3\mathbf{k})$ with the same notation as in the previous paragraph. Expanding the product, we find :

$$
\begin{aligned}
\mathbf{t\,q} = {}& (q_0 + q_1\mathbf{i} + q_2\mathbf{j} + q_3\mathbf{k}) + \\
& \tfrac{1}{2}\epsilon\,(-(q_1 t_1 + q_2 t_2 + q_3 t_3) + (q_0 t_1 + q_2 t_3 - q_3 t_2)\mathbf{i} \\
& + (t_2 q_0 - t_3 q_1 + t_1 q_3)\mathbf{j} + (t_3 q_0 + t_2 q_1 - t_1 q_2)\mathbf{k})
\end{aligned}
$$

Let us denote : $d_0 = \frac{1}{2}(q_1 t_1 + q_2 t_2 + q_3 t_3)$, $d_1 = \frac{1}{2}(-t_1 q_0 - t_3 q_2 + t_2 q_3)$, $d_2 = \frac{1}{2}(-t_2 q_0 + t_3 q_1 - t_1 q_3)$, $d_3 = \frac{1}{2}(-t_3 q_0 - t_2 q_1 + t_1 q_2)$ so that $\mathbf{t\,q} = \mathbf{q} - \epsilon\,(d_0 + d_1\mathbf{i} + d_2\mathbf{j} + d_3\mathbf{k}) = \mathbf{q} - \epsilon\,\mathbf{d}$. Let $\mathbf{s} = t_1\mathbf{i} + t_2\mathbf{j} + t_3\mathbf{k}$, so that $\mathbf{d} = \mathbf{s\,q}$ and

$$
q_0 d_0 + q_1 d_1 + q_2 d_2 + q_3 d_3 = (\mathbf{d}|\bar{\mathbf{q}}) = \langle \mathbf{d}\bar{\mathbf{q}} \rangle_0 = \langle \mathbf{s}\mathbf{q}\bar{\mathbf{q}} \rangle_0 = \langle \mathbf{s} \rangle_0 (\mathbf{q}|\bar{\mathbf{q}}) = 0
$$

Conversely, given $\mathbf{q} \in \mathbb{H}$ and $\mathbf{d} \in \mathbb{H}$ such that $(\bar{\mathbf{q}}|\mathbf{q}) = 1$ and $(\mathbf{d}|\bar{\mathbf{q}}) = 0$, $\psi_{\mathbf{q}+\epsilon\mathbf{d}}$ represents a displacement of $\mathbb{E}_4$ (for $\mathbf{s} = \mathbf{d}\,\bar{\mathbf{q}}$ is in $\langle \mathbf{i}, \mathbf{j}, \mathbf{k} \rangle$). If $\bar{\mathbf{q}}\mathbf{q} = 1$, we easily check that the inverse of $\mathbf{q} + \epsilon\mathbf{d}$ is $\bar{\mathbf{q}} + \epsilon\,\bar{\mathbf{d}}$, for we have $\mathbf{d}\,\bar{\mathbf{q}} + \mathbf{q}\,\bar{\mathbf{d}} = 0$. Expanding the product : $\overline{(\mathbf{t\,q})\tilde{x}(\mathbf{t\,q})^{-1}}$, where $x = e_0 + x_1 e_1 + x_2 e_2 + x_3 e_3$, we find the matrix $[\Psi]$ (expressed relative to (e_0, e_1, e_2, e_3)) :

$$
\begin{pmatrix}
1 & 0 & 0 & 0 \\
2(d_0 q_1 - d_3 q_2 + d_2 q_3 - d_1 q_0) & q_0^2 + q_1^2 - q_2^2 - q_3^2 & 2q_1 q_2 - 2q_0 q_3 & 2q_1 q_3 + 2q_2 q_0 \\
2(d_3 q_1 + d_0 q_2 - d_1 q_3 - d_2 q_0) & 2q_1 q_2 + 2q_0 q_3 & q_0^2 - q_1^2 + q_2^2 - q_3^2 & 2q_2 q_3 - 2q_0 q_1 \\
2(d_1 q_2 - d_2 q_1 + d_0 q_3 - d_3 q_0) & 2q_1 q_3 - 2q_2 q_0 & 2q_2 q_3 + 2q_0 q_1 & q_0^2 - q_1^2 - q_2^2 + q_3^2
\end{pmatrix}
$$

Example : we want to compose the rotation throught the angle π, about the e_1 with the translation of vector (t_1, t_2, t_3). We find $q_1 = 1, q_0 = q_2 = q_3 = 0$, and $d_0 = \frac{1}{2}t_1, d_2 = \frac{1}{2}t_3, d_3 = -\frac{1}{2}t_2$. Thus, for $x = e_0 + x_1 e_1 + x_2 e_2 + x_3 e_3$, we get $\psi_{\mathbf{tq}}(x) = (x_1 + t_1)e_1 - (x_2 + t_2)e_2 - (x_3 + t_3)e_3 + e_0$.

2.3. SPACE OF SPHERES

Let us denote by S the space of spheres, in which the sphere in the 3-dimensional affine space $\mathbb{A}^3$ with equation $u_0(x^2 + y^2 + z^2) - 2u_1 x - 2u_2 y - 2u_3 z + u_4 = 0$, represented by the point with coordinates $(u_0 : u_1 : u_2 : u_3 : u_4)$ in the four-dimensional projective space $\mathbb{P}^4$. We will denote by ω the special sphere $(0 : 0 : 0 : 0 : 1)$, and by $\dot{S}$ the normalized sphere $(1 : v_1 : v_2 : v_3 : v_4)$. The set of sphere-points (spheres with null radius), is defined by the quadratic form:

$$Q(x) = u_1^2 + u_2^2 + u_3^2 - u_0 u_4$$

This quadratic form Q induces a scalar product between spheres such that: for S (resp. S') defined by $(u_0 : u_1 : u_2 : u_3 : u_4)$ (resp. $(u_0' : u_1' : u_2' : u_3' : u_4')$) , the scalar product is: $(S|S') = u_1 u_1' + u_2 u_2' + u_3 u_3' - \frac{1}{2}(u_0 u_4' + u_4 u_0')$. We can embed $\mathbb{E}_4$ in S as follows: If we denote by $(1, x, y, z)$ the coordinates of a point P in $\mathbb{E}_4$, then the coordinates of the normalized sphere-point $\dot{P}$ associated to P will be $(1 : x : y : z : x^2 + y^2 + z^2)$.

A lot of geometric properties can be described with this formalism (for more details, see (Mourrain and Stolfi, 1994)). One of them is the following : *For all sphere S, and for all point P, $P \in S$ iff $(\dot{P}|\dot{S}) = 0$.*

Let us construct the Clifford algebra on $C(\mathbb{E}_5, Q)$ with this quadratic form. Let $(e_0, \ldots, e_4)$ be the canonical basis of $\mathbb{E}_5$. In the hyperplane H : $(u_4 = 0)$, the quadratic form is $Q(u_0, u_1, u_2, u_3) = u_1^2 + u_2^2 + u_3^2$. It is degenerate. Identifying H with $\mathbb{E}_4$, we find the dual quaternions. So, with the same notations as before, we know that we can describe the isometries of $O(\mathbb{E})$ with applications ψ_g with g belonging to $\mathbb{H} + \epsilon \mathbb{H}$. We have the following properties:

- For any $g \in \mathbb{H} + \epsilon \mathbb{H}$ and any sphere S, $\psi_g(S)$ is sphere, image of S by the displacement associated with g.
  Consider $S \in S$. It can be written $\dot{S} = \dot{C} - R^2 \omega$, where C is the sphere-point associated with the center C of the sphere, and R the radius of this sphere.

$$\psi_g(\dot{S}) = \psi_g(\dot{C}) - R^2 \psi_g(\omega)$$

Let us compute $\psi_g(\omega)$: for all $g \in \mathbb{H} + \epsilon \mathbb{H}$, we have $g\omega = \omega g$, so $g\omega g^{-1} = \omega$.

The center $\dot{C}$ is the point of intersection of the line (ω, C) and the quadric Q. So using linearity, $\psi_g(\dot{C}) \in (\psi_g(\omega), \psi_g(C)) \cap \psi_g(Q)$, then $\psi_g(\dot{C}) \in (\omega, \psi_g(C)) \cap Q$ (for ψ_g is an isometry $\psi_g(Q) = Q$). Thus $\psi_q(\dot{C}) = \psi_q(C)$ and $\psi_q(\dot{S}) = \overline{\psi_q(C)} - R^2\omega$ is the sphere centered in $\psi_g(C)$, of radius R.

- Let $A, B, C, D \in \mathbb{A}^3$ be 4 *non-coplanar points, the sphere S passing through these 4 points is* $\overline{\dot{A} \wedge \dot{B} \wedge \dot{C} \wedge \dot{D}}$.

For any sphere S, we have $|\dot{A}, \dot{B}, \dot{C}, \dot{D}, S| = -(\overline{\dot{A} \wedge \dot{B} \wedge \dot{C} \wedge \dot{D}}|S)$, so that $\overline{\dot{A} \wedge \dot{B} \wedge \dot{C} \wedge \dot{D}}$ is "orthogonal" to $\dot{A}, \ldots, \dot{D}$. It is the sphere passing through these four points.

- Let A, B, C, D be 4 *non-coplanar points of* $\mathbb{A}^3$. *Then the radius R of the sphere passing through these 4 points is given by :*

$$R^2 = \frac{1}{4}\frac{(\dot{A} \wedge \cdots \wedge \dot{D}|\dot{A} \wedge \cdots \wedge \dot{D})}{|A, \ldots, D|^2}.$$

Any sphere S can be decomposed as $S = \lambda(\dot{C} - R^2\omega)$. As $(\dot{C}|\omega) = -\frac{1}{2}$ and $(\dot{C}|\dot{C}) = (\omega|\omega) = 0$, we have $(S|\omega) = -\frac{\lambda}{2}$ and $(S|S) = \lambda^2 R^2$. So that $R^2 = \frac{(S|S)}{4(S|\omega)^2}$. Note that $(S|\omega) = |\dot{A}, \ldots, \dot{D}, \omega| = |A, \ldots, D|$ and $(S|S) = (\dot{A} \wedge \dot{B} \wedge \dot{C} \wedge \dot{D}|\dot{A} \wedge \dot{B} \wedge \dot{C} \wedge \dot{D})$, which yields the previous formula.

Using the same notation as in the paragraph on dual quaternions, for the composition of the rotation defined by $q = q_0 + q_1\mathbf{i} + q_2\mathbf{j} + q_3\mathbf{k}$, and the translation of vector (t_1, t_2, t_3), we get the following matrix (obtained in the basis $(e_0, \ldots, e_4)$) :

$$\begin{pmatrix} [\mathbf{\Psi}] & 0 \\ 0 & 1 \end{pmatrix}$$

where $[\mathbf{\Psi}]$ is the matrix defined in section (2.2).

Application to the parallel robot.

Consider six fixed points $(X_i)_{1 \leq i \leq 6}$ (of a fixed solid $\mathcal{S}_X$) and six other points Z_i, attached to a moving solid $\mathcal{S}_Z$. The articulations between the two solids $\mathcal{S}_X$ and $\mathcal{S}_Z$ are extensible bars $(X_i, Z_i)_{1 \leq i \leq 6}$ with spherical joints. Let us fix the length of these bars. The problem consists is to find know how many (real) positions of the platform there are for these fixed lengths.

Let g be an element of the Clifford algebra representing the displacement $[R, T]$ applied to the platform. Then each point Z_i must be on a sphere S_i centered in X_i of radius d_i. In the space of spheres described previously, we find the conditions :

$$(\psi_g(\dot{Z}_i)|S_i) = (\overline{(\mathbf{q} + \epsilon\mathbf{d})\tilde{\dot{Z}}_i(\bar{\mathbf{q}} + \epsilon\bar{\mathbf{d}})}|S_i) = 0$$

where $\dot{Z}_i$ represents the sphere-point associated to the point Z_i. This can be rewritten as

$$\overline{(\mathbf{q}\,\tilde{\dot{Z}}_i\bar{\mathbf{q}}|S_i)} + \overline{(\epsilon\,(\mathbf{d}\tilde{\dot{Z}}_i\bar{\mathbf{q}} - \mathbf{q}\tilde{\dot{Z}}_i\bar{\mathbf{d}})|S)} = [\mathbf{q}]^t\,A_i\,[\mathbf{q}] + [\mathbf{q}]^t\,B_i\,[\mathbf{d}] = 0$$

where $[\mathbf{q}], [\mathbf{d}]$ are the vector of coordinates of $\mathbf{q}, \mathbf{d}$ in the basis $(1, \mathbf{i}, \mathbf{j}, \mathbf{k})$ of $\mathbb{H}$, A_i, B_i are 4×4 matrices with coefficients in $\mathbb{K}$ and B_i is antisymmetric ($[\mathbf{q}]^t\,B_i\,[\mathbf{d}] = -[\mathbf{d}]^t\,B_i\,[\mathbf{q}]$). Let $[\mathbf{x}] = [\mathbf{q}] + \epsilon[\mathbf{d}]$ and $[\mathbf{x}'] = [\mathbf{q}] + \epsilon[\mathbf{d}]$. They are vectors of $(\mathbb{K} + \epsilon\,\mathbb{K})^4$ and satisfy the equations $[\mathbf{x}]^t(\epsilon\,A_i + B_i)[\mathbf{x}'] = 0$ and $[\mathbf{x}]^t[\mathbf{x}] = (\mathbf{q} + \epsilon\,\mathbf{d}|\bar{\mathbf{q}} + \epsilon\,\bar{\mathbf{d}}) = 1$, $[\mathbf{x}']^t[\mathbf{x}'] = (\mathbf{q} - \epsilon\,\mathbf{d}|\bar{\mathbf{q}} - \epsilon\,\bar{\mathbf{d}}) = 1$. This is the formulation used in (Wampler, 1994) and yields another (simple) proof that the number of solutions in $(\mathbb{K} + \epsilon\,\mathbb{K})^4$ is 40. See also (Husty, 1994) where these dual quaternions are also underlying and where there is an example of resolution.

3. Conclusion

In this paper we present a bief but general framework for computation in 3D-Geometry, based on the structure of Clifford Algebra. The formulation that we use is not the usual one but it allows us to handle points, lines, planes, spheres and displacements in a same homogeneous environment. These objects are treated in a natural and symbolic way, as variables with special rewriting rules for normalization of expressions. With this formalism, geometric properties correspond to simple relations so that heavy computations with coordinates can be avoided. This is illustrated by computations in the space of spheres and by the direct kinematic problem of a parallel robot where this formalism is used to derive a "good" polynomial system.

References

M. Barnabei, A. Brini, and G.C. Rota. On the exterior calculus of invariant theory. *J. of Algebra*, 96:p 120–160, 1985.

A. Crumeyrolle. *Orthogonal and Simplectic Clifford Algebra*. Kluwer Academic Plublishers, 1990.

A.W.M. Dress and T.F. Havel. Distance geometry and Geometric algebra. *Foundations of Physics*, 23(10):1357–1374, October 1991.

D. Hestenes. *Space Time Algebra*. Gordon and Breach, 1987.

M.L. Husty. An algorithm for solving the direct kinematics of Stewart-Gough-type platforms. Technical Report TR-CIM-94-7, Université McGill, Montréal, June 1994.

S. Lang. *Algebra*. Addison-Wesley, 1980.

B. Mourrain and N. Stolfi. *Invariants methods in Discrete and Computational Geometry*, chapter in Computational Symbolic Geometry. Kluwer acad. pub., 1994. (to appear).

B. Mourrain. New aspects of geometrical calculus with invariants. *Advances in Mathematics*, 1994. to appear.

C.W. Wampler. Forward displacement analysis of general six-in-parallel sps (stewart) platform manipulators using soma coordinates. submitted to Mech. Mach. Theory, 1994.

ELIMINATION METHODS FOR SPATIAL SYNTHESIS

J. NIELSEN AND B. ROTH

Design Division, Mechanical Engineering Department
Stanford University
`nielsen@leland.stanford.edu, roth@flamingo.stanford.edu`

Abstract. Many spatial, dimensional, position-synthesis problems lead to finite sets of solutions. In this paper, four such problems are presented. Two of these problems are solved using multihomogeneous resultant theory. The other two problems are solved using elimination strategy to obtain a solution based on rank reduction of a rectangular matrix of the form $(A - \lambda B)$. In the latter case, the appearance of linearly dependent equations during the elimination process is fully analyzed.

1. Introduction

In dimensional position-synthesis problems, we seek the dimensions of linkages which guide a moving body through a set of specified positions. The position-synthesis problem can also be used to solve other syntheses, such as the function generation and point-angle problems. There are many different types of planar and spatial linkages, all of which lead to different position-synthesis problems.

Recently, Innocenti gave a solution to the position-synthesis problem for the sphere-sphere binary link [1]. This simple solution eliminated in one step all but one of the unknowns from the constraint equations for the sphere-sphere link. Some recent advances in multihomogeneous resultant theory [2] predict this solution, and in fact predict solutions to certain other spatial synthesis problems.

Unfortunately, not all spatial synthesis problems lead to constraint equations which fall under multihomogeneous resultant theory. However, a similar elimination strategy can, for some such problems, lead to a solution based on rank reduction of a rectangular matrix of the form $(A - \lambda B)$.

J.-P. Merlet and B. Ravani (eds.), Computational Kinematics, 51–62.

2. Multihomogeneous Resultant Solutions

Given a multihomogeneous system of polynomial equations, there is a sufficient condition for the existence of a multigraded resultant of the Sylvester type [2]. This resultant is a determinental function, of the polynomial coefficients, whose vanishing indicates that the equations may be satisfied by a common system of non-trivial values of the variables. It is thus modeled after the classical Sylvester resultant of two homogeneous equations [3].

The condition can be expressed as follows: suppose we are given a set of polynomials which are homogeneous of degree d_k in each group of l_k variables, where there are r such groups and $k \in \{1, \ldots, r\}$. If $l_k = 1$ or $d_k = 1$ for all k, then the Sylvester type resultant exists, and in fact there are at least $r!$ different coefficient matrices representing this resultant. The actual construction of these Sylvester type resultants is discussed in [2].

Some binary links lead to sets of design equations for position synthesis which can be viewed as multihomogeneous and satisfying the above condition. After suppression of a variable (that is, considering one of the variables to be part of the coefficient for each term), the resultant may be constructed from the remaining multihomogeneous set. The necessary and sufficient condition for the existence of a solution set, then, is the vanishing of the determinant of the coefficients, which will be a univariate polynomial in the suppressed variable.

2.1. SLIDER-SLIDER-SPHERE DYAD

Consider a moving Cartesian coordinate system σ, representing the moving body, within the fixed system Σ. The equations for kinematic position synthesis of the slider-slider-sphere dyad are developed in [4]. If $P(x_1, x_2, x_3)$ is a point in σ, then the coordinates of that point in Σ when the moving body is in its jth finitely separated position will be denoted by $\left(X_{1_j}, X_{2_j}, X_{3_j}\right)$.

We will use the screw displacement representation to specify the arbitrary positions of the moving body. Assuming that Σ and σ are coincident in the first position, then $X_{1_1} = x_1$, $X_{2_1} = x_2$, and $X_{3_1} = x_3$. In terms of the rotation angle ϕ_j, the translation distance d_j, the unit screw direction $s_j\left(s_{1_j}, s_{2_j}, s_{3_j}\right)$, and a vector from the origin of Σ to the screw axis $S_j\left(S_{1_j}, S_{2_j}, S_{3_j}\right)$, the values of $X_{1_j}, X_{2_j}, X_{3_j}, j = 2, 3, \ldots$ are given by [5]:

$$\begin{bmatrix} X_{1_j} \\ X_{2_j} \\ X_{3_j} \end{bmatrix} = [R_j] \begin{bmatrix} x_1 \\ x_2 \\ x_3 \end{bmatrix} + \begin{bmatrix} d_{1_j} \\ d_{2_j} \\ d_{3_j} \end{bmatrix} \qquad (1)$$

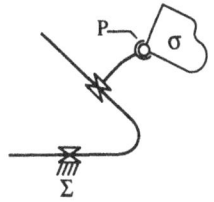

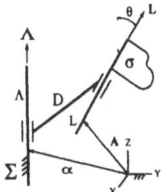

Figure 1. Slider-Slider-Sphere Dyad   *Figure 2.* Cylinder-Cylinder Binary Link

where the elements of $[R_j]$ are r_{ik_j}, and

$$
\begin{aligned}
r_{11_j} &= \left(s_{1_j}{}^2 - 1\right)\left(1 - \cos\phi_j\right) + 1 \\
r_{12_j} &= s_{1_j}s_{2_j}\left(1 - \cos\phi_j\right) - s_{3_j}\sin\phi_j \\
r_{13_j} &= s_{1_j}s_{3_j}\left(1 - \cos\phi_j\right) + s_{2_j}\sin\phi_j \\
d_{1_j} &= d_j s_{1_j} - S_{1_j}\left(r_{11_j} - 1\right) - S_{2_j}r_{12_j} - S_{3_j}r_{13_j} \\
r_{21_j} &= s_{2_j}s_{1_j}\left(1 - \cos\phi_j\right) + s_{3_j}\sin\phi_j, \quad \text{etc.}
\end{aligned} \tag{2}
$$

The slider-slider-sphere dyad, shown in Figure 1, constrains a point P located at the spherical joint to move in a plane parallel to the axes of the slider joints. The constraint equations for point $P(x_1, x_2, x_3)$ to lie on a plane for six arbitrary positions, then, are

$$
\left(X_{1_j} - y_1\right)l + \left(X_{2_j} - y_2\right)m + \left(X_{3_j} - y_3\right)n = 0 \qquad j = 1, 2, \dots, 6 \tag{3}
$$

where (l, m, n) are the direction numbers of the normal direction to the plane, and (y_1, y_2, y_3) are the coordinates of a point in the plane.

We divide (3) by l, and make the substitutions $b_1 = m/l$ and $b_2 = n/l$. Then by introducing (1), and subtracting the first equation from the remaining five, (3) may be reduced to

$$
f_{j1} + f_{j2}b_1 + f_{j3}b_2 = 0 \qquad j = 2, 3, \dots, 6 \tag{4}
$$

Here the f_{j_i} are linear functions of x_1, x_2, and x_3. Now by suppressing the variable b_2, and introducing the homogenizing variables x_0 and b_0, equation (4) may be written as

$$
\sum_{p=0,1, q=0,1,2,3} d_{pqj}b_p x_q = 0 \qquad j = 2, 3, \dots, 6 \tag{5}
$$

where the d_{pqj} are linear functions of b_2. This represents five multihomogeneous equations, in which the first group of variables is $\{b_0, b_1\}$, and the second group is $\{x_0, x_1, x_2, x_3\}$. The degree of both groups is one. It is therefore possible to construct the Sylvester type multigraded resultant.

TABLE 1.   Arbitrary Finitely Separated Positions

Disp	Axis cosines			Axis location			Screw parameters	
1 to j	s_{1_j}	s_{2_j}	s_{3_j}	S_{1_j}	S_{2_j}	S_{3_j}	ϕ_j(deg)	d_j
1 to 2	-0 5281	0 1947	0 8265	1 1173	-0 2281	0 7676	-106 2	0 837
1 to 3	0 7344	0 0505	0 6769	-0 6119	1 8700	0 5242	-49 1	0 961
1 to 4	-0 0508	0 1749	0 9833	-0 9809	1 3178	-0 2851	97 4	1 380
1 to 5	-0 4518	0 0691	0 8894	1 7809	-1 6889	1 0358	-9 2	-0 622
1 to 6	0 2911	0 2280	0 9291	0 7220	1 6061	-0 6202	13 3	1 828
1 to 7	0 8483	-0 4231	-0 3183	-1 1559	1 9577	-5 6839	43 0	0 576

TABLE 2. **Slider-Slider-Sphere Numerical Example.** *Each number is represented by a real and imaginary part. For complex numbers, one out of each pair of complex conjugate solutions is included*

b_1	b_2	x_1	x_2	x_3
(17 0280, 0 0000)	(43 6151,0 0000)	(-24 6996, 0 0000)	(4 7448, 0 0000)	(40 4675, 0 0000)
(0 2892, 0 0000)	(-2 8012,0 0000)	(16 5814, 0 0000)	(0 9501, 0 0000)	(22 7059, 0 0000)
(-3 1175,-0 4937)	(-5 6324,5 8188)	(-6 4364, 12 2066)	(5 5732,12 4732)	(36 8176,-8 1848)
(0 0458,-0 9274)	(0 2944,0 2772)	(-28 4010,-11 5330)	(-10 0961,29 5558)	(-7 6200,-0 4181)
(0 0574, 0 8555)	(-0 5451,0 1973)	(11 8040, 1 7488)	(-2 7496, 9 3272)	(-3 7270,-8 5057)
(-1 3530, 0 1303)	(-0 3511,0 0172)	(-1 3579, 0 0301)	(0 5136, -0 2370)	(-2 1005,-0 5548)

To construct the resultant, we multiply the five equations in (5) by the four variables x_r, where $r = 0, 1, 2, 3$. The resulting 20 equations may be written as $M_1 z_1 = 0$ where M_1 is a 20×20 matrix linear in b_2 and z_1 is a vector whose 20 components are "power products" (unsuppressed variable terms) of the form $b_p x_q x_r$, where $p = 0, 1; q, r = 0, 1, 2, 3$.

Because one entry of z_1 is non-trivial ($b_0 x_0^2 = 1$), the determinant of M_1 must vanish in order to admit a non-trivial solution. Expansion of this determinant shows that it is a tenth-degree polynomial in b_2. For each root of this polynomial, the remaining unknowns can be found by solving $M_1 z_1 = 0$ for z_1, after substituting the proper values of b_2.

As an example, the first five rows of Table 1 specify six arbitrary positions for the moving body. The synthesis problem was solved for this numerical example: the ten solutions are given in Table 2.

The multihomogeneous resultant theory promises $2! = 2$ different resultants, since there are two groups of variables. The second resultant can be obtained by multiplying the five equations in (5) by the terms $\{b_1^3, b_1^2, b_1, 1\}$. This also leads to a 20×20 matrix, and the same ten solutions.

2.2. CYLINDER-CYLINDER BINARY LINK

A cylinder-cylinder binary link, as shown in Figure 2, constrains a line L in the moving system σ so that the distance D and angle θ between L and the fixed axis Λ remain constant. If $\mathbf{L}(l, m, n)$ is a unit vector parallel to L

in σ, then that vector in Σ when the moving body is in its jth position will be denoted $\mathbf{L}_j$ (l_j, m_j, n_j). Using the screw displacement representation, $\mathbf{L}_j = [R_j] \mathbf{L}$ where $[R_j]$ is as defined in (1) and (2). Let $\boldsymbol{\Lambda}$ (λ, μ, ν) be a unit vector parallel to Λ, and $\mathbf{A}$ (a, b, c) and $\boldsymbol{\alpha}$ (α, β, γ) be position vectors of points on L and Λ respectively. Then the constraint equations for five positions are [4]

$$\boldsymbol{\Lambda} \cdot \mathbf{L}_j = \cos\theta = \text{const} \qquad j = 1, 2, \ldots, 5 \qquad (6)$$

$$(\boldsymbol{\Lambda} \times \mathbf{L}_j) \cdot (\mathbf{A}_j - \boldsymbol{\alpha}) = D\sin\theta = \text{const} \qquad j = 1, 2, \ldots, 5 \qquad (7)$$

where

$$\mathbf{A}_j = [R_j] \begin{bmatrix} a \\ b \\ c \end{bmatrix} + \begin{bmatrix} d_{1_j} \\ d_{2_j} \\ d_{3_j} \end{bmatrix} \qquad (8)$$

This system can be decoupled. We divide (6) by $l\lambda$. Then by making the substitutions $b_1 = m/l$, $b_2 = n/l$, $c_1 = \mu/\lambda$, $c_2 = \nu/\lambda$, introducing homogenizing variables b_0 and c_0, and subtracting the first equation from the remaining four, (6) may be transformed to

$$\sum_{p=0,1,2, q=0,1,2} d_{pqj} b_p c_q = 0 \qquad j = 2, 3, 4, 5 \qquad (9)$$

where the d_{pqj} are constant. This represents four equations in four unknowns; they may be solved independent of (7). To solve (9), we suppress the unknown b_2, changing (9) to

$$\sum_{p=0,1, q=0,1,2} \bar{d}_{pqj} b_p c_q = 0 \qquad j = 2, 3, 4, 5 \qquad (10)$$

where the $\bar{d}_{pqj}$ are linear in b_2. This represents four multihomogeneous equations, where the first group of variables is $\{b_0, b_1\}$, and the second group is $\{c_0, c_1, c_2\}$. Since the degree of both groups is one, we may again construct a Sylvester type resultant.

To construct the resultant, we multiply (10) by the three variables c_r where $r = 0, 1, 2$. This results in 12 equations which may be written as $M_2 \mathbf{z}_2 = 0$ where M_2 is a 12×12 matrix linear in b_2, and $\mathbf{z}_2$ contains power products of the form $b_p c_q c_r$, where $p = 0, 1; q, r = 0, 1, 2$. Setting the determinant to zero gives a sixth degree polynomial in b_2. After finding b_1, c_1, and c_2 for a given value of b_2, equation (7) is linear in the remaining unknowns ($\mathbf{A}$ and $\boldsymbol{\alpha}$) and can be easily solved.

The system was solved numerically for the five positions given by the first four rows of Table 1. The results are shown in Table 3. A different

TABLE 3. Cylinder-Cylinder Numerical Example. *Only one out of each pair of complex conjugate solutions is included.*

b_1	b_2	c_1	c_2
(-0.4590, 0.0000)	(3.2124,0.0000)	(-0.1805, 0.0000)	(-1.6901, 0.00000)
(0.1055, 0.0000)	(-1.2499,0.0000)	(458.7605, 0.0000)	(1479.1400, 0.0000)
(0.3709, 0.9381)	(-2.6452,1.1799)	(-1.8203, 1.6182)	(-5.0940,-0.5935)
(-0.0160,-0.9513)	(0.2682,0.0906)	(0.1022,-0.8960)	(0.1514, 0.3966)

Sylvester type resultant may be obtained by multiplying (10) by the power products $\{b_1^2, b_1, 1\}$.

3. Rank Reduction Solutions

Many kinematic synthesis problems encountered in practice will not meet the sufficient condition for the existence of a Sylvester type resultant. This does not necessarily indicate that such a resultant does not exist; if it does exist, though, it cannot be constructed by the methods introduced in [2]. Nevertheless, some of the same ideas may be adopted and extended, such as splitting the unknowns into groups then multiplying by power products from only one group to perform the elimination.

Difficulties are encountered during such an elimination which were not seen in the Sylvester type resultant cases. Most notably, linear dependencies appear between the equations produced during elimination. We now show that in the synthesis of both the revolute-slider-sphere dyad and the cylinder-sphere binary link, a full understanding of these linear dependencies can lead to a solution based on rank reduction of a matrix $(A - \lambda B)$.

3.1. REVOLUTE-SLIDER-SPHERE DYAD

The revolute-slider-sphere dyad constrains a point P located at the spherical joint to move on a one-sheet hyperboloid of revolution which shares its axis with the revolute joint and whose generator through P is parallel to the sliding direction. The constraint equations for positions synthesis are developed in [4]: for seven positions we have

$$
\left(X_{1_j} - a\right)^2 + \left(X_{2_j} - b\right)^2 + \left(X_{3_j} - c\right)^2
$$

$$
- \left[l\left(X_{1_j} - a\right) + m\left(X_{2_j} - b\right) + n\left(X_{3_j} - c\right)\right]^2 \left[1 + \frac{\alpha^2}{\beta^2}\right]
$$

$$
- \left\{k^2 - 2k\left[l\left(X_{1_j} - a\right) + m\left(X_{2_j} - b\right) + n\left(X_{3_j} - c\right)\right]\right\} \frac{\alpha^2}{\beta^2} = \alpha^2
$$

$$
j = 1, 2, \ldots, 7 \qquad (11)
$$

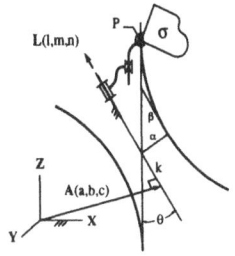

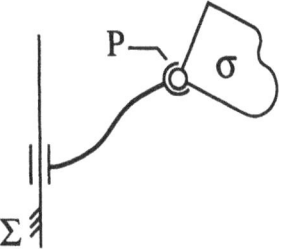

Figure 3. Revolute-Slider-Sphere Dyad   *Figure 4.* Cylinder-Sphere Dyad

Here (l, m, n) are the direction cosines of the hyperbola axis $\mathbf{L}$, $\mathbf{A}\,(a, b, c)$ is a vector to a point on that axis, k is the distance from the hyperbola center to the point $A\,(a, b, c)$, α is the distance between the axis and a generator, and α/β is the tangent of the angle θ between the axis and a generator. These quantities are shown in Figure 3.

Assuming (l, m, n) to be specified, we choose to make $\mathbf{A}$ perpendicular to $\mathbf{L}$. Thus

$$al + bm + cn = 0 \tag{12}$$

which simplifies (11). Now subtracting the first equation from the remaining six, and using (12) to find a in terms of b and c, we are left with equations of the form

$$f_{j_1}b + f_{j_2}c + s_j k\frac{\alpha^2}{\beta^2} + t_j^2\left(1 + \frac{\alpha^2}{\beta^2}\right) + f_{j_3} = 0 \qquad j = 2, 3, \ldots, 7 \tag{13}$$

Here f_{j_1}, f_{j_2}, f_{j_3}, and s_j are linear in x_1, x_2, and x_3; t_j^2 is quadratic in x_1, x_2, and x_3. If we specify $\frac{\alpha^2}{\beta^2}$, suppress b, and introduce the homogenizing variable x_0, then (13) may be written as

$$c \sum_{p=0,1,2,3} d_{1pj}x_p + k \sum_{p=0,1,2,3} d_{2pj}x_p + \sum_{p,q=0,1,2,3} d_{3pqj}x_p x_q = 0$$

$$j = 2, 3, \ldots, 7 \tag{14}$$

As a first try at elimination, we multiply the six equations in (14) by all terms of the form $x_r x_s x_t x_u$ where $r, s, t, u = 0, 1, 2, 3$. There are 35 such terms; the number of combinations of n things taken m at a time is $\binom{n+m-1}{m}$, and here we have $\binom{4+4-1}{4} = 35$. This gives a total of 210 equations, which contain power products of the form

$$c x_p x_r x_s x_t x_u$$
$$k x_p x_r x_s x_t x_u$$
$$x_p x_q x_r x_s x_t x_u \qquad p, q, r, s, t, u = 0, 1, 2, 3 \tag{15}$$

There are 196 such power products. Thus we may write $M_3\mathbf{z}_3 = 0$, where M_3 is a 210×196 matrix which is linear in b, and $\mathbf{z}_3$ is a vector containing the power products in (15).

Non-trivial solutions exist if M_3 is rank deficient for specific values of b; in this case M_3 is in fact rank deficient for all values of the suppressed variable b, and we cannot determine b from M_3. To see this, write any four equations from (14) as

$$g_i = c\hat{g}_i + k\bar{g}_i + \tilde{g}_i = 0 \qquad i = 1, 2, 3, 4 \tag{16}$$

where the $\hat{g}_i$'s and $\bar{g}_i$'s are linear in $\{x_0, x_1, x_2, x_3\}$, and the $\tilde{g}_i$'s are quadratic in these variables. Substitution of (16) shows the following to be true:

$$
\begin{aligned}
h_{1234} = {} & [\bar{g}_3\hat{g}_4\tilde{g}_2 - \hat{g}_3\bar{g}_4\tilde{g}_2 + \hat{g}_2\bar{g}_4\tilde{g}_3 - \bar{g}_2\hat{g}_4\tilde{g}_3 + \hat{g}_3\bar{g}_2\tilde{g}_4 - \bar{g}_3\hat{g}_2\tilde{g}_4]\, g_1 \\
& - [\bar{g}_4\hat{g}_1\tilde{g}_3 - \hat{g}_4\bar{g}_1\tilde{g}_3 + \hat{g}_3\bar{g}_1\tilde{g}_4 - \bar{g}_3\hat{g}_1\tilde{g}_4 + \hat{g}_4\bar{g}_3\tilde{g}_1 - \bar{g}_4\hat{g}_3\tilde{g}_1]\, g_2 \\
& + [\bar{g}_1\hat{g}_2\tilde{g}_4 - \hat{g}_1\bar{g}_2\tilde{g}_4 + \hat{g}_4\bar{g}_2\tilde{g}_1 - \bar{g}_4\hat{g}_2\tilde{g}_1 + \hat{g}_1\bar{g}_4\tilde{g}_2 - \bar{g}_1\hat{g}_4\tilde{g}_2]\, g_3 \\
& - [\bar{g}_2\hat{g}_3\tilde{g}_1 - \hat{g}_2\bar{g}_3\tilde{g}_1 + \hat{g}_1\bar{g}_3\tilde{g}_2 - \bar{g}_1\hat{g}_3\tilde{g}_2 + \hat{g}_2\bar{g}_1\tilde{g}_3 - \bar{g}_2\hat{g}_1\tilde{g}_3]\, g_4 = 0
\end{aligned}
\tag{17}
$$

If we substitute the actual values for only the bracketed quantities above, then (17) represents a linear relation among terms of the form $x_r x_s x_t x_u g_i$; $i = 1, 2, 3, 4$; $r, s, t, u = 0, 1, 2, 3$. This, however, is precisely the form of the 210 equations used to create the matrix M_3. So, for any value of the suppressed variable b, (17) represents one linear relation between the rows of M_3.

Since we choose the four equations in (16) from the six in (14), we actually have $\binom{6}{4} = 15$ possible linear relations from M_3. Thus M_3 has rank at most $210 - 15 = 195$, and since M_3 has 196 columns it is rank deficient for all values of the suppressed variable b. Numerical tests confirm this result.

To obtain an equation set which has full rank for arbitrary value of the suppressed variable, multiply the six equations in (14) by all terms of the form $x_r x_s x_t x_u x_v$ where $r, s, t, u, v = 0, 1, 2, 3$. This yields 336 equations which may be written as $M_4\mathbf{z}_4 = 0$, where M_4 is linear in the suppressed variable b and $\mathbf{z}_4$ is a vector containing all power products of the form

$$
\begin{aligned}
& cx_p x_r x_s x_t x_u x_v \\
& kx_p x_r x_s x_t x_u x_v \\
& x_p x_q x_r x_s x_t x_u x_v \qquad p, q, r, s, t, u, v = 0, 1, 2, 3
\end{aligned}
\tag{18}
$$

There are 288 such power products, so M_4 is a 336×288 matrix.

Numerical tests show that M_4 has full rank for arbitrary value of b. At first glance, it appears that the previously developed theory again predicts rank deficiency. Since the rows of M_4 represent equations of the form

$x_r x_s x_t x_u x_v g_i$; $i = 1, 2, 3, 4$; $r, s, t, u, v = 0, 1, 2, 3$, we would expect to be able to multiply each of the linear relations in (17) by x_0, x_1, x_2, and x_3. This would yield 60 linear relations. In doing this, however, we encounter linear relations among those 60 linear relations. Choosing any five of the six equations in (14) and using our previous notation, we have the five linear relations

$$h_{1234}, h_{1235}, h_{1245}, h_{1345}, h_{2345} \tag{19}$$

Substitution of (16) and (17), though, shows the following to be true:

$$\hat{g}_5 h_{1234} - \hat{g}_4 h_{1235} + \hat{g}_3 h_{1245} - \hat{g}_2 h_{1345} + \hat{g}_1 h_{2345} = 0$$
$$\bar{g}_5 h_{1234} - \bar{g}_4 h_{1235} + \bar{g}_3 h_{1245} - \bar{g}_2 h_{1345} + \bar{g}_1 h_{2345} = 0 \tag{20}$$

Note that (20) represents linear relations among terms of the form $x_p h_{ijkl}$; $p = 0, 1, 2, 3$; $i, j, k, l \in \{1, 2, 3, 4, 5\}$. Thus, these are linear relations among the 60 linear relations described above. There are two such relations for each way we can choose five of the six equations in (14); a total of 12 linear relations.

So we predict $60 - 12 = 48$ independent linear relations among the rows of M_4. If no other linear relations exist, then M_4 has rank $336 - 48 = 288$ for arbitrary value of the suppressed variable b; numerical tests show that the rank is indeed 288. This is full rank, since M_4 has 288 columns. We now seek those values of b for which the rank of M_4 is reduced; here M_4 can be divided into two matrices G and H such that $M_4 = (G - bH)$. In [6], Thompson and Weil give an algorithm to find all such values. This algorithm, whether applied symbolically or numerically, leads to a square generalized eigenvalue problem whose solutions correspond to the rank-reducing values of $(G - bH)$. For this problem, the resulting square generalized eigenvalue problem is 42×42. The use of eigenvalues for solving polynomial systems was proposed previously in [7].

To find the remaining unknowns corresponding to each values of b, we can solve $M_4 \mathbf{z}_4 = 0$ for $\mathbf{z}_4$ after substituting in the proper value of b. Note that this can be approached as a full rank least squares problem, but it is computationally less expensive to find a basis for the row space of M_4 and simply solve a linear system after substituting the proper value of b. One may find such a basis by examining the linear relations among the rows of M_4, but this is non-trivial since care must be taken to avoid choosing rows which cause the column rank to drop. The basis need only be chosen once, and it is valid for all problems of this type.

Table 1 specifies seven arbitrary positions for the moving body; the 42 solutions corresponding to these positions were found numerically using the foregoing procedure. The results are shown in Table 4.

TABLE 4. **Revolute-Slider-Sphere Numerical Example.** *Only one out of each pair of complex conjugate solutions is included* $(\frac{\alpha^2}{\beta^2} = 0\,54, l = 0.40, m = 0\,23, n = 0\,89)$

b	c	k	x_1	x_2	x_3
(1 58, 0 00)	(0 43, 0 00)	(-4 92, 0 00)	(0 17, 0 00)	(-2 65, 0 00)	(-4 35, 0 00)
(2 40, 0 00)	(0 44, 0 00)	(-4 47, 0 00)	(-0 93, 0 00)	(-1 55, 0 00)	(-3 81, 0 00)
(-2 48, 0 00)	(0 95, 0 00)	(7 95, 0 00)	(0 26, 0 00)	(-6 29, 0 00)	(9 46, 0 00)
(-3 89, 0 00)	(-1 19, 0 00)	(-155 00, 0 00)	(55 46, 0 00)	(-29 51, 0 00)	(-94 75, 0 00)
(-7 52, 0 00)	(-0 86, 0 00)	(-451 45, 0 00)	(64 39, 0 00)	(-62 77, 0 00)	(-231 24, 0 00)
(-38 82, 0 00)	(-264 30, 0 00)	(320 97, 0 00)	(-7 76, 0 00)	(43 02, 0 00)	(87 34, 0 00)
(0 71, -0 13)	(0 09, -0 05)	(0 99, -0 82)	(-0 54, 0 65)	(0 11, -1 26)	(0 60, -0 63)
(-0 15, -0 08)	(1 30, -0 31)	(1 98, 0 65)	(-1 72, 0 75)	(-0 61, 0 07)	(3 41, -0 10)
(0 43, -0 75)	(0 39, 0 72)	(2 53, 1 00)	(-0 52, -0 30)	(0 36, -0 86)	(2 02, 1 94)
(0 38, -0 90)	(-0 21, 0 40)	(-6 47, -1 23)	(-3 59, -5 20)	(-0 27, -0 01)	(-5 76, -0 11)
(-1 11, -0 34)	(1 31, -1 27)	(2 89, -7 99)	(-8 77, -12 27)	(2 94, -3 60)	(1 27, -5 06)
(1 78, -1 13)	(0 03, -0 25)	(-2 33, 4 00)	(-1 90, -1 27)	(-0 74, 2 59)	(-1 32, 3 09)
(0 39, -2 01)	(0 02, 0 29)	(4 91, -0 25)	(0 20, 0 98)	(3 55, -3 81)	(2 70, 0 35)
(3 24, -0 68)	(0 16, 1 98)	(0 91, 1 20)	(-2 96, -1 56)	(2 34, -1 61)	(-0 10, 3 88)
(-1 44, -2 72)	(-2 96, 0 36)	(29 84, 31 93)	(3 35, -1 23)	(-5 40, 1 23)	(16 70, 19 42)
(3 49, -4 01)	(-2 53, -0 22)	(-2 34, 0 86)	(1 90, 3 47)	(4 83, -3 35)	(-4 26, 0 25)
(3 00, -4 58)	(0 39, -0 79)	(-6 08, -1 18)	(-1 18, 0 18)	(-1 10, -1 30)	(-6 08, -0 81)
(2 73, -6 41)	(-3 72, -2 75)	(68 56, -59 56)	(14 70, 16 31)	(12 85, -3 05)	(11 70, -42 32)
(-4 56, -4 96)	(-0 94, 2 03)	(80 66, 35 89)	(-31 33, 21 56)	(26 19, 18 01)	(34 60, 1 38)
(-6 92, -5 01)	(11 29, 2 59)	(-85 15, -128 92)	(-77 10, -56 99)	(-30 25, -39 59)	(7 49, -36 31)
(15 97, -2 95)	(5 72, 3 27)	(494 28, -5 54)	(-92 29, -2 01)	(206 57, -27 76)	(187 98, 3 24)
(-7 98, -14 00)	(7 77, 15 85)	(-11 27, -61 73)	(8 26, -10 87)	(-2 64, -4 06)	(21 55, -5 92)
(-42 54, -24 22)	(21 33, 0 38)	(-57 35, -55 65)	(-18 47, 16 69)	(-20 59, -15 01)	(-11 55, 28 01)
(-9 67, -51 77)	(29 37, 3 02)	(17 98, 29 78)	(16 02, 24 21)	(-21 34, 10 45)	(-2 43, -15 17)

3.2. CYLINDER-SPHERE BINARY LINK

A cylinder-sphere binary link (Figure 4) is equivalent to the special case of the revolute-slider-sphere link where $\alpha/\beta = 0$. So for six positions, the constraint equations are obtained directly from (13)

$$f_{j1}b + f_{j2}c + t_j^2 + f_{j3} = 0 \qquad j = 2, 3, \ldots, 6 \qquad (21)$$

Suppressing b and introducing homogenizing variable x_0, (21) may be written as

$$c \sum_{p=0,1,2,3} d_{1pj}x_p + \sum_{p,q=0,1,2,3} d_{2pqj}x_p x_q = 0 \qquad j = 2, 3, \ldots, 6 \qquad (22)$$

If we multiply (22) by terms of the form $x_r x_s x_t$, where $r, s, t = 0, 1, 2, 3$, then we obtain 100 equations containing only 91 power products. Again, though, there are linear relations among the equations. To see this, write any three equations from (22) as

$$g_i = c\hat{g}_i + \tilde{g}_i = 0 \qquad i = 1, 2, 3 \qquad (23)$$

where the g_i's are as previously defined. By substitution of (23), we see that

$$h_{123} = [\hat{g}_2\tilde{g}_3 - \hat{g}_3\tilde{g}_2]\,g_1 + [\hat{g}_3\tilde{g}_1 - \hat{g}_1\tilde{g}_3]\,g_2 + [\hat{g}_1\tilde{g}_2 - \hat{g}_2\tilde{g}_1]\,g_3 = 0 \qquad (24)$$

So for every way to choose three equations from (22), there is one linear relation. This gives a total of 10 linear relations, and leads to a rank-deficiency problem.

If we multiply (22) by terms of the form $x_r x_s x_t x_u$, where $r, s, t, u = 0, 1, 2, 3$, then we have 175 equations containing 140 power products. We expect to get 40 linear relations by multiplying each of the 10 equations like (24) by x_0, x_1, x_2, and x_3. But again there are linear relations among these 40; choosing any four of the equations in (22), we have the four linear relations $h_{123}, h_{124}, h_{134}$, and h_{234}. Substitution of (23) and (24) shows that

$$\hat{g}_4 h_{123} - \hat{g}_3 h_{124} + \hat{g}_2 h_{134} - \hat{g}_1 h_{234} = 0 \qquad (25)$$

So there is one linear relation among the expected 40 for each way we can choose four equations out of the five in (22). Thus, our theory predicts that there are 40-5=35 independent linear relations; numerical tests confirm that these are the only linear relations. We may write the 175 equations obtained above as $M_5 z_5 = 0$, where M_5 is a 175×140 matrix, which has full rank for arbitrary b since there are 35 independent linear relations among the rows of M_5. The power products in z_5 are of the form

$$c x_p x_r x_s x_t x_u$$
$$x_p x_q x_r x_s x_t x_u \qquad p, q, r, s, t, u = 0, 1, 2, 3 \qquad (26)$$

We find those values of the suppressed variable which reduce the rank of M_5, using the same algorithm as for the revolute-slider-sphere case. This leads to a 26×26 generalized eigenvalue problem, giving 26 possible values for b. Back-substitution to find the remaining unknowns is accomplished as in the revolute-slider-sphere case.

The first five rows of Table 1 specify six arbitrary positions for the moving body; the 26 solutions corresponding to these positions were found numerically. The results are shown in Table 5.

4. Conclusion

Despite the size and complexity of the eliminations, the procedures described in this paper introduced no extraneous roots. After solving for all unknowns in each numerical example, all roots, including complex roots, led to solutions which satisfied the original sets of equations. The procedures described herein therefore constitute a reliable means by which these four, dimensional, position-synthesis problems may be solved.

Multihomogeneous resultant theory has proven useful for certain synthesis problems. The rank reduction solutions presented here represent extensions of some of the elimination methods prescribed by multihomogeneous

TABLE 5. **Cylinder-Sphere Numerical Example.** *Only one out of each pair of complex conjugate solutions is included* $(l = 0.40, m = 0.23, n = 0.89)$

b	c	x_1	x_2	x_3
(0 7651, 0 0000)	(-2 6130, 0 0000)	(-36 5639, 0 0000)	(-2 1921, 0 0000)	(-10 4294, 0 0000)
(0 3279, 0 0000)	(-0 2514, 0 0000)	(1 4479, 0 0000)	(-2 5713, 0 0000)	(3 5292, 0 0000)
(0 2204, 0 0000)	(-0 7770, 0 0000)	(-1 3013, 0 0000)	(1 1668, 0 0000)	(-2 7876, 0 0000)
(0 0264, 0 0000)	(-0 1695, 0 0000)	(1 4500, 0 0000)	(-3 2740, 0 0000)	(3 9661, 0 0000)
(3 2535, 0 0000)	(-1 2072, 0 0000)	(13 5242, 0 0000)	(32 5185, 0 0000)	(-8 8145, 0 0000)
(7 0262, 0 0000)	(-2 3712, 0 0000)	(14 1068, 0 0000)	(50 5853, 0 0000)	(-15 4864, 0 0000)
(0 9137, -0 2675)	(0 1273, -0 0718)	(-1 2665, 0 3008)	(1 3725, -0 4976)	(0 7182, -0 3100)
(0 5399, -0 6177)	(0 0113, 0 4889)	(-0 8329, -0 3343)	(0 4942, -0 4359)	(0 0527, 1 2602)
(1 4226, -0 2997)	(0 0707, -0 1203)	(-1 4990, 1 1254)	(2 2197, -0 4213)	(0 7217, 0 5614)
(1 6371, -0 8689)	(-0 1624, 0 3944)	(-0 2517, 1 0215)	(1 9061, -0 8632)	(1 0836, 2 8007)
(2 3066, -2 7292)	(1 5844, 0 4153)	(0 6761, 8 0265)	(-6 0103, -4 4220)	(-5 3810,-10 3704)
(-1 7529, -1 8971)	(1 0596, 0 2025)	(0 0110, -1 2959)	(-0 9860, -4 7608)	(8 1200, -1 8698)
(1 0319, -3 5824)	(-0 5000,-0 5837)	(-1 0201, -0 1145)	(0 6247, -1 2773)	(-3 8957, -2 0223)
(4 5086, -0 6786)	(-0 6827, 1 7435)	(-3 5830, -3 1533)	(3 9193, -2 4686)	(-2 4681, 0 8318)
(-6 7685, -1 7916)	(13 1328, 7 8876)	(-41 9087,183 5588)	(2 3801, 51 4071)	(57 1207, 3 1381)
(-33 6188,-14 9874)	(1 3105,21 7136)	(33 0932, 37 7862)	(31 1384,-26 3885)	(-21 1942, 0 9908)

resultant theory, and they involve a novel use of rank reducing algorithms for rectangular matrices. Such algorithms, coupled with a firm understanding of the origin of linear relations, could prove useful in the solution of many other nonlinear sets of equations.

5. Acknowledgment

The financial support of the National Science Foundation is gratefully acknowledged.

References

1. C. Innocenti. Polynomial solution of the spatial Burmester problem. In G. R. Pennock, editor, *Mechanism Synthesis and Analysis*, volume DE-70, pages 161–166, New York, 1994. ASME.
2. B. Sturmfels and A. Zelevinsky. Multigraded resultants of Sylvester type. *J. Algebra*, 163:115–127, 1994.
3. G. Salmon. *Lessons Introductory to the Modern Higher Algebra*, pages 79–83. Chelsea Publishing Company, fifth edition, 1964.
4. P. Chen and B. Roth. Design equations for the finitely and infinitesimally separated position synthesis of binary links and combined link chains. *Journal of Engineering for Industry, Trans. ASME, Series B*, 91(1):209–219, February 1969.
5. O. Bottema and B. Roth. *Theoretical Kinematics*, pages 56–62. North Holland, 1979. (reprinted by Dover Publications, NY, 1990).
6. G. Thompson and R. Weil. Reducing the rank of $(A - \lambda B)$. *Proc. Amer. Math. Soc.*, 26:548–554, December 1970.
7. M. Ghazvini. Reducing the inverse kinematics of manipulators to the solution of a generalized eigenproblem. In J. Angeles, G. Hommel, and P. Kovacs, editors, *Computational Kinematics*, pages 15–26. Kluwer Academic Publishers, 1993.

THE APPLICATION OF FINITE DISPLACEMENT SCREWS TO DRAWING CONSTRAINTS

I.A. PARKIN
Basser Department of Computer Science
University of Sydney, N.S.W. 2006, Australia

Abstract. Recent investigations show that *finite displacement screws* form *linearly combined* sets when they describe incompletely specified displacements of a body. These results are considered in the context of *geometric constraints* as they are used in computer-aided drawing or design. A constraint — requiring that one or more of the geometric elements *point, plane* or *line*, embedded in a body, must remain in coincidence with other such elements which are fixed — is defined by a set of screws which specify all finite displacements which are then available to the body. When so represented, constraints may be manipulated in a manner which is both simple and uniform, and which, for the chosen geometric elements, is linear.

1. Introduction

Of importance in enlarging the expressive power of software for *computer-aided design* is the provision of *geometric constraints* whereby a user may specify, for example, that a nominated point or line in one body must coincide with a particular point, or must lie in a particular plane, of another body. Such geometric (or kinematic) constraints serve to define, and are effectively specified by, the set of *finite displacements* which are available to one of the bodies in its movements relative to the other. With a view to developing a uniform method for the computational manipulation of constraints, this paper examines the manner in which such sets interact under certain types of constraint combination.

This work is in the general area of such studies as (Angeles, 1986; Ravani *et al.*, 1993) but particularly exploits recent results (Parkin, 1992; Huang *et al.*, 1994; Hunt *et al.*, 1995) which show that certain *finite displacement*

63

J.-P. Merlet and B. Ravani (eds.), Computational Kinematics, 63–72.

screws (defined in Section 2, *et seq.*) form a *linear combined* set when they describe an incompletely specified displacement of a body. All screws of such a set can be expressed as *dual* linear combinations of three basis screws.

We specify a constraint by requiring that two of the geometric elements *point, plane* and *line* — one embedded in each of two rigid bodies — should remain in coincidence. We enumerate the nine linear combination sets of screws obtained (three of which are duplicates under interchange of the bodies) when such elements, taken two at a time, are constrained to coincide. For each coincidence, the rule for constructing these sets from the displacement freedoms intrinsic to the geometric elements is uniformly found to be the *screw triangle rule* (Bottema *et al.*, 1990) given in Equation 2.

In distinction, the imposition of multiple constraints always amounts to finding the intersection of two linearly combined sets, a matter which is readily solved in a linear manner.

2. Definition of a Screw

We write the general *screw* as a 3-vector of dual numbers, $\hat{\mathbf{S}} \equiv \mathbf{S} + \varepsilon\, \mathbf{S_p}$, in which the real 3-vectors $\mathbf{S}$ and $\mathbf{S_p}$ are respectively the *direction* and *moment* parts of the screw, and ε is the *quasi-scalar* with the property $\varepsilon^2 = 0$. The *magnitude, pitch* and *origin radius* of the screw are

$$|\mathbf{S}|\ ,\quad p = \frac{\mathbf{S} \bullet \mathbf{S_p}}{\mathbf{S}^2}\ ,\quad \mathbf{R} = \frac{\mathbf{S} \times \mathbf{S_p}}{\mathbf{S}^2}\ ,$$

respectively and, in terms of these, the screw may alternatively be written $\hat{\mathbf{S}} = |\mathbf{S}|\,(1 + \varepsilon p)\,\hat{\mathbf{s}}$, where $\hat{\mathbf{s}}$ is the *unit line*, of unit magnitude and zero pitch, of the screw. Whenever, throughout this paper, we refer to the *normalised* instance of a screw $\hat{\mathbf{S}}$, we shall mean this unit line, written with the corresponding lower case letter, like $\hat{\mathbf{s}}$; and we shall write $\mathbf{s}$ for the unit direction 3-vector of that screw. For any two screws $\hat{\mathbf{S}}_1 \equiv \mathbf{S}_1 + \varepsilon\, \mathbf{S_{p_1}}$ and $\hat{\mathbf{S}}_2 \equiv \mathbf{S}_2 + \varepsilon\, \mathbf{S_{p_2}}$, the scalar product $\hat{\mathbf{S}}_1 \bullet \hat{\mathbf{S}}_2$ and cross product $\hat{\mathbf{S}}_1 \times \hat{\mathbf{S}}_2$ are defined over their dual elements in the same manner as they are defined over the real elements of real 3-vectors. The scalar product yields a dual number

$$\hat{\mathbf{S}}_1 \bullet \hat{\mathbf{S}}_2 = \mathbf{S}_1 \bullet \mathbf{S}_2 + \varepsilon\, \hat{\mathbf{S}}_1 \otimes \hat{\mathbf{S}}_2 \quad where \quad \hat{\mathbf{S}}_1 \otimes \hat{\mathbf{S}}_2 \equiv \mathbf{S}_1 \bullet \mathbf{S_{p_2}} + \mathbf{S_{p_1}} \bullet \mathbf{S}_2\ ,$$

in which the binary operation $\hat{\mathbf{S}}_1 \otimes \hat{\mathbf{S}}_2$ measures the *mutual moment* of the screws. In the situation $\mathbf{S}_1 \bullet \mathbf{S}_2 = 0$ where two screws are mutually perpendicular, their mutual moment vanishes if, and only if, the screws intersect. Thus the condition $\hat{\mathbf{S}}_1 \bullet \hat{\mathbf{S}}_2 = 0$, in which the screws are *orthogonal*, implies that each intersects the other at right angles. We note without proof that the cross product screw $\hat{\mathbf{S}}_1 \times \hat{\mathbf{S}}_2$ is sited in the *common perpendicular line* of $\hat{\mathbf{S}}_1$ and $\hat{\mathbf{S}}_2$. If, in particular, $\hat{\mathbf{s}}_1$ and $\hat{\mathbf{s}}_2$ are orthogonal lines, then $\hat{\mathbf{s}}_1 \times \hat{\mathbf{s}}_2$ is also a line and is orthogonal to both $\hat{\mathbf{s}}_1$ and $\hat{\mathbf{s}}_2$.

3. The Screw Axis of a Finite Displacement

When a rigid body of arbitrary shape suffers a finite displacement between an initial and a final location, there is a directed line $\hat{\mathbf{s}}$ — the *screw axis* — such that the displacement could have been achieved by means of a translation parallel to that line and by a rotation about it. The line $\hat{\mathbf{s}}$ is unique except in the special case that the displacement consists of a pure translation, when *any* line parallel with the translation will serve.

We follow Hunt (Hunt, 1987) in defining the *cardinal motion* for the given displacement as that which achieves the final location from the initial location by means of translation through a positive distance 2σ parallel to the direction of $\hat{\mathbf{s}}$ (thereby fixing the direction of $\hat{\mathbf{s}}$) and by rotation through an angle 2θ, $-\pi < 2\theta \leq \pi$, in a right-handed sense about that direction.

4. Specification of a Finite Displacement Screw

Adopting the standard interpretation of a *dual angle*, thus

$$\hat{\theta} \equiv \theta + \varepsilon\,\sigma \ , \quad \sin\hat{\theta} \equiv \sin\theta + \varepsilon\,\sigma\cos\theta \ , \quad \cos\hat{\theta} \equiv \cos\theta - \varepsilon\,\sigma\sin\theta$$

we use the elements of the cardinal motion, namely the axis $\hat{\mathbf{s}}$, the *half-translation* σ, and the *half-rotation* θ, $-\pi/2 < \theta \leq \pi/2$, to define the (directed) *finite displacement screw* which represents the displacement, viz.

$$\hat{\mathbf{S}} \equiv \sin\hat{\theta}\,\hat{\mathbf{s}} = \sin\theta\,(1 + \varepsilon\,P_S)\,\hat{\mathbf{s}} \ , \quad P_S \equiv \frac{\sigma}{\tan\theta} \ , \tag{1}$$

of amplitude $\sin\theta$ and pitch P_S whose line is $\hat{\mathbf{s}}$.

When a sequence $\hat{\mathbf{S}}_1$, $\hat{\mathbf{S}}_2$, $\hat{\mathbf{S}}_3$, ..., of such finite displacements is applied to a rigid body, their operations can not, in general, be *commuted* without causing change to the resultant; but they may be freely *associated* in pairs. As shown in (Parkin, 1994), the resultant of applying such a pair, first $\hat{\mathbf{S}}_1 \equiv \sin\hat{\theta}_1\,\hat{\mathbf{s}}_1$ and then $\hat{\mathbf{S}}_2 \equiv \sin\hat{\theta}_2\,\hat{\mathbf{s}}_2$, is given by the screw

$$\hat{\mathbf{S}} \equiv (\hat{\mathbf{S}}_1,\ \hat{\mathbf{S}}_2) = \cos\hat{\theta}_2\,\hat{\mathbf{S}}_1 + \cos\hat{\theta}_1\,\hat{\mathbf{S}}_2 - \hat{\mathbf{S}}_1 \times \hat{\mathbf{S}}_2 \ . \tag{2}$$

This is a screw formulation of the *screw triangle rule* (Bottema *et al.*, 1990).

5. Reference Frames

In saying that a body has adopted a particular location we shall imply that we know the coordinates, in some *ground frame*, of three normalised mutually orthogonal axial lines $\hat{\mathbf{x}}$, $\hat{\mathbf{y}}$, and $\hat{\mathbf{z}}$ which provide the x-, y-, and z-axes of a reference frame embedded within the body. So defined, the lines necessarily intersect at right angles in an origin point, and satisfy

$$\hat{\mathbf{x}} \bullet \hat{\mathbf{x}} = \hat{\mathbf{y}} \bullet \hat{\mathbf{y}} = \hat{\mathbf{z}} \bullet \hat{\mathbf{z}} = 1 \quad and \quad \hat{\mathbf{x}} \bullet \hat{\mathbf{y}} = \hat{\mathbf{y}} \bullet \hat{\mathbf{z}} = \hat{\mathbf{z}} \bullet \hat{\mathbf{x}} = 0 . \tag{3}$$

A general screw, written $\hat{\mathbf{S}}$ in ground frame coordinates, takes the form

$$\hat{\mathbf{S}}' = \begin{bmatrix} \hat{\mathbf{x}} \bullet \hat{\mathbf{S}} \\ \hat{\mathbf{y}} \bullet \hat{\mathbf{S}} \\ \hat{\mathbf{z}} \bullet \hat{\mathbf{S}} \end{bmatrix} = \begin{bmatrix} \hat{\mathbf{x}}^{\mathrm{T}} \\ \hat{\mathbf{y}}^{\mathrm{T}} \\ \hat{\mathbf{z}}^{\mathrm{T}} \end{bmatrix} \hat{\mathbf{S}} , \tag{4}$$

in the coordinates of the frame defined by $\hat{\mathbf{x}}$, $\hat{\mathbf{y}}$, and $\hat{\mathbf{z}}$. Equations 3 imply that the 3×3 *dual matrix* on the right is an *orthogonal* matrix.

6. The Unconstrained Body

The finite displacement screws available to an unconstrained body lie on every line in space, the screws $\hat{\mathbf{S}}$ which occupy any such line presenting every possible combination of permitted magnitude ($0 \leq |\hat{\mathbf{S}}| \leq 1$) and pitch ($-\infty \leq P_S \leq \infty$). Adopting any three normalised mutually orthogonal lines $\hat{\mathbf{x}}$, $\hat{\mathbf{y}}$, and $\hat{\mathbf{z}}$ as forming a fixed frame, we can represent any one of this six-fold infinity of screws by the dual linear combination

$$\hat{\mathbf{S}} = \begin{bmatrix} \hat{\mathbf{x}} & \hat{\mathbf{y}} & \hat{\mathbf{z}} \end{bmatrix} \begin{bmatrix} \hat{L} \\ \hat{M} \\ \hat{N} \end{bmatrix} = \begin{bmatrix} \hat{\mathbf{x}} & \hat{\mathbf{y}} & \hat{\mathbf{z}} \end{bmatrix} \begin{bmatrix} L + \varepsilon P \\ M + \varepsilon Q \\ N + \varepsilon R \end{bmatrix} , \quad L^2 + M^2 + N^2 \leq 1 ,$$

in which, subject to the inequality, the dual values $\hat{L}$, $\hat{M}$, $\hat{N}$ are arbitrary.

7. The Constrained Body

In following sections we find the finite displacement screws available to a body which is constrained by the fact that an embedded point, line, or plane, must remain in coincidence with another such element which is fixed.

In each case considered, certain sets of available screws are obvious, being intrinsic to symmetries of the contributing geometric elements. Often, however, these sets do not exhaust the available screws. If a constructed set is to comprise *all* finite displacements which maintain a particular coincidence, it must include those which result from sequential application of any pair of its member displacements. Thus, though proofs of closure are omitted, the development often proceeds through application of Equation 2.

The three geometric elements, point, line, and plane, admit of nine pairwise combinations. Sections 8 through 10 cover the three cases in which the elements are of the same type. Of the six combinations remaining, three are duplicates of others under interchange of the roles of being *fixed* and being *mobile*. These duplicated situations are considered in Sections 11 through 13. The reader may confirm that the interchange of roles is achieved by reversing the sequence of application of the screws $\hat{S}_1$ and $\hat{S}_2$ defined in those sections; and that, given the arbitrary nature of the parameters applying, the screw set is in each case the same as that for its duplicate.

8. A Point Constrained to lie in a Point

We distinguish the *fixed point* and the *mobile point*. It is obvious that the mobile point (and any body in which it is embedded) may rotate through an arbitrary angle 2θ about any line which passes through the fixed point; but it may not translate, so, along any such line, the translation distance 2σ vanishes. Adopting axes $\hat{x}$, $\hat{y}$, and $\hat{z}$ which intersect in an origin at the fixed point, we write the general normalised line through that point as

$$\hat{s} = l\,\hat{x} + m\,\hat{y} + n\,\hat{z} \ , \ \ l^2 + m^2 + n^2 = 1 \ ,$$

in which the coordinates l, m, and n are purely real. Following the definition given in Equation 1, we then construct the finite displacement screw

$$\hat{S} \equiv \sin\hat{\theta}\,\hat{s} = \sin\theta\,\hat{s} \ , \ \ \theta \ \ arbitrary \ , \ \ \sigma = 0 \ .$$

So the screw for the general finite displacement which leaves the mobile point invariant at the fixed point has the form of a linear combination

$$\hat{S} = \begin{bmatrix} \hat{x} & \hat{y} & \hat{z} \end{bmatrix} \begin{bmatrix} L \\ M \\ N \end{bmatrix} = \sin\theta \begin{bmatrix} \hat{x} & \hat{y} & \hat{z} \end{bmatrix} \begin{bmatrix} l \\ m \\ n \end{bmatrix} \ , \ \ L^2 + M^2 + N^2 \leq 1$$

$$(5)$$

in which the potentially dual coefficients are, in fact, purely real. Subject only to the inequality just written, these real values may be arbitrarily chosen.

9. A Line Constrained to lie in a Line

We distinguish the *mobile line* and the *fixed line*, which latter we adopt as the axis $\hat{x}$ with the axes $\hat{y}$ and $\hat{z}$ generally disposed. The freedom of the mobile line to slide and rotate arbitrarily in $\hat{x}$ is described completely by the screw of arbitrary magnitude and pitch which lies on that line, namely

$$\hat{S} \equiv \sin\hat{\theta}\,\hat{x} \ , \ \ \theta \ \ arbitrary \ , \ \ \sigma \ \ arbitrary \ . \tag{6}$$

10. A Plane Constrained to lie in a Plane

Arbitrarily distinguishing the *fixed plane* and the *mobile plane*, we adopt $\hat{\mathbf{x}}$ and $\hat{\mathbf{y}}$ to be axes lying in the fixed plane with $\hat{\mathbf{z}}$ in its normal direction. Obvious freedoms of the mobile plane are, firstly,

$$\hat{\mathbf{S}}_1 \equiv \sin\hat{\theta}_1\,\hat{\mathbf{s}}_1 = \varepsilon\,\sigma_1\,(l\,\hat{\mathbf{x}} + m\,\hat{\mathbf{y}})\ ,\quad l^2+m^2=1\ ,\quad \theta_1=0\ ,\quad \sigma_1\ \ arbitrary\ ,$$

which describes translation through the arbitrary distance $2\sigma_1$ in the fixed plane in a direction at arbitrary angle $\tan^{-1} m/l$ from $\hat{\mathbf{x}}$; and, secondly,

$$\hat{\mathbf{S}}_2 \equiv \sin\hat{\theta}_2\,\hat{\mathbf{z}} = \sin\theta_2\,\hat{\mathbf{z}}\ ,\quad \theta_2\ \ arbitrary\ ,\quad \sigma_2=0\ ,$$

which describes a pure rotation through the arbitrary angle $2\theta_2$ about the normal axis $\hat{\mathbf{z}}$. Although we may not do so in general, we can in this case write the totality of available screws as the simple union of these sets, viz.

$$\left.\begin{array}{l}\hat{\mathbf{S}} \equiv \sin\hat{\theta}\,\hat{\mathbf{s}} = \sin\theta_2\,\hat{\mathbf{z}} + \varepsilon\,\sigma_1\,(l\,\hat{\mathbf{x}} + m\,\hat{\mathbf{y}})\ ,\quad l^2+m^2=1\ ,\\[4pt] \theta=\theta_2\ \ arbitrary\ ,\quad \sigma=\sigma_2=0\ ,\quad \sigma_1\ \ arbitrary\ ,\end{array}\right\} \tag{7}$$

which screws are parallel to the plane normal axis $\hat{\mathbf{z}}$. To confirm this step we use Equation 2 for the resultant of applying first $\hat{\mathbf{S}}_1$ and then $\hat{\mathbf{S}}_2$. Thus

$$\begin{aligned}\hat{\mathbf{S}} &= \cos\theta_2\,\varepsilon\,\sigma_1\,(l\,\hat{\mathbf{x}}+m\,\hat{\mathbf{y}}) + \sin\theta_2\,\hat{\mathbf{z}} + \sin\theta_2\,\varepsilon\,\sigma_1\,(l\,\hat{\mathbf{y}}-m\,\hat{\mathbf{x}})\\[4pt] &= \sin\theta_2\,\hat{\mathbf{z}} + \varepsilon\,\sigma_1\,\{\cos(\theta_2+\tan^{-1} m/l)\,\hat{\mathbf{x}} + \sin(\theta_2+\tan^{-1} m/l)\,\hat{\mathbf{y}}\}\ ,\end{aligned}$$

which, to the extent that l and m may be arbitrarily chosen, has the same form as Equation 7. Equation 2 shows that applying those displacements in the opposite sequence has no effect other than to express this result in terms of angular differences rather than angular sums. Thus, in brief,

$$\hat{\mathbf{S}} = \begin{bmatrix} \hat{\mathbf{x}} & \hat{\mathbf{y}} & \hat{\mathbf{z}} \end{bmatrix} \begin{bmatrix} \hat{L} \\ \hat{M} \\ \hat{N} \end{bmatrix} = \begin{bmatrix} \hat{\mathbf{x}} & \hat{\mathbf{y}} & \hat{\mathbf{z}} \end{bmatrix} \begin{bmatrix} \varepsilon\,L \\ \varepsilon\,M \\ N \end{bmatrix}\ , \tag{8}$$

in which the direction numbers $\hat{L}$ and $\hat{M}$ are imaginary, and $\hat{N}$ is real.

11. A Point Constrained to lie in a Line

We place the origin of the $\hat{\mathbf{x}}\hat{\mathbf{y}}\hat{\mathbf{z}}$-frame at the given *mobile point*, with $\hat{\mathbf{x}}$ lying on the *fixed line* and with $\hat{\mathbf{y}}$ and $\hat{\mathbf{z}}$ generally disposed. For convenience in evaluation we attribute all rotations about the line $\hat{\mathbf{x}}$ to the rotational freedoms provided by the given point, so that the line itself provides only a sliding freedom. Obvious freedoms are, firstly, by Equation 5,

$$\left.\begin{array}{l}\hat{\mathbf{S}}_1 \equiv \sin\hat{\theta}_1\,\hat{\mathbf{s}}_1 = L\,\hat{\mathbf{x}} + M\,\hat{\mathbf{y}} + N\,\hat{\mathbf{z}}\ ,\quad L^2+M^2+N^2\leq 1\ ,\\[4pt] \theta_1\ \ arbitrary\ ,\quad \sigma_1=0\ ,\quad \sin\hat{\theta}_1=\sin\theta_1=\pm\sqrt{L^2+M^2+N^2}\ ,\end{array}\right\} \tag{9}$$

which describes a pure rotation through the arbitrary angle $2\theta_1$ about an arbitrarily chosen line through the given point and, secondly,

$$\hat{S}_2 \equiv \sin\hat{\theta}_2\,\hat{x} = \varepsilon\,\sigma_2\,\hat{x}\;,\quad \theta_2 = 0\;,\quad \sigma_2\;\; arbitrary\;,$$

which describes a pure translation through the arbitrary distance $2\sigma_2$ along the given line $\hat{x}$. Using Equation 2 we find the general available displacement, being the resultant screw arising from applying first $\hat{S}_1$ and then $\hat{S}_2$, to be the dual linear combination

$$\hat{S} = \sin\theta_1\,\begin{bmatrix}\hat{x} & \hat{y} & \hat{z}\end{bmatrix}\begin{bmatrix} l & + & \varepsilon\,\sigma_2\cot\theta_1 \\ m & - & \varepsilon\,\sigma_2 n \\ n & + & \varepsilon\,\sigma_2 m \end{bmatrix}\;,\quad l^2+m^2+n^2=1\;. \quad (10)$$

12. A Point Constrained to lie in a Plane

We place the origin of the $\hat{x}\hat{y}\hat{z}$-frame at the *mobile point* with $\hat{x}$ and $\hat{y}$ lying in the *fixed plane* and $\hat{z}$ in normal direction. For convenience we attribute all rotations about $\hat{z}$ to the rotational freedoms of the point, so the plane provides only sliding freedoms. Obvious freedoms are: firstly, those inherent in the point itself, as defined for the screw $\hat{S}_1$ in Equation 9; and, secondly,

$$\hat{S}_2 \equiv \sin\hat{\theta}_2\,\hat{s}_2 = \varepsilon\,\sigma_2\,(l_2\,\hat{x}+m_2\,\hat{y})\;,\quad l_2^2+m_2^2=1\;,\quad \theta_2=0\;,\quad \sigma_2\;\; arbitrary\;,$$

which describes translation through the arbitrary distance $2\sigma_2$ along any line which lies in the $\hat{x}\hat{y}$-plane and passes through the origin at arbitrary angle $\tan^{-1} m_2/\,l_2$ from $\hat{x}$. Using Equation 2, we find the general available displacement, arising from applying first $\hat{S}_1$ and then $\hat{S}_2$, to be

$$\hat{S} = \sin\theta_1\,\begin{bmatrix}\hat{x} & \hat{y} & \hat{z}\end{bmatrix}\begin{bmatrix} l & + \varepsilon\,\sigma_2(\cot\theta_1\,l_2+n\,m_2) \\ m & + \varepsilon\,\sigma_2(\cot\theta_1\,m_2-n\,l_2) \\ n & + \varepsilon\,\sigma_2(m\,l_2-l\,m_2) \end{bmatrix}\;,\quad l^2+m^2+n^2=1\;.$$
$$(11)$$

13. A Plane Constrained to pass through a Line

We adopt fixed axes $\hat{x}$, $\hat{y}$, and $\hat{z}$ such that $\hat{x}$ lies along the *fixed line* and $\hat{z}$ is a normal line of the given location of the *mobile plane*. Available freedoms are: firstly, by Equation 7,

$$\hat{S}_1 \equiv \sin\hat{\theta}_1\,\hat{s}_1 = \sin\theta_1\,\hat{z} + \varepsilon\,\sigma_1\,(l\,\hat{x}+m\,\hat{y})\;,\quad l^2+m^2=1\;,\quad \theta_1,\,\sigma_1\;\; arbitrary\;,$$

which describes an arbitrary translation within the $\hat{x}\hat{y}$-plane through the distance $2\sigma_1$ at angle $\tan^{-1} m/\,l$ from the axis $\hat{x}$, together with an arbitrary rotation through angle $2\theta_1$ about the normal line $\hat{z}$; and, secondly,

$$\hat{S}_2 \equiv \sin\hat{\theta}_2\,\hat{x} = \sin\theta_2\,\hat{x}\;,\quad \theta_2\;\; arbitrary\;,\quad \sigma_2=0\;,$$

which describes a pure rotation through the arbitrary angle $2\theta_2$ about the given line $\hat{\mathbf{x}}$. Using Equation 2 we find the general available displacement arising from applying first $\hat{\mathbf{S}}_1$ and then $\hat{\mathbf{S}}_2$, to be the dual linear combination

$$\hat{\mathbf{S}} = [\ \hat{\mathbf{x}}\ \ \hat{\mathbf{y}}\ \ \hat{\mathbf{z}}\ ]\begin{bmatrix} \cos\theta_1\sin\theta_2 & + & \varepsilon\,\sigma_1\,l\cos\theta_2 \\ -\sin\theta_1\sin\theta_2 & + & \varepsilon\,\sigma_1\,m\cos\theta_2 \\ \sin\theta_1\cos\theta_2 & + & \varepsilon\,\sigma_1\,m\sin\theta_2 \end{bmatrix}\ ,\ \ l^2+m^2=1\ . \quad (12)$$

14. Compounding Constraints

When two constraints, each represented by a set of finite displacement screws, are both made to apply to the mobility of a body, the only screws which remain available are those which lie in the intersection of the sets.

Limitations of space do not permit demonstration of a general solution method. For simplicity in the following illustration, we consider situations in which the directions of corresponding axial basis screws of the contributing screw sets may validly be made parallel.

15. A Body Constrained to be fixed at Two Points

Twice making reference to Equation 5, we write the identity of the common screw set in the ground frame as

$$\hat{\mathbf{S}} = [\ \hat{\mathbf{x}}_1\ \ \hat{\mathbf{y}}_1\ \ \hat{\mathbf{z}}_1\ ]\begin{bmatrix} L_1 \\ M_1 \\ N_1 \end{bmatrix} = [\ \hat{\mathbf{x}}_2\ \ \hat{\mathbf{y}}_2\ \ \hat{\mathbf{z}}_2\ ]\begin{bmatrix} L_2 \\ M_2 \\ N_2 \end{bmatrix}\ ,\ \ \begin{array}{l} L_1^2+M_1^2+N_1^2\leq 1 \\ L_2^2+M_2^2+N_2^2\leq 1 \end{array}$$

in which, subject to these inequalities, the real coefficients L_1, M_1, N_1 and L_2, M_2, N_2 may be arbitrarily chosen. This last fact directly implies that we are free to choose the axial directions of the reference frames.

We choose corresponding axes to be parallel but otherwise arbitrary. On pre-multiplying by the inverse of the former orthogonal matrix, so the equation is expressed in the coordinates of the $\hat{\mathbf{x}}_1\hat{\mathbf{y}}_1\hat{\mathbf{z}}_1$-frame, we obtain

$$\begin{bmatrix} 1 & 0 & 0 \\ 0 & 1 & 0 \\ 0 & 0 & 1 \end{bmatrix}\begin{bmatrix} L_1 \\ M_1 \\ N_1 \end{bmatrix} = \begin{bmatrix} 1 & \varepsilon\,\hat{\mathbf{x}}_1\otimes\hat{\mathbf{y}}_2 & \varepsilon\,\hat{\mathbf{x}}_1\otimes\hat{\mathbf{z}}_2 \\ \varepsilon\,\hat{\mathbf{y}}_1\otimes\hat{\mathbf{x}}_2 & 1 & \varepsilon\,\hat{\mathbf{y}}_1\otimes\hat{\mathbf{z}}_2 \\ \varepsilon\,\hat{\mathbf{z}}_1\otimes\hat{\mathbf{x}}_2 & \varepsilon\,\hat{\mathbf{z}}_1\otimes\hat{\mathbf{y}}_2 & 1 \end{bmatrix}\begin{bmatrix} L_2 \\ M_2 \\ N_2 \end{bmatrix}\ ,$$

Since the coefficients are real, any solution screw $\hat{\mathbf{S}}$ must be of zero pitch and must pass through the origin of the $\hat{\mathbf{x}}_1\hat{\mathbf{y}}_1\hat{\mathbf{z}}_1$-frame. The real parts of this equation state the obvious: that corresponding direction numbers must be equal in the parallel frames. On further expressing the parallelism of corresponding axes, the imaginary-part equation can be written

$$
\begin{bmatrix} 0 \\ 0 \\ 0 \end{bmatrix} = \begin{bmatrix} 0 & \hat{\mathbf{x}}_1 \otimes \hat{\mathbf{y}}_2 & \hat{\mathbf{x}}_1 \otimes \hat{\mathbf{z}}_2 \\ -\hat{\mathbf{x}}_1 \otimes \hat{\mathbf{y}}_2 & 0 & \hat{\mathbf{y}}_1 \otimes \hat{\mathbf{z}}_2 \\ -\hat{\mathbf{x}}_1 \otimes \hat{\mathbf{z}}_2 & -\hat{\mathbf{y}}_1 \otimes \hat{\mathbf{z}}_2 & 0 \end{bmatrix} \begin{bmatrix} L_2 \\ M_2 \\ N_2 \end{bmatrix} = \mathbf{D} \times \begin{bmatrix} L_2 \\ M_2 \\ N_2 \end{bmatrix} ,
$$

in which the multiplying matrix, if non-null, is skew-symmetric so that its product with any vector can be re-expressed, as shown on the right, as a vector cross product containing the vector

$$
\mathbf{D} = \begin{bmatrix} \hat{\mathbf{y}}_1 \otimes \hat{\mathbf{z}}_2 \\ \hat{\mathbf{z}}_1 \otimes \hat{\mathbf{x}}_2 \\ \hat{\mathbf{x}}_1 \otimes \hat{\mathbf{y}}_2 \end{bmatrix} = \begin{bmatrix} d_x \\ d_y \\ d_z \end{bmatrix} ,
$$

which measures the vector distance from the origin of the $\hat{\mathbf{x}}_1 \hat{\mathbf{y}}_1 \hat{\mathbf{z}}_1$-frame to the origin of the $\hat{\mathbf{x}}_2 \hat{\mathbf{y}}_2 \hat{\mathbf{z}}_2$-frame. The vanishing of that cross product effectively states that any solution screw $\hat{\mathbf{S}}$ must lie parallel with the vector $\mathbf{D}$. So we discover, as expected, that the solution screws

$$
\hat{\mathbf{S}} = \frac{\sin \theta_1}{|\mathbf{D}|} \begin{bmatrix} \hat{\mathbf{x}}_1 & \hat{\mathbf{y}}_1 & \hat{\mathbf{z}}_1 \end{bmatrix} \mathbf{D} = \frac{\sin \theta_1}{|\mathbf{D}|} \begin{bmatrix} \hat{\mathbf{x}}_1 & \hat{\mathbf{y}}_1 & \hat{\mathbf{z}}_1 \end{bmatrix} \begin{bmatrix} \hat{\mathbf{y}}_1 \otimes \hat{\mathbf{z}}_2 \\ \hat{\mathbf{z}}_1 \otimes \hat{\mathbf{x}}_2 \\ \hat{\mathbf{x}}_1 \otimes \hat{\mathbf{y}}_2 \end{bmatrix} , \quad (13)
$$

lie on the line that joins the given points and, while permitting no translation (since the pitch vanishes), provide arbitrary rotation about that line.

16. A Body Constrained to be fixed at Three Points

We intersect sets of screws for one of the given points and a line $\hat{\mathbf{S}}_2$ determined as above for the other points. With origin at the point, we choose $\hat{\mathbf{x}}_1$ to be parallel to $\hat{\mathbf{S}}_2$, with $\hat{\mathbf{y}}_1$ and $\hat{\mathbf{z}}_1$ arbitrarily disposed. We write the identity of the common screw set in the ground frame as

$$
\hat{\mathbf{S}} = \begin{bmatrix} \hat{\mathbf{x}}_1 & \hat{\mathbf{y}}_1 & \hat{\mathbf{z}}_1 \end{bmatrix} \begin{bmatrix} L_1 \\ M_1 \\ N_1 \end{bmatrix} = \hat{\mathbf{S}}_2 = |\mathbf{S}_2| \hat{\mathbf{s}}_2 , \quad L_1^2 + M_1^2 + N_1^2 \leq 1 .
$$

On multiplying by the inverse of the orthogonal matrix we obtain

$$
\begin{bmatrix} 1 & 0 & 0 \\ 0 & 1 & 0 \\ 0 & 0 & 1 \end{bmatrix} \begin{bmatrix} L_1 \\ M_1 \\ N_1 \end{bmatrix} - |\mathbf{S}_2| \begin{bmatrix} 1 \\ \varepsilon \hat{\mathbf{y}}_1 \otimes \hat{\mathbf{s}}_2 \\ \varepsilon \hat{\mathbf{z}}_1 \otimes \hat{\mathbf{s}}_2 \end{bmatrix} = 0 ,
$$

which has the real solutions required on the left only if the imaginary quantities vanish; that is, as expected, only if the line $\hat{\mathbf{S}}_2$ passes through the origin. In any other situation the only available solution is the null screw, which carries the implication that the body is completely constrained.

17. Conclusion

We have exemplified a methodolgy in which geometric constraints on the mobility of solid bodies may be manipulated in a uniform manner — in terms of sets of finite displacement screws which can be expressed as dual linear combinations of an orthogonal *basis* of normalised axial lines.

Starting with the point, the line, and the plane as geometric elements, Sections 8 through 10 have shown that whenever a mobile element (and any body within which it is embedded) is constrained to lie in another fixed element *of the same type*, the available screws may be constructed as obvious dual linear combinations of certain symmetry lines. Sections 11 through 13, where elements *of different types* are involved, have then shown that the available screws may be constructed by combining the freedoms of each type through the agency of the *screw triangle rule* of Equation 2. It is in the nature of this rule that if the sets to be combined are each expressed as a dual linear combination of certain orthogonal lines, then so is the resultant set.

Finally, in Sections 14 through 16, where the evaluation of intersection sets was discussed, it has been seen that the solution coefficients must again apply in linear combinations of the adopted axial lines. It can be seen that the uniformity of this form of representation of constraints is preserved throughout.

References

J. Angeles. Automatic computation of the screw parameters of rigid body motions. Part I: Finitely-separated positions. *Trans. ASME* (*J. Dynamic Systems, Measurement and Control*) **108**, pp. 32–38 (1986).

B. Ravani and Q.J. Ge. Computations of spatial displacements from geometric features. *Trans. ASME* (*J. Mechanical Design*) **115**, pp. 95–102 (1993).

I.A. Parkin. A third conformation with the screw systems: finite twist displacements of a directed line and point. *Mechanism and Machine Theory* **27**, pp. 177–188 (1992).

C. Huang and B. Roth. Analytic expressions for the finite screw systems. *Mechanism and Machine Theory* **29**, pp. 207–222 (1994).

K.H. Hunt and I.A. Parkin. Finite displacements of points, planes and lines *via* screw theory. *Mechanism and Machine Theory* **30**, pp. 177–192 (1995).

O. Bottema and B. Roth. *Theoretical kinematics*. North-Holland Publishing Company, Amsterdam (1979). Reprinted Dover, New York (1990).

K.H. Hunt. Manipulating a body through a finite displacement. *Proc. Seventh World Congress on the Theory of Machines and Mechanisms, Sevilla, Spain*, pp. 187–91 (1987).

I.A. Parkin. Zero magnitude screws in the 3-system of finite displacement screws for a pair of revolute joints. *Advances in robot kinematics and computational geometry* (eds. J. Lenarcic and B. Ravani), Kluwer Academic Publishers, pp. 401–410 (1994).

REAL ROOT COUNTING FOR SOME ROBOTICS PROBLEMS

FABRICE ROUILLIER
IRMAR Université de Rennes I
Avenue des Buttes de Coesme, 35042 Rennes cedex, France

Abstract. We propose two algorithms to compute the number of real roots of zero-dimensional systems, using effective algebraic methods. To compare their behaviour on practical examples, we apply these methods to systems that describe some robotics problems (e.g. direct kinematic problem of parallel manipulators).

1. Introduction

Let Z be a domain, K its Fraction field, $S = \{f_1, f_2, \cdots, f_s\}$ a system of polynomial equations in $Z[X_1, \cdots, X_n]$ with a finite set of distinct roots $\mathcal{X} = \{\alpha_1, \ldots, \alpha_d\}$ of respective multiplicities $\{m_1, \ldots, m_d\}$ and , I the ideal generated by S .

In such cases, $A = \frac{K[X_1, \cdots, X_n]}{I}$ is a finite dimensional vector space of dimension $D = \sum_{i=1}^{d} m_i$.

Referring to (Rouillier , 1995), we assume in this paper the existence of efficient algorithms that compute the following (from a Groebner basis) :

- A Z-basis $\mathcal{B} = \omega_1, \ldots, \omega_D$ of A
- The multiplication table of A (with respect to $\mathcal{B}$), defined by a $D \times D$ matrix $MT = (MT[i,j])_{1 \leq i,j \leq D}$, where $MT[i,j]$ is the column-vector whose coordinates are the coefficients of $\omega_i \omega_j$ with respect to $\mathcal{B}$ (e.g. $\omega_i \omega_j = \sum_{k=1}^{D} (MT[i,j])_k \cdot \omega_k$).

We study two different strategies in order to count the number of real roots :

- Hermite's method, which computes a quadratic form whose signature gives the number of real roots of the system.

J.-P. Merlet and B. Ravani (eds.), Computational Kinematics, 73–82.
© 1995 *Kluwer Academic Publishers.*

- The Generalized Shape Lemma , which computes an univariate polynomial that has the same number of real roots as the system . The number of distinct real roots can then be computed using a Sturm-Habicht sequence.

Since the reduction of a quadratic form needs $O(n^3)$ basic arithmetic operations and Sturm-Habicht sequences needs $O(n^2)$ operations, the second method is asymptotically better. In practice however, the first method behaves better in many cases.

2. Hermite's method

For every $h \in A$, we define :

- The linear homomorphism of multiplication by h:

$$m_h : \quad A \quad \to \quad A$$
$$q \quad \longmapsto \quad h \cdot q$$

- The matrix M_h of m_h with respect to $\mathcal{B}$
- The h-trace symmetric bilinear form (or simply trace if $h = 1$) :

$$Tr_h : \quad A \times A \quad \to \quad K$$
$$(f, g) \quad \longmapsto \quad Trace(fgh)$$

where $Trace(fgh)$ is the trace of M_{fgh}.
- Hermite's quadratic form :

$$Q_h : \quad A \quad \to \quad K$$
$$f \quad \longmapsto \quad Tr_h(f, f)$$

Let C (resp. R) be the algebraic closure (resp. real closure) of Z, the following theorem relates the rank and signature of Q_h to the number of zeros of S in C^n or R^n (see (Petersen & al., 1993)) :

Theorem 1 *Let S be a zero-dimensional system of $Z[X_1, \ldots, X_n]$ with a finite set of distinct zeros $\mathcal{X}$, C (resp. R) the algebraic closure (resp. real closure) of Z. Then :*

- $rank(Q_h) = \sharp\{\delta \in C^n \bigcap \mathcal{X} , h(\delta) \neq 0\}$
- $signature(Q_h) = \sharp\{\delta \in R^n \bigcap \mathcal{X}, h(\delta) > 0\} - \sharp\{\delta \in R^n \bigcap \mathcal{X}, h(\delta) < 0\}$

Remark 1 *The second formula allows us to produce an algorithm "à la Ben Or - Kozen -Reif" to deal with polynomial inequalities $(= 0 , > 0 , < 0)$ over $\mathcal{X} \bigcap R$ (see (Ben-Or & al., 1986)).*

Applying the previous theorem with $h = 1$, we obtain an algorithm to compute the number of distinct real or complex roots of a zero-dimensional system :

Corollary 1 *With the notations of theorem 1,*

$- \ rank(Q_1) = \sharp\{\delta \in C^n \bigcap \mathcal{X}\}$
$- \ signature(Q_1) = \sharp\{\delta \in R^n \bigcap \mathcal{X}\}$

2.1. COMPUTING HERMITE'S QUADRATIC FORM

Let the multiplication table MT of A with respect to a linear basis $\mathcal{B} = \{\omega_1, \ldots, \omega_D\}$ be given and for $P \in A$, let $Vect(P)$ be the column-vector whose coordinates are the coefficients of P with respect to $\mathcal{B}$.

With these notations, the matrix $Q_{h,\mathcal{B}} = (Q_{h,\mathcal{B}}[\imath, \jmath])_{1 \leq \imath, \jmath \leq D}$ of Q_h with respect to the basis $\mathcal{B}$ is defined by :

$$Q_{h,\mathcal{B}}[\imath, \jmath] = Q_h(\omega_\imath, \omega_\jmath) = Trace(\omega_\imath \omega_\jmath h) = \sum_{k=1}^{D} Vect(\omega_\imath \omega_\jmath \omega_k h)_k$$

A naive algorithm would require us to compute all the products $\omega_\imath \omega_\jmath \omega_k$ (resp. $\omega_\imath \omega_\jmath \omega_k \omega_l$) to get $Q_{1,\mathcal{B}}$ (resp. $Q_{h,\mathcal{B}}$). Even if we can do these computations using the strategy proposed in (Rouillier , 1995), the method would be inefficient because of a dramatic growth of the number of monomials involved . In order to improve this, we use the linearity of the mapping trace, as much as possible :

Lemma 1 *If we denote by $Vtr(h)$ the column-vector*

$$[Trace(h\omega_1), \ldots, Trace(h\omega_D)]^T$$

then :

$- \ Vtr(h) = (Q_{1,\mathcal{B}})^T \cdot Vect(h)$
$- \ \text{for all } 1 \leq \imath, \jmath \leq D \ , \ Q_{h,\mathcal{B}}[\imath, \jmath] = MT[\imath, \jmath] \cdot Vtr(h)$

Proof : Let $h = \sum_{k=1}^{D} a_k \omega_k$. Then, for all $1 \leq \imath \leq D$,

$$Trace(h\omega_\imath) = \sum_{k=1}^{D} a_k Trace(\omega_k \omega_\imath)$$

and so $Vtr(h) = (Q_{1,\mathcal{B}})^T \cdot Vect(h)$.

Let $\omega_\imath \omega_\jmath = \sum_{l=1}^{D} a_l^{(\imath,\jmath)} \omega_l$, then :

$$\begin{aligned} Q_{h,\mathcal{B}}[\imath, \jmath] &= \sum_{k=1}^{D} Vect(\omega_\imath \omega_\jmath \omega_k h)_k = \sum_{k=1}^{D} \sum_{l=1}^{D} a_l^{(\imath,\jmath)} Vect(\omega_l \omega_k h)_k \\ &= \sum_{l=1}^{D} a_l^{(\imath,\jmath)} \sum_{k=1}^{D} Vect(\omega_l \omega_k h)_k = \sum_{l=1}^{D} a_l^{(\imath,\jmath)} Trace(h\omega_l) \\ &= Vect(\omega_\imath \omega_\jmath) \cdot Vtr(h) \end{aligned}$$

$\square$

When studying systems with coefficients in Z, one obtains for Hermite's quadratic form a matrix with coefficients in K, but using a simple transformation, one can assume that the matrix of $Q_{h,\mathcal{B}}$ has coefficients in Z.

The author's recent algorithm (see (Rouillier, 1994)) generalizes the Bareiss identities in order to reduce $Q_{h,B}$ by doing all the computations in Z with $O(D^3)$ basic operations, and allows a good control of the size of the intermediate results, when Z is the domain of integers (e.g $O(D(t+log(D)))$ if t is the maximum size of the coefficients of $Q_{h,B}$).

We can now describe a first algorithm for computing the number of distinct real roots of a given zero-dimensional system :

Algorithm I

Input : A Groebner basis of S for any admissible monomial ordering and the associated multiplication table MT.

- **step 1** : Compute $Vtr(1) = [\sum_{i=1}^{D}(MT[1,i])_i, \ldots, \sum_{i=1}^{D}(MT[D,i])_i]^T$
- **step 2** : Compute

$$Q_{1,B}[j,i] = Q_{1,B}[i,j] = MT[i,j] \cdot Vtr(1) , \ 1 \le i \le j \le D$$

- **step 3** : Compute the signature of $Q_{1,B}$.

3. Generalized Shape lemma

The idea of the Generalized Shape lemma (see (Alonso & al., 1994)) is to express the solutions of a polynomial system as rational functions in the roots of a univariate polynomial.

Let $\mathcal{Y} = \{\beta_1, \ldots, \beta_D\}$ be the set of the roots (not necessary distinct) of S.

Let t be a separating element of A i.e. such that for every $x \ne y$ in $\mathcal{Y}$, $t(x) \ne t(y)$ and let v be any element of A.

Consider the polynomials

$$f(t,T) = \prod_{y \in \mathcal{Y}} (T - t(y))$$

$$g(v,t,T) = \sum_{x \in \mathcal{Y}} v(x) \prod_{y \in \mathcal{Y}, y \ne x} (T - t(y))$$

If β is a zero of S of multiplicity m, then $f(t, t(\beta)) = 0$, and we have

$$v(\beta) = \frac{g^{(m-1)}(v,t,t(\beta))}{g^{(m-1)}(t,t(\beta))} = \left(\frac{\frac{\partial^{m-1} g(v,t,T)}{\partial T^{m-1}}}{\frac{\partial^{m-1} g(t,T)}{\partial T^{m-1}}} \right)_{T=t(\beta)}$$

where $g(t,T) = f'(t,T) = g(1,t,T)$.

Proposition 1 *Let $S = \{f_1, \ldots, f_s\}$ be a zero-dimensional system of polynomials of $Z[X_1, \ldots, X_n]$, and $\mathcal{X} = \{\alpha_1, \ldots \alpha_d\}$ the distinct solutions of S with respective multiplicities $\{m_1, \ldots, m_d\}$.*

There exist a separating element t in A and polynomials

$$f(t,T) \, , \, g(t,T) \, , \, g_1(t,T), \ldots, g_n(t,T)$$

so that :

- *The roots of $f(t,T)$ are exactly $\{t(\alpha_1), \ldots, t(\alpha_d)\}$ with respective multiplicities $\{m_1, \ldots, m_d\}$*
- *If β is a zero of S of multiplicity m,*

$$X_i(\beta) = \frac{g_i^{(m-1)}(v,t,t(\beta))}{g^{(m-1)}(t,t(\beta))}$$

Proof : The existence of a separating element is given by the following lemma :

Lemma 2 *If $\mathcal{X}$ contains less than d points, then at least one among the $u_i = X_1 + iX_2 + \ldots + i^{n-1}X_n$ for $0 \leq i \leq (n-1)\frac{d(d-1)}{2}$ is a separating element.*

Proof : Consider a couple $(x,y) = ((x_1, \ldots, x_n), (y_1, \ldots, y_n))$ of distinct points of $\mathcal{X}$, and let $l(x,y)$ be the number of index i such that $u_i(x) = u_i(y)$. Since the polynomial $(x_1 - y_1) + \ldots + (x_n - y_n)t^{n-1}$ has no more than $k-1$ distinct roots (it is not indentically null because $x \neq y$), $l(x,y)$ is less than $k-1$. Since the total number of couples of distinct points of $\mathcal{X}$ is less than $\frac{d(d-1)}{2}$, this complete the proof. □

Given a separating element t, we complete the proof of the proposition by taking $g_i(t,T) = g(X_i,t,T)$, $i = 1 \ldots n$ □

According to the previous proposition, the number of distinct real roots of a given system can be easily computed from the Generalized Shape Lemma :

Corollary 2 *Let t be a separating element of A. The number of distinct real roots of S is exactly the number of real roots of $f(t,T)$. It can also be found by computing the Sturm-Habicht sequence of $f(t,T)$*

We will not discuss here the problem of finding a separating element, but, given $f(t,T)$, $g(t,T)$, $g_1(t,T), \ldots, g_n(t,T)$ for any t (randomly chosen) , we assume that there exists a simple method to check if t is a separating element or not (see (Rouillier , 1995)).

An algorithm that computes the number of distinct real roots of a given zero dimensional system can now be described :

Algorithm II

Input : A Groebner basis of S for any admissible monomial ordering and the associated multiplication table MT.

- **step 1** : Take any t among

$$\{X_1 + \imath X_2 + \ldots + \imath^{n-1} X_n \ , \ \imath = 0 \ldots (n-1)\frac{D(D-1)}{2}\}$$

- **step 2** : Compute $f(t,T)$, $g(t,T)$, $g_1(t,T), \ldots, g_n(t,T)$
- **step 3** : Check if t is a separating element, and if not go to **step 1**.
- **step 4** : Compute the Sturm-Habicht sequence of $f(t,T)$.

3 1   COMPUTING GSL USING TRACES AND SYMMETRIC FUNCTIONS

The method we propose for computing $f(t,T) = \prod_{y \in \mathcal{Y}}(T - t(y))$ uses the classical notion of elementary symmetric functions and their connection with Newton sums.

Notation 1 *We denote by :*

- $S_\imath(t, \mathcal{Y}) = \sum_{\mathcal{I} \subset \mathcal{Y} \ \sharp \mathcal{I} = \imath} \prod_{y \in \mathcal{I}} t(y)$ *the $\imath^{th}$ elementary symmetric function associated with* $\{t(\beta_1), \ldots, t(\beta_D)\}$.
- $N_\imath(t, \mathcal{Y}) = \sum_{y \in \mathcal{Y}} t(y)^\imath$ *the $\imath^{th}$ elementary Newton sum associated with* $\{t(\beta_1), \ldots, t(\beta_D)\}$.

According to this notation, the classical relations that link elementary symmetric functions to elementary Newton sums can be written as follows :

$$(D - \imath)S_\imath(t, \mathcal{Y}) = \sum_{\jmath=0}^{\imath}(-1)^\jmath N_\jmath(t, \mathcal{Y})S_{\imath-\jmath}(t, \mathcal{Y})$$

with the convention $S_0(t, \mathcal{Y}) = 1$.

Let M_t the matrix of multiplication by a polynomial t in A. Since the eigenvalues of M_t are the scalars : $\{t(y) \ , \ y \in \mathcal{Y}\}$ (see (Petersen & al., 1993)), the set $\{N_\jmath(t, \mathcal{Y})\}$ can be computed using the relation :

$$Trace(t^\imath) = N_\imath(t, \mathcal{Y})$$

Since $S_\imath(t, \mathcal{Y})$ is the coefficient of $T^{D-\imath}$ in the polynomial $\prod_{y \in \mathcal{Y}}(T - t(y))$, $f(t,T)$ can be easily computed by using the traces $Trace(t^\imath)$, $\imath = 1, \ldots D$.

Expanding the polynomial $g(v, t, T)$ (of degree $D - 1$), we note that the coefficient of $T^{D-\imath-1}$ in $g(v, t, T)$ is

$$(-1)^\imath \sum_{y \in \mathcal{Y}} v(y)S_\imath(t, \mathcal{Y} \setminus \{y\})$$

We also extend the notion of elementary symmetric function and elementary Newton sums :

Definition 1 *Given two polynomials t and $v \in K[X_1, \ldots, X_n]$ and $\mathcal{Y} \in C^n$,*

- $S_\imath(v, t, \mathcal{Y}) = \sum_{y \in \mathcal{Y}} v(y) S_\imath(t, \mathcal{Y} \setminus \{y\})$
- $N_\imath(v, t, \mathcal{Y}) = \sum_{y \in \mathcal{Y}} v(y) t(y)^\imath$

Lemma 3 *Generalized elementary Newton sums and classical elementary symmetric functions can be linked using the following formula : For $0 \leq k < \imath$,*

$$\sum_{y \in \mathcal{Y}} v(y) t(y)^k S_{\imath-k}(t, \mathcal{Y} \setminus \{y\}) = N_k(v, t, \mathcal{Y}) S_{\imath-k}(t, \mathcal{Y}) - \sum_{y \in \mathcal{Y}} v(y) t(y)^{k+1} S_{\imath-k-1}(t, \mathcal{Y} \setminus \{y\})$$

Proof : For all k, $0 \leq k < \imath$,

$$
\begin{aligned}
N_k(v, t, \mathcal{Y}) S_{\imath-k}(t, \mathcal{Y}) &= \left(\sum_{y \in \mathcal{Y}} v(y) t(y)^k \right) \left(\sum_{I \subset \mathcal{Y}, \, \sharp I = \imath-k} \prod_{z \in I} t(z) \right) \\
&= \sum_{y \in \mathcal{Y}} \sum_{I \subset \mathcal{Y}, \, \sharp I = \imath-k} v(y) t(y)^k \prod_{z \in I} t(z) \\
&= \sum_{y \in \mathcal{Y}} (\sum_{I \subset \mathcal{Y}, \, \sharp I = \imath-k} v(y) t(y)^k \prod_{z \in I, \, z \neq y} t(z) \\
&\quad + \sum_{I \subset \mathcal{Y}, \, \sharp I = \imath-k-1} v(y) t(y)^{k+1} \prod_{z \in I, \, z \neq y} t(z)) \\
&= \sum_{y \in \mathcal{Y}} v(y) t(y)^k S_{\imath-k}(t, \mathcal{Y} \setminus \{y\}) + \\
&\quad \sum_{y \in \mathcal{Y}} v(y) t(y)^{k+1} S_{\imath-k-1}(t, \mathcal{Y} \setminus \{y\})
\end{aligned}
$$

□

We now extend the relation between Newton sums and symmetric functions :

Proposition 2

$$S_\imath(v, t, \mathcal{Y}) = \sum_{j=0}^{\imath} (-1)^j N_j(v, t, \mathcal{Y}) S_{\imath-j}(t, \mathcal{Y})$$

Proof According to lemma 3, we have :

$$
\begin{aligned}
S_\imath(v, t, \mathcal{Y}) &= \sum_{y \in \mathcal{Y}} v(y) S_\imath(t, \mathcal{Y} \setminus \{y\}) \\
&= N_0(v, t, \mathcal{Y}) S_\imath(t, \mathcal{Y}) - \sum_{y \in \mathcal{Y}} v(y) t(y) S_{\imath-1}(t, \mathcal{Y} \setminus \{y\})
\end{aligned}
$$
B

Using the same argument, we obtain by induction :

$$
\begin{aligned}
S_\imath(v, t, \mathcal{Y}) &= \sum_{j=0}^{\imath-1} (-1)^j N_j(v, t, \mathcal{Y}) S_{\imath-j}(t, \mathcal{Y}) \\
&\quad + (-1)^\imath \sum_{y \in \mathcal{Y}} v(y) t(y)^\imath S_0(t, \mathcal{Y} \setminus \{y\})
\end{aligned}
$$

Since $S_0(t, \mathcal{Y} \setminus \{y\}) = 1$ the proof is complete. □

Since $Trace(vt^\imath) = N_\imath(v, t, \mathcal{Y})$ (for $y \in \mathcal{Y}$, the scalars $vt^\imath(y)$ are the eigenvalues of $M_{vt^\imath}$) $g(v, t, T)$ can be deduced from $f(t, T)$ and the traces $Trace(vt^\imath)$, $\imath = 1 \ldots (D-1)$.

3.2. COMPUTING GENERALIZED NEWTON SUMS

According to the previous part, there is an easy way to compute the generalized Shape Lemma from the traces : $Trace(vt^i)$, $i = 1, \ldots, D - 1$ and $Trace(t^i)$, $i = 1, \ldots, D$.

Assume that all the products $Vect(\omega_i\omega_j)$ are known. A straightforward algorithm computes all the needed products vt^i and, using the previous results, the expressions $Trace(vt^i)$.

Procedure Compute-Traces-I
   Input :
      MT
      t /* a separating element */
   Output :
      $Newton_t$, a vector of dimension $D + 1$ so that $Newton_t[i] = N_i(t, \mathcal{Y})$
      Tr_t, a $n \cdot D$ matrix so that $Tr_t[i, j] = N_j(X_i, t, \mathcal{Y})$
Begin
   $Vtr := [Trace(\omega_1), \ldots, Trace(\omega_D)]^T$
   $tmp := [1, 0, \ldots, 0]$
   For j from 0 to $D - 1$ do
      $Newton_t[j] := tmp \cdot Vtr$
      For i from 1 to n do
         $Tr_t[i, j] := (M_{X_i} \cdot tmp) \cdot Vtr$
      If $j < D - 1$
         $tmp := M_t \cdot tmp$
   $Newton_t[D] := tmp \cdot Vtr$

End

Proof of the algorithm Since $Vect(P) \cdot Vtr = Trace(P)$, the proof of the algorithm is obvious.   □

The following procedure is an optimization of the previous one, significantly decreasing the number of basic operations :

Procedure Compute-Traces
   /* Same Input and Output as Compute-Traces-I */
Begin
   $tmp := [Trace(\omega_1), \ldots, Trace(\omega_D)]^T$
   $Newton_t[0] := D$
   For j from 0 to $D - 1$ do
      For i from 1 to n do
         $Tr_t[i, j] := Vect(X_i) \cdot tmp$
      $Newton_t[j + 1] := Vect(t) \cdot tmp$
      $tmp := (M_t)^T \cdot tmp$
End

Proof of the algorithm If we notice that

$$Vect(P) \cdot [Trace(Q\omega_1), \ldots, Trace(Q\omega_D)] = Trace(PQ)$$

then, after the jth step in the principal loop :

$$tmp = [Trace(u^j\omega_1), \ldots, Trace(u^j\omega_D)]^T$$

and so, $\begin{cases} Newton_t[i] = N_i(t, \mathcal{Y}) \, , \ i = 1, \ldots, n \\ Tr_t[i,j] = N_j(X_i, t, \mathcal{Y}) \end{cases}$

$\square$

4. Benchmarks

In this section, we use examples from robotics (e.g. direct kinematic problem of a parallel robot, see (Ditrit & al., 1995), (Faugère and Lazard, 1994), (Merlet, 1993), ...) in order to compare the two methods described before.

Legend :

- GSL : computation of the Generalized Shape Lemma.
- St-Habicht : Sturm-Habicht's algorithm (applied on the univariate polynomial given by GSL).
- Hermite : Hermite's algorithm.
- Roots :

  - R : number of distinct real roots

  - C : number of distinct complex roots

- T. : computation time in seconds on a Sun-Sparc 10Mhz/128Mo, using the PoSSo-Library.
- L. : binary length of the largest coefficient that appears in the result.

We assume that GSL and Hermite algorithms take as input a Groebner basis (computed with respect to the *Degre Reverse Lexicographic* monomial ordering) and the associated multiplication table.

	Alg. II				Alg. I		Roots	
	GSL		St-Habicht		Hermite			
Name	T.	L.	T.	L.	T.	L.	R	C
kin1 (Ditrit)	7	8	63	444	22	132	8	40
kin2 (Merlet)	247	87	591	1226	457	1647	16	36
planN3I (Faugère)	25	32	198	936	179	654	0	40
spatial1 (Faugère)	51	30	44	815	15	590	0	16
spatial2 (Innocenti)	1236	77	4803	3514	5015	2944	24	40
spat2A2I (Faugère)	212	38	626	1839	316	894	0	32
spat66 (Faugère)	1935	40	2006	2852	7955	2348	2	40

5. Conclusion

Even if we do not take care of the preprocessing that is needed in order to compute an univariate representation of the systems, Hermite's method is more efficient in most cases. Since Sturm - Habicht sequences need $O(n^2)$ basic operations and the quadratic form reduction needs $O(n^3)$ operations, this result is due to a better control of the size of the coefficients in the algorithm that reduces the quadratic forms.

References

Alonso, M.E., Becker, E., Roy, M.F., Wörmann, T.(1994) Zero's, Multiplicities and Idempotents for Zero Dimensional Systems, *MEGA 94* **to be published**.

Ben-Or, M., Kozen, D., and Reif, J. (1986) Complexity of Elementary Algebra and Geometry *Journal of Computation and Systems Sciences* **Vol. 32 p. 251-264**.

Ditrit, O., Petitot, M., Walter, E. (1995) Guaranteed numerical solution of sets of nonlinear equations by Hansen's algorithm, implementation and applications *Proceedings of the PoSSo Workshop on Software* **p. 19-34**.

Faugère, J.C. (1994) Résolution des systèmes d'équations polynômiales *Doctoral Thesis*.

Faugère, J.C., Gianni, P., Lazard, D. and Mora, T. (1994) Efficient Computation of Zero-dimensional Gröbner Basis by Change of Ordering. *J. Symbolic Computation*

Faugère, J.C. and Lazard, D. (1994) The combinatorial classes of parallel manipulators *Mechanism and Machine Theory*.

Merlet, J-P. (1990) Les Robots Parallèles *Hermès, Paris*.

Pedersen, P. Roy, M.F, Szpirglas, A. (1993) Counting Real Zeros in the Multivariate Case *Computational Algebraic Geometry, Frédéric Eyssette, André Galligo (editors), Birkhäuser* **p. 203-223**.

Rouillier, F. (1994) Formules de Bareiss et réduction de formes quadratiques *Notes aux Comptes Rendus de l'Académie des Sciences* **To be published**.

Rouillier, F. (1995) PoSSo-RealSolving (Zero-dimensional systems of polynomials) *Proceedings of the PoSSo Workshop on Software* **p. 125-151**.

A SPATIAL CONSTRAINT PROBLEM

C. M. HOFFMANN
Purdue University
Department of Computer Sciences
West Lafayette, IN 47907-1398, USA

AND

P. J. VERMEER
Washington and Lee University
Department of Computer Science
Lexington, VA 24450, USA

Abstract. Three-dimensional geometric constraint solving is a rapidly developing field, with applications in areas such as kinematics, molecular modeling, surveying, and geometric theorem proving. While two-dimensional constraint solving has been studied extensively, there remain many open questions in the arena of three-dimensional problems. In this paper, we continue the development of our previous work on configuring a set of points and planes in three-space so that the configuration satisfies a given system of constraints. The constraint system considered consists of six geometric elements and pairwise constraints between triples of the elements. We first review the basic techniques developed in our earlier work germane to the current problem and explain how the problem we consider in this paper occurs. We then demonstrate how to solve the case of a geometric constraint system with four points and two planes.

1. Introduction

The spatial geometric constraint problem consists of a set of geometric entities, and a prescription of geometric constraints between the elements. The goal is to find all placements of the geometric entities which satisfy the given constraints. Two-dimensional constraint solving has been studied extensively, yet many open questions remain for the spatial problem.

J.-P. Merlet and B. Ravani (eds.), Computational Kinematics, 83–92.
© 1995 *Kluwer Academic Publishers.*

A problem is *well-constrained* if it has a finite number of solutions, while a problem with an infinite number of solutions is *underconstrained*. A problem is *overconstrained* if one constraint can be deleted yet the problem still has a finite number of solutions. An overconstrained problem may have a solution when the additional constraints are consistent with previous constraints, but often overconstrained problems have no solution.

Applications of constraint solving to kinematics include determining whether a mechanism is over- or underconstrained and its degrees of freedom. Further, by arbitrarily fixing underconstrained sketches, instance configurations of mechanisms can be computed. In the case of complex mechanisms, this may lead to reasonable kinematic simulations. Conversely, some techniques used in kinematics can also be used in constraint solving.

In this paper, we continue to develop our previous work on configuring a set of points and planes in three-space so that a given system of constraints is satisfied. The problem considered has six geometric elements and pairwise constrains four triples of the elements. After a brief review of basic techniques and problem context, we demonstrate how to solve the case of a certain geometric constraint system with four points and two planes.

2. General Solution Technique

Given a set of geometric elements and certain constraints between them, there are two basic strategies for solving the constraint problem. An *instance solver* uses the explicit values of the given constraints to determine the possible geometric configurations which satisfy the constraints. A *generic solver* determines whether the given geometric elements can be placed independent of the particular values assigned to the constraints. That is, the constraints have a symbolic rather than numerical value. The geometric elements are placed only after a decision has been made about whether or not the problem is generically well-constrained.

There are a variety of ways to implement each of these two strategies. Numerical constraint solvers first translate the constraints into a system of algebraic equations. This system is then solved using an iterative technique such as the Newton-Raphson method. Examples of numerical solvers include Sketchpad (Sut63) and ThingLab (Bor81). Symbolic constraint solvers also begin by setting up a system of algebraic equations. However, general symbolic computations are applied first to simplify the system, before solving it numerically. Methods such as Gröbner bases (Bos85) or Wu-Ritt (Wu86) techniques can be applied to find symbolic expressions for the solutions. For example, Kondo uses the symbolic computation for adding and deleting constraints from a given system (Kon92). Logical inference and term rewriting applies general logical reasoning techniques to the constraint solving problem. This approach has been taken by Aldefeld (Ald88) and Bruderlin (Bru90). For a deeper literature review see (Fud93; DH95).

2.1. GRAPH-BASED CONSTRUCTION SEQUENCES

Our approach to constraint solving is graph-based. Graph-based algorithms for solving geometric constraint problems have a graph analysis phase and a construction phase. First, a graph representation of the problem is constructed, where the nodes of the graph correspond to geometric entities, and an edge corresponds to a constraint between entities. A graph analysis determines whether the problem is well-constrained and a construction sequence. The graph analysis is more discriminating than Gruebler's or Reuleaux's criteria (Bar93; Phi84). It does not, however, account for degeneracies such as three planes intersecting in a line rather than a point due to specific constraint values. If the graph is (generically) well-constrained, this phase also determines a sequence of steps for solving the problem.

The second phase of the graph method takes the construction sequence determined from the first phase and performs the necessary construction steps to actually place the geometric elements. Since the first phase does not depend on the values of the constraints but only on the number and type of constraints between the geometric elements, we have a generic method of constraint solving. The actual values of the constraints only considered in the second phase when the construction steps are carried out.

One way to handle the analysis phase of the graph-based method looks for collections of geometric elements whose members can be placed with respect to one another based on constraints between them. These collections are then placed relative to one another, thus forming new, larger collections of elements, until all constraints have been processed and the locations of all the elements are known. We propose to use this recursive method to analyze the constraint graph for the three dimensional problems. This approach extends the two-dimensional constraint solving method developed in (BFH+94; Fud93). In the following sections we give a very brief overview of the method; further details and examples can be found in (DH95).

2.2. GEOMETRIC ENTITIES AND CONSTRAINTS

In the following, we restrict to considering only points and planes in $\mathcal{R}^3$. A point is represented simply by its coordinate triple (x, y, z). A plane is represented by its unit normal (n_x, n_y, n_z) and the distance from the origin to the plane d. This simplifies matters because each of the geometric entities has three degrees of freedom, thus allowing a uniform handling of the constraints and geometries throughout the solution. The constraints allowed are distance between two points, signed distance between a point and a plane, and angle between two planes. When an angle is given between two planes, it is assumed to be the angle between the sides of each plane positive with respect to the plane normals.

2.3. GRAPH ANALYSIS

The first step of the construction is to form *clusters* of geometric elements which are placed in fixed postions with respect to one another. Because each geometric element has three degrees of freedom, placing a new element requires that it be constrained by three known elements. Thus to begin a cluster, a set of three pairwise constrained nodes is necessary. These three geometric elements are placed into a standard position and the resulting configuration is fixed up to a rigid motion in space. Subsequently, a node is added to the cluster if it is incident to three nodes already in the cluster.

When no further nodes can be added to the cluster, the edges representing constraints between nodes in the cluster are deleted from the original graph, as well as all isolated nodes. Cluster formation is then applied recursively to this subgraph. This cluster forming and subsequent graph pruning is carried out as long as possible. Because a cluster start requires three pairwise constrained elements, there may eventually be unused constraints in the remaining subgraph, yet no new cluster can be started. In this case, any remaining constraint and its two incident nodes forms a *degenerate cluster*.

For cluster formation we need to find a generic placement for the first three elements which are pairwise constrained within a cluster, and we need to place a geometric element from three known elements. Because we use signed distances and angles, well-defined generic configurations exist. For the details see (DH95).

2.4. CLUSTER MERGING

Once initial clusters have been formed, clusters which have geometric elements in common can be placed relative to one another. Each cluster is a rigid body and has in general six degrees of freedom, three rotational and three translational. Exceptions include degenerate clusters such as a plane with an incident point. Here, one degree of freedom is lost by symmetry. To fix a cluster in space, we must determine how to place the shared elements of the cluster with respect to the other clusters sharing them. Three *separate* geometric entities are needed to fix the cluster. Once these elements are known, the rigid-body motion to position the cluster can be determined using the techniques of (RG93). Degenerate clusters share only two geometric elements and can be positioned from them because of cluster symmetry.

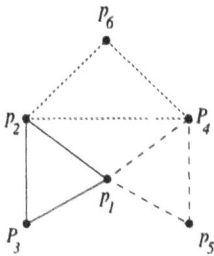

Figure 1. Constraint graph for Case 1 of four points and two planes in four clusters

Now, if two clusters A and B share two geometric elements, then the clusters are overconstrained, because the relative position of the shared elements is determined independently in both cluster. Therefore, the three shared elements in the cluster must belong to three *separate* other clusters. (For degenerate clusters, the two elements in the cluster are each shared with a different cluster.)

3. Merging Four Full Clusters

We consider merging four clusters, each with three geometric elements shared with the other clusters. Six geometric elements must be placed relative to each other. When the geometric elements are all points, the problem can be solved as a Stewart platform problem(NWM90). In (DH95) the case of five points and one plane is handled. We now show how to solve the constraint system when two of the elements are planes. There are two cases.

3.1. CASE 1

The first case we consider is when the two planes are not in the same cluster. The four clusters for the problem are (p_1, p_2, P_3), (p_1, P_4, p_5), (p_2, P_4, p_6), and (P_3, p_5, p_6). Note that this arrangement satisfies the criterion that each element of a given cluster is in exactly one of the other three clusters. The constraints within each cluster are the distances between the two points in each cluster, and the (signed) distance from each point to the plane in the cluster. A graphical representation of these clusters is shown in Figure 1, with the clusters distinguished by the type of line of the constraint edges. Each edge in the graph represents a distance constraint. A diagram of the geometry of this situation is shown in Figure 2. The open circles in P_3 represent the points in P_3 which satisfy the distance constraints between P_3 and the points p_2, p_5, and p_6, and analogously for the open circles in P_4. The elements of each cluster are connected by the same type of line.

Suppose that the distance between any two geometric elements g_i and

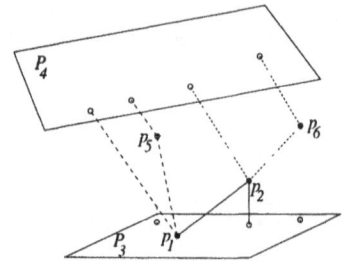

Figure 2. Four points and two planes with distance constraints only

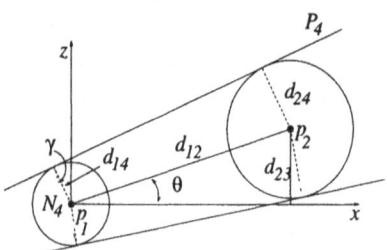

Figure 3. Constructing the plane P_4

g_j, where g is a point or a plane, is given by d_{ij}. If we assume the distances between a point and the plane are signed, then we can modify the problem so that the point p_1 lies in the plane P_3. If we can find a configuration where p_1 lies in P_3 and the respective distances between each of the three points p_2, p_5, and p_6 and P_3 are reduced by the given distance from p_1 to P_3, the actual configuration which satisfies the given constraints can be found by offsetting P_3 in the found configuration by d_{13}.

We place p_1 at the origin, make P_3 the xy-plane, and place p_2 at $(l_1, 0, d_{23})$, where $l_1^2 + d_{23}^2 = d_{12}^2$. The distance constraints between p_1 and P_4 and between p_2 and P_4 force P_4 to be tangent both to the sphere S_{14} centered at p_1 with radius d_{14} and to the sphere S_{24} centered at p_2 with radius d_{24}. If the two distances have the same sign, P_4 must be tangent to the cone containing and tangent to both spheres as shown in Figure 3. If the signs are opposite, the two spheres are contained in and tangent to a different cone, and the spheres are in opposite half-cones of the cone.

Note that the direction of the normal to P_4, $N_4 = (n_x, n_y, n_z)$ must be on a circle within the sphere S_{14}. The projection to the cross-section of this circle is shown in dotted line in Figure 3. To obtain a unit normal, we begin with a circle of radius $\cos \gamma$ in the xz-plane, with center $(-\sin \gamma, 0, 0)$, where γ is the apical angle of the cone. This circle is then rotated by θ about the y-axis, where θ is the angle between the line $p_1 p_2$ and the x-axis. The result of these operations is

$$N_4 : (\cos\gamma\sin\theta\sin u - \sin\gamma\cos\theta, \cos\gamma\cos u, \cos\gamma\cos\theta\sin u + \sin\gamma\sin\theta)$$

This means that the plane P_4 can be written in terms of the variable u as
$P_4 : N_4 \cdot (x, y, z) = d_{14}$.

Now, p_5 lies in a plane parallel to P_3 and also on a sphere of radius d_{15} centered at p_1. Thus it must lie on the circle C_5 given by

$$C_5 : (r_5\cos v, r_5\,\sin v, d_{35})$$

Similarly p_6 lies in a plane parallel to P_3 and also on a sphere of radius d_{26} centered at p_2. Since the distance from p_1 to the projection of p_2 into P_3 is l_1 (from the earlier positioning of p_2), p_6 must lie on the circle C_6 given by

$$C_6 : (l_1 + r_6\,\cos w, r_6\sin w, d_{36})$$

The parameters l_1, $\cos\theta$, $\sin\theta$, $\cos\gamma$, $\sin\gamma$, r_5, and r_6 are all dependent only on the distances given between elements of the clusters. Specifically, we have the following relationships:

$$
\begin{aligned}
l_1 &= \sqrt{d_{12}^2 - d_{23}^2} & l_2 &= \sqrt{d_{12}^2 - (d_{24} - d_{14})^2} \\
\cos\theta &= l_1/d_{12} & \cos\gamma &= l_2/d_{12} \\
\sin\theta &= d_{23}/d_{12} & \sin\gamma &= (d_{24} - d_{14})/d_{12} \\
r_5 &= \sqrt{d_{15}^2 - d_{35}^2} & r_6 &= \sqrt{d_{26}^2 - (d_{36} - d_{23})^2}
\end{aligned}
\tag{1}
$$

We now can use the remaining distance constraints to set up three equations in the unknowns u, v, and w which when solved will give the configurations. The remaining constraints are the distance between p_5 and P_4, the distance between p_6 and P_4, and the distance between p_5 and p_6. These constraints are translated into the following trigonometric equations:

$$N_4 \cdot p_5 + d_{45} = d_{12} \tag{2}$$

$$N_4 \cdot p_6 + d_{46} = d_{12} \tag{3}$$

$$(r_5\cos v - r_6\cos w - l_1)^2 + (r_5\sin v - r_6\sin w)^2 +$$
$$(d_{35} - d_{36})^2 = d_{56}^2 \tag{4}$$

Making the standard substitutions $\cos u = \frac{1-q_4^2}{1+q_4^2}$ $\sin u = \frac{2q_4}{1+q_4^2}$, $\cos v = \frac{1-q_5^2}{1+q_5^2}$ $\sin v = \frac{2q_5}{1+q_5^2}$, $\cos w = \frac{1-q_6^2}{1+q_6^2}$, and $\sin w = \frac{2q_6}{1+q_6^2}$, we obtain three equations with the following structure:

$$(A_1\,q_5^2 + A_2\,q_5 + A_3)\,q_4^2 + (A_4\,q_5^2 + A_6)\,q_4 + (A_7\,q_5^2 + A_8\,q_5 + A9) = 0 \tag{5}$$

$$(B_1\,q_6^2 + B_2\,q_6 + B_3)\,q_4^2 + (B_4\,q_6^2 + B_6)\,q_4 + (B_7\,q_6^2 + B_8\,q_6 + B9) = 0 \tag{6}$$

$$(D_1\,q_5^2 + D_3)\,q_6^2 + D_5\,q_5\,q_6 + (D_7\,q_5^2 + D_9) = 0 \tag{7}$$

Here the coefficients A_i, B_j, and D_k are functions only of the constants l_1, $\cos\theta$, $\sin\theta$, $\cos\gamma$, $\sin\gamma$, r_5, d_{35}, d_{36}, d_{45}, d_{46}, and d_{56}. This is exactly the same structure as the system of equations which arose in the case of five points and one plane, and its solution is detailed in (DH95). Note that the system has 16 solutions, in complex projective space.

3.2. CASE 2

The second case occurs when the two planes are both in one of the clusters, and another cluster has only points. Specifically, the clusters now are (p_1, p_2, P_3), (p_1, p_4, p_5), (p_2, p_4, P_6), and (P_3, p_5, P_6). The constraints are distances between any point-point or point-plane pairs within each cluster, and an angle constraint α_{36} between the two planes in the fourth cluster.

As before, we assume that p_1 lies in P_3, and we begin by positioning p_1 as the origin, P_3 as the xy-plane, and p_2 as the point $(l_1, 0, d_{23})$, where $l_1^2 + d_{23}^2 = d_{12}^2$. Based on the distance constraints between the point p_4 and the points p_1 and p_2, we can determine the circle C_4 on which p_4 must lie, which is the intersection of the sphere S_{14} centered at p_1 with radius d_{14} and the sphere S_{24} centered at p_2 with radius d_{24}. Assume that the center of this circle is l_4 units from p_1 along the line $p_1 p_2$, and that the radius of the intersection circle is r_4. Then the intersection circle can be found by rotating the circle of radius r_4 centered at $(l_4, 0, 0)$ about the y-axis by θ degrees, where θ is the angle between the line $p_1 p_2$ and the x-axis. This results in the following equation for C_4:

$$C_4 : (l_4 \cos\theta + r_4 \sin\theta \sin u, \; r_4 \cos u, \; -l_4 \sin\theta + r_4 \cos\theta \sin u)$$

The values of $\cos\theta$ and $\sin\theta$ are computed as in Equations 1, and the values of l_4 and r_4 are

$$l_4 = \frac{d_{24}^2 - d_{14}^2 - d_{12}^2}{2d_{12}} \qquad r_4 = \sqrt{d_{14}^2 - l_4^2}$$

The circle of possible points for p_5 is identical to the previous case:

$$C_5 : (r_5 \cos v, \; r_5 \sin v, \; d_{35})$$

The plane P_6 is completely determined by its normal and its distance from the origin. Note, however, that d_{16} is not one of the known constraints. Thus we refer to this distance as the variable l_{16}. Now the normal of P_6 can be computed as a function of the angle between P_6 and P_3, and some parameter w as

$$N_6 : (\cos\alpha_{36} \cos w, \; \cos\alpha_{16} \sin w, \; \sin\alpha_{16})$$

Since we do know the distance from p_2 to P_6, and p_2 is fixed, we can express l_{16} in terms of this constraint as

$$l_{16} = p_2 \cdot N_6 + d_{26}$$

The three equations which express the remaining constraints are

$$(l_4 \cos \theta + r_4 \sin \theta \sin u - r_5 \cos v)^2 + (r_4 \cos u - r_5 \sin v)^2 +$$
$$(-l_4 \sin \theta + r_4 \cos \theta \sin u - d_{35})^2 = d_{45}^2$$
$$C_4 \cdot N_6 + d_{46} = l_{16}$$
$$C_5 \cdot N_6 + d_{56} = l_{16}$$

When standard rational substitutions are made for the trigonometric functions, the resulting three equations have form identical to Equations 5, 6, and 7.

3.3. NUMERICAL EXAMPLE

To verify the above process, we consider a numerical example with one predetermined solution The initial configuration is

$$
\begin{array}{llll}
p_1 &=& (0,0,0) & p_2 &=& (3,0,4) & P_3 &=& xy-\text{plane} \\
P_4 &=& N : (3/5, 4/5, 0); d : 5 & p_5 &=& (2,2,2) & p_6 &=& (6,1,1)
\end{array}
$$

This means the input to the problem is four clusters with the following distance constraints:

Cluster 1	Cluster 2	Cluster 3	Cluster 4
$d_{13} = 0$	$d_{14} = 5$	$d_{24} = -16/5$	$d_{56} = 3\sqrt{2}$
$d_{23} = 4$	$d_{15} = 2\sqrt{3}$	$d_{46} = -3/5$	$d_{35} = 2$
$d_{12} = 5$	$d_{45} = -11/5$	$d_{26} = \sqrt{19}$	$d_{36} = 1$

Note the signed distances between P_3 and P_4, and the points with distance constraints with respect to the two planes. From these distances the parameters of the three equations are computed:

$$
\begin{array}{lllll}
l_1 &=& 3, & l_2 &=& 4/5\sqrt{34}, & \cos \theta &=& 3/5, & \cos \gamma &=& 4\sqrt{34}/25, \\
\sin \theta &=& 4/5, & \sin \gamma &=& 9/25, & r_5 &=& 2\sqrt{2}, & r_6 &=& \sqrt{10}
\end{array}
$$

Substituting these values into Eqs. (5), (6), and (7) and following the solution procedure detailed in (DH95), we found six solutions for q_5^2, two of which were positive. The corresponding values of q_5, $\cos v$, $\sin v$, $\cos u$, $\sin u$, $\cos w$, and $\sin w$ are shown in Table 3.3, rounded to six digits. Evaluating Eqs. (2), (3), and (4) at these points yielded the following results for the normal N_4 of P_4 and for the points p_5 and p_6 on the circles C_5, and C_6, respectively, rounded to six digits:

Solution 1

$N_4 = (-0.937703,.237830, -0.253277)$
$p_5 = (2.67722, 0.9124, 2.0)$
$p_6 = (3.63658, -3.09754, 1.0)$

Solution 2

$N_4 = (-0.937703, -0.237830, -0.253277)$
$p_5 = (2.67722, -0.9124, 2.0)$
$p_6 = (3.63658, 3.09754, 1.0)$

Solution 3

$N_4 = (-0.6,-0.8, 0.0)$
$p_5 = (2.0, 2.0, 2.0)$
$p_6 = (6.0, 1.0, 1.0)$

Solution 4

$N_4 = (-0.6, 0.8, 0.0)$
$p_5 = (2.0, -2.0, 2.0)$
$p_6 = (6.0, -1.0, 1.0)$

TABLE 1. Real solutions to the numerical example

q_5	$\cos v$	$\sin v$	$\cos u$	$\sin u$	$\cos w$	$\sin w$
0.165721	0.946541	0.322582	.254922	-.966962	.201306	-.979528
-0.165721	0.946541	-0.322582	-.254922	-.966962	.201306	.979528
0.414214	0.707107	0.707107	-.857493	-.514496	.948683	.316228
-0.414214	0.707107	-0.707107	.857493	-.514496	.948683	-.316228

Solution 3 is the predetermined solution, which corroborates the correctness of the solution procedure.

References

B. Aldefeld. Variation of geometries based on a geometric-reasoning method. *CAD*, 20(3):117–126, April 1988.

L. O. Barton. *Mechanism Analysis*. Marcel Dekker Inc., 1993.

W. Bouma, I. Fudos, C. Hoffmann, J. Cai, and R. Paige. A geometric constraint solver. *CAD*, 1994.

A. H. Borning. The programming language aspects of ThingLab, a constraint oriented simulation laboratory. *ACM TOPLAS*, 3(4):353–387, 1981.

N.K. Bose, editor. *Multidimensional Systems Theory*, pages 184–232. D. Reidel Publishing Co., 1985.

B. Bruderlin. Symbolic computer geometry for computer aided geometric design. In *Advances in Design and Manufacturing Systems*. NSF, 1990.

D.-Z. Du and F. Hwang, editors *Computing in Euclidean Geometry*, pages 266–298. World Scientific Publishing Co Pte Ltd, 2 edition, 1995.

I. Fudos and C. Hoffmann. Correctness of a geometric constraint solver. Technical Report CSD-TR-93-076, Purdue University, 1993.

I. Fudos. Editable representations for 2d geometric design. Master's thesis, Purdue University, 1993.

K. Kondo. Algebraic method for manipulation of dimensional relationships in geometric models. *CAD*, 24(3):141–147, March 1992.

P. Nanua, K. J. Waldron, and V. Murthy. Direct kinematic solution of a stewart platform. *IEEE Transactions on Robotics and Automation*, 6(4):438–444, Aug. 1990.

J. Phillips. *Freedom in Machinery*, volume 1. Cambridge University Press, 1984.

B. Ravani and Q. J. Ge. Computation of spatial displacements from geometric features. *J. Mech. Design*, 115:95–102, March 1993.

I. E. Sutherland. Sketchpad: A man-machine graphical communication system. In *Proceedings–Spring Joint Computer Conference*, pages 329–346, 1963.

W.-T. Wu. Principles of mechanical theorem proving. *J. of Automated Reasoning*, 2:221–252, 1986.

MOTOR TENSOR CALCULUS

K. Wohlhart
Institute for Mechanics
Graz University of Technology
A 8010 Graz, Kopernikusgasse 24 Austria

Summary This paper reviews the motor calculus as it was presented by Richard von Mises [1] and extends it to a general motor tensor calculus on the basis of unit motors. R. v. Mises defined a scalar and a motorial product of two motors without using Clifford's duality unit ε ($\varepsilon^2 = 0$) and introduced motor dyads. In complete analogy to the common tensor analysis, we define motor tensors of any order and show how the motorial product of two motors can be converted into a product of a motor multiplied by second order motor tensor, the cross motor tensor. With this motor tensor, whose matrix is a special function of the motor elements, and the unit motor dyad, the transfer of motor equations into matrix equations, or the decomposition of a motor equation into its six scalar equations, becomes a simple and straightforward procedure. A consistent notation for motor and motor dyads, which are used in the forthcoming English translation of v. Mises motor calculus [2], facilitates the algebra.

1. Introduction

In 1924 Richard von Mises, then Professor and Director of the Institute for Applied Mathematics at the University in Berlin and editor of the „Zeitschrift für Angewandte Mathematik und Mechanik" (which he founded in 1921), published in quick succession two long articles with the title: „Motorrechnung, ein neues Hilfsmittel in Mechanik" (Motor Calculus, a New Device in Mechanics). R. v. Mises adopted the term "motor" (coined originally by Clifford [3] for screws with which a magnitude is associated) in the sense given to it by E. Study [4], who defined the motor as an ordered pair of straight lines. It was v. Mises' declared intention to create a motor calculus without Clifford's duality unit, which was used by Study and many others, following Clifford's line of thinking. Mises' motor calculus was immediately recognized by the most prominent scientists as an important contribution to mechanics. Shortly after its publication it was treated and used in the Handbuch der Physik [5], which then had great influence on mechanics. However the work of v. Mises remained almost unknown to the community of engineers. Although motor calculus was used in Theoretical Mechanics now and then [6], [7], it never entered any of the many German textbooks on Applied Mechanics. One can only speculate about the reason for this poor acceptance by the engineers. Was it the clumsy notation? v. Mises uses bold capital Gothic and Greek letters for motors and motor dyads, which needs an artist to write them neatly. Was it the tricky overcross multiplication rule for the vectorial components of a motor, which makes the decomposition of a motor equation into a set of scalar equations cumbersome and mistake-prone? R. v. Mises' „Motorrechnung" was reprinted, yet not translated into English in the "Selected

J.-P. Merlet and B. Ravani (eds.), Computational Kinematics, 93–102.
© 1995 *Kluwer Academic Publishers.*

Papers of Richard von Mises" [8], which appeared in 1963 in two large volumes. This might be the reason why even now English speaking scientists are mostly unaware of this part of v. Mises' work. From time to time redefinitions of v. Mises' scalar motor product [9],[10] have emerged in the literature. The main advantage of motor calculus (in the sense of v. Mises as well in the sense of Clifford) is that it allows one to write geometric, kinematic and dynamic equations in the most transparent and compact form. The information density of its formulas is the highest possible, and motor calculus thus contributes greatly to the " economy of thought" (E. Mach). At the end of the fifties, engineers started to search for methods appropriate for solving the complex problems of robot kinematics. The dissertation by M. Keler [11], the paper by A.T. Yang and F. Freudenstein [12] and the booklet by F.M. Dimentberg [13] were trail-blazing, and subsequently "dual methods" became somewhat fashionable. As long as geometry and kinematics were concerned, these methods turned out to be extremely successful and they stabilized the Clifford-direction of the motor calculus. In dynamics the dual methods remained less successful. Neither the inertia binor introduced by S.G. Kislitzin [14], nor the inertia operator recently defined by M. Shoham and V. Brodsky [15], allow one to write the equations of motion of a rigid body in a pure dual form. In both cases dual vectors do not enter the equation as a whole, but only vector parts of them. The best result in dual dynamics is due to A.T. Yang [16], who gave the equation of the motion of a rigid body as formally identical to the Euler equation. But only v. Mises motor calculus permits writing this equation in the most compact style, i.e., formally identical to that of point mechanics. Is Clifford's duality unit, which is so fruitful in geometry and kinematics, an impediment in dynamics?

2. Motors, Scalar, Motorial and Dyadic Products of Motors

Regardless of whether a motor $\hat{\mathbf{f}} \Leftrightarrow \{\mathbf{f}, \mathbf{f}_A\}$, as an object, is represented as an ordered pair of straight lines (E. Study), or as an oriented screw with a magnitude (W. K. Clifford), it assigns to any point A in space two vectors $\mathbf{f}$ and $\mathbf{f}_A$, one constant and the other one depending on the position of A. There exists a one to one correspondence between the motor $\hat{\mathbf{f}}$ and the vectors $\mathbf{f}$ and $\mathbf{f}_A$: $\hat{\mathbf{f}} \Leftrightarrow \{\mathbf{f}, \mathbf{f}_A\}$. Therefore it can be said that the pair of vectors $\mathbf{f}$ and $\mathbf{f}_A$ is the representation of the motor $\hat{\mathbf{f}}$ at point A. If two point representations of a motor $\hat{\mathbf{f}}$ are given:

$$\hat{\mathbf{f}} \Leftrightarrow \{\mathbf{f}, \mathbf{f}_A\} \text{ and } \hat{\mathbf{f}} \Leftrightarrow \{\mathbf{f}, \mathbf{f}_B\}, \tag{1}$$

then
$$\mathbf{f}_B = \mathbf{f}_A + \mathbf{r}_{AB} \times \mathbf{f}, \tag{2}$$

with the vector $\mathbf{r}_{AB}$ leading from point B to point A. R. v. Mises defines the scalar, the motorial and the dyadic products of two motors $\hat{\mathbf{a}}$ and $\hat{\mathbf{b}}$ as follows. The scalar motor product yields a constant:

$$\hat{\mathbf{a}} \circ \hat{\mathbf{b}} = \mathbf{a} \circ \mathbf{b}_A + \mathbf{a}_A \circ \mathbf{b} = c. \tag{3}$$

The motorial product is a motor:

$$\hat{\mathbf{a}} \times \hat{\mathbf{b}} = \hat{\mathbf{c}}, \quad \hat{\mathbf{c}} \Leftrightarrow (\mathbf{c}, \mathbf{c}_A) = (\mathbf{a} \times \mathbf{b}, \ \mathbf{a} \times \mathbf{b}_A + \mathbf{a}_A \times \mathbf{b}). \tag{4}$$

The dyadic product:
$$\widehat{C} = \widehat{a} \otimes \widehat{b} \equiv \widehat{a}\widehat{b} \qquad (5)$$

represents a new entity which assigns to any motor $\widehat{f}$ the motor $c\widehat{a}$ or $d\widehat{b}$ by:

$$\widehat{A} \circ \widehat{f} = \widehat{a}\widehat{b} \circ \widehat{f} = c\widehat{a} \quad \text{or} \quad \widehat{f} \circ \widehat{A} = \widehat{f} \circ \widehat{a}\widehat{b} = d\widehat{b}, \qquad (6)$$

where the multiplication of a motor $\widehat{a}$ by a scalar means multiplication of the vector parts $\mathbf{a}$ and $\mathbf{a}_A$ by this scalar:

$$c\widehat{a} \Leftrightarrow \{c\mathbf{a}, c\mathbf{a}_A\}. \qquad (7)$$

It is easy to prove with Eq. (2) that the scalar product as well as the motorial product of two motors is independent of the point of representation of the motors:

$$\widehat{a} \circ \widehat{b} = \mathbf{a} \circ \mathbf{b}_A + \mathbf{a}_A \circ \mathbf{b} = c, \text{ and } \quad \widehat{a} \circ \widehat{b} = \mathbf{a} \circ \mathbf{b}_B + \mathbf{a}_B \circ \mathbf{b} = c \qquad (8)$$

$$\widehat{c} \Leftrightarrow \{\mathbf{c}, \mathbf{c}_A\} = \{\mathbf{a} \times \mathbf{b}, \ \mathbf{a} \times \mathbf{b}_A + \mathbf{a}_A \times \mathbf{b}\} \text{ and } \widehat{c} \Leftrightarrow \{\mathbf{c}, \mathbf{c}_B\} = \{\mathbf{a} \times \mathbf{b}, \ \mathbf{a} \times \mathbf{b}_B + \mathbf{a}_B \times \mathbf{b}\}. \qquad (9)$$

Both products have their own mechanical significance: If $\widehat{f}$ denotes the force motor and $\widehat{s}$ the speed motor of the rigid body, then $\widehat{f} \circ \widehat{s}$ gives the power, and $\widehat{s} \times \widehat{p}$ measures the time rate of the motor $\widehat{p}$ if fixed to the body moving with the speed motor $\widehat{s}$.

From
$$\widehat{f} \Leftrightarrow \{\mathbf{F}, \mathbf{M}_A\} \text{ and } \widehat{s} \Leftrightarrow \{\omega, \mathbf{v}_A\} \qquad (10)$$

it follows that
$$\widehat{f} \circ \widehat{s} = \mathbf{F} \circ \mathbf{v}_A + \mathbf{M}_A \circ \omega = P. \qquad (11)$$

If the motor $\widehat{p}$ if fixed to a body moving with the speed motor $\widehat{s}$, then the vectorial parts $\mathbf{p}$ and $\mathbf{p}_A$ change within the time span dt to

$$\mathbf{p}^l = \mathbf{p} + dt\omega \times \mathbf{p} \quad \text{and} \quad \mathbf{p}_{A^l}^l = \mathbf{p}_A + dt\omega \times \mathbf{p}_A.$$

From
$$\mathbf{p}_A^l = \mathbf{p}_{A^l}^l + \mathbf{r}_{A^lA} \times \mathbf{p}^l \quad \text{and} \quad \mathbf{r}_{A^lA} = dt\mathbf{v}_A \qquad \text{it follows that}$$

$$d\widehat{p} = dt\widehat{s} \times \widehat{p} \Leftrightarrow \{\mathbf{p}^l - \mathbf{p}, \mathbf{p}_A^l - \mathbf{p}_A\} \text{ or } \frac{d\widehat{p}}{dt} = \widehat{s} \times \widehat{p} \Leftrightarrow \{\omega \times \mathbf{p}, \ \omega \times \mathbf{p}_A + \mathbf{v}_A \times \mathbf{p}\}. \qquad (12)$$

3. General Motor Tensors

A unit motor $\widehat{n}$ is a motor for which one of the following two point representations is valid:

$$\widehat{n} \Leftrightarrow \{\mathbf{n}, \mathbf{n}_A\} \text{ with } \mathbf{n} \circ \mathbf{n} = 1 \text{ and } \mathbf{n} \circ \mathbf{n}_A = 0, \text{ or } \widehat{n} \Leftrightarrow \{0, \mathbf{n}\} \text{ with } \mathbf{n} \circ \mathbf{n} = 1. \qquad (13)$$

Together with the vector $\mathbf{r}_{AO}$, a cartesian coordinate system, consisting of an origin A and three mutually orthogonal unit vectors $\mathbf{n}_1, \mathbf{n}_2$ and $\mathbf{n}_3$ can be interpreted as six unit motors related to point O

$$\widehat{n}_\alpha \Leftrightarrow \{\mathbf{n}_\alpha, \mathbf{r}_{AO} \times \mathbf{n}_\alpha\} \text{ for } \alpha = 1,2,3 \text{ and } \widehat{n}_\alpha \Leftrightarrow \{0, \mathbf{n}_{\alpha-3}\} \text{ for } \alpha = 4,5,6. \qquad (14)$$

A motor $\widehat{a}$ can be represented as a linear combination of these six unit motors $\widehat{n}_\alpha$:

$$\hat{\mathbf{a}} = a_1\hat{\mathbf{n}}_1 + a_2\hat{\mathbf{n}}_2 + a_3\hat{\mathbf{n}}_3 + a_4\hat{\mathbf{n}}_4 + a_5\hat{\mathbf{n}}_5 + a_6\hat{\mathbf{n}}_6 \equiv a_\alpha\hat{\mathbf{n}}_\alpha, \tag{15}$$

where the first three motor coordinates a_1, a_2, a_3 are identical to the vector coordinates of the first vector part $\mathbf{a}$, and the second three motor coordinates a_4, a_5, a_6 are identical to the vector coordinates of the second vector part $\mathbf{a}_A$, both measured in the cartesian coordinate system $(A|\mathbf{n}_1, \mathbf{n}_2, \mathbf{n}_3)$. With the matrices

$$\underline{a}^{\mathrm{T}} = \boxed{\begin{array}{cccccc} a_1 & a_2 & a_3 & a_4 & a_5 & a_6 \end{array}} \text{ and } \underline{\hat{\mathbf{n}}}^{\mathrm{T}} = \boxed{\begin{array}{cccccc} \hat{\mathbf{n}}_1 & \hat{\mathbf{n}}_2 & \hat{\mathbf{n}}_3 & \hat{\mathbf{n}}_4 & \hat{\mathbf{n}}_5 & \hat{\mathbf{n}}_6 \end{array}} \tag{16}$$

one can write for the motor $\hat{\mathbf{a}}$: $\qquad \hat{\mathbf{a}} = \underline{a}^{\mathrm{T}}\underline{\mathbf{n}} = \underline{\mathbf{n}}^{\mathrm{T}}\underline{a}.$ \hfill (17)

A motor tensor of order n $\hat{\mathbf{A}}^{(n)} = A_{\alpha_1\alpha_2\ldots\alpha_{n-1}\alpha_n}\hat{\mathbf{n}}_{\alpha_1}\hat{\mathbf{n}}_{\alpha_2}\ldots\hat{\mathbf{n}}_{\alpha_{n-1}}\hat{\mathbf{n}}_{\alpha_n}$ \hfill (18)

can be defined as an entity which assigns to a motor $\hat{\mathbf{b}}$ a motor tensor $\hat{\mathbf{C}}^{(n-1)}$ of order $n-1$ by the inner product:

$$\hat{\mathbf{A}}^{(n)} \circ \hat{\mathbf{b}} = A_{\alpha_1\alpha_2\ldots\alpha_{n-1}\alpha_n}\hat{\mathbf{n}}_{\alpha_1}\hat{\mathbf{n}}_{\alpha_2}\ldots\hat{\mathbf{n}}_{\alpha_{n-1}}\hat{\mathbf{n}}_{\alpha_n} \circ \hat{\mathbf{b}},$$

which gives with (3) $\quad \hat{\mathbf{A}}^{(n)} \circ \hat{\mathbf{b}} = C_{\beta_1\beta_2\ldots\beta_{n-1}}\hat{\mathbf{n}}_{\beta_1}\hat{\mathbf{n}}_{\beta_2}\ldots\hat{\mathbf{n}}_{\beta_{n-1}} = \hat{\mathbf{C}}^{(n-1)}.$ \hfill (19)

The motor tensor of second order $\hat{\mathbf{A}}^{(2)} \equiv \hat{\mathbf{A}}$ can be written in the forms:

$$\hat{\mathbf{A}} = A_{\alpha\beta}\hat{\mathbf{n}}_\alpha\hat{\mathbf{n}}_\beta = \underline{\hat{\mathbf{n}}}^{\mathrm{T}}\underline{A}\underline{\hat{\mathbf{n}}}, \text{ with the } 6\times 6 \text{ matrix } \quad \underline{A} = \|A_{\alpha\beta}\|. \tag{20}$$

Post– or premultiplication of the motor $\hat{\mathbf{b}}$ with the motor tensor $\hat{\mathbf{A}}$ lead to different motors.

From $\qquad\qquad\qquad\qquad \hat{\mathbf{n}}_\alpha \circ \hat{\mathbf{n}}_\beta = \delta_{\alpha\beta}$ \hfill (21)

we obtain $\qquad \hat{\mathbf{A}} \circ \hat{\mathbf{b}} = A_{\alpha\beta}\hat{\mathbf{n}}_\alpha\hat{\mathbf{n}}_\beta \circ b_\gamma\hat{\mathbf{n}}_\gamma = A_{\alpha\beta}\delta_{\beta\gamma}b_\gamma\hat{\mathbf{n}}_\alpha = c_\alpha\hat{\mathbf{n}}_\alpha = \hat{\mathbf{c}}$ \hfill (22)

or $\qquad \hat{\mathbf{b}} \circ \hat{\mathbf{A}} = b_\alpha\hat{\mathbf{n}}_\alpha \circ A_{\beta\gamma}\hat{\mathbf{n}}_\beta\hat{\mathbf{n}}_\gamma = b_\alpha\delta_{\alpha\beta}A_{\beta\gamma}\hat{\mathbf{n}}_\gamma = d_\alpha\hat{\mathbf{n}}_\alpha = \hat{\mathbf{d}}.$ \hfill (23)

With the matrix $\underline{\delta}$, which collects the scalar products $\hat{\mathbf{n}}_\alpha \circ \hat{\mathbf{n}}_\beta$ of the unit motors (16) of the cartesian coordinate system:

$$\underline{\delta} = \underline{\delta}^{\mathrm{T}} = \|\delta_{\alpha\beta}\| = \underline{\hat{\mathbf{n}}} \circ \underline{\hat{\mathbf{n}}}^{\mathrm{T}} = \begin{array}{|cccccc|} \hline 0 & 0 & 0 & 1 & 0 & 0 \\ 0 & 0 & 0 & 0 & 1 & 0 \\ 0 & 0 & 0 & 0 & 0 & 1 \\ 1 & 0 & 0 & 0 & 0 & 0 \\ 0 & 1 & 0 & 0 & 0 & 0 \\ 0 & 0 & 1 & 0 & 0 & 0 \\ \hline \end{array}, \tag{24}$$

the formulas (22) and (23) can also be written in matrix form:

$$\hat{A} \circ \hat{b} = \underline{\hat{n}}^T \underline{A}\hat{n} \circ \underline{\hat{n}}^T \underline{b} = \underline{\hat{n}}^T \underline{A}\delta \underline{b} = \underline{\hat{n}}^T \underline{c} = \hat{c} \quad \text{and} \quad \hat{b} \circ \hat{A} = \underline{b}^T \underline{\hat{n}} \circ \underline{\hat{n}}^T \underline{A}\hat{n} = \underline{b}^T \underline{\delta A}\hat{n} = \underline{d}^T \underline{\hat{n}} = \hat{d}.$$

For later use we define the antisymmetric 6×6 matrix $\underline{\hat{N}}$ here whose elements $\hat{n}_\alpha \times \hat{n}_\beta$ are the motorial products of the unit motors (14) of the cartesian coordinate system. According to definition (4) we obtain:

$$\underline{\hat{N}} = -\underline{\hat{N}}^T = \| \hat{n}_\alpha \times \hat{n}_\beta \| = \underline{\hat{n}} \times \underline{\hat{n}}^T = \begin{vmatrix} 0 & \hat{n}_3 & -\hat{n}_2 & 0 & \hat{n}_6 & -\hat{n}_5 \\ -\hat{n}_3 & 0 & \hat{n}_1 & -\hat{n}_6 & 0 & \hat{n}_4 \\ \hat{n}_2 & -\hat{n}_1 & 0 & \hat{n}_5 & -\hat{n}_4 & 0 \\ 0 & \hat{n}_6 & -\hat{n}_5 & 0 & 0 & 0 \\ -\hat{n}_6 & 0 & \hat{n}_4 & 0 & 0 & 0 \\ \hat{n}_5 & -\hat{n}_4 & 0 & 0 & 0 & 0 \end{vmatrix}. \tag{25}$$

4. The Cross Motor Tensor

According to (4) the motorial product of two motors is a new motor. Therefore we ask whether this product can be converted into a scalar product of one of its motors multiplied by a motor tensor of second order. The question is whether it is possible to identify special motor tensors $\hat{A}$ or $\hat{B}$ for which

$$\hat{c} = \hat{a} \times \hat{b} = \hat{A} \circ \hat{b} = \hat{a} \circ \hat{B}. \tag{26}$$

Evidently the coordinates $A_{\alpha\beta}$ of $\hat{A}$ must be functions of the coordinates a_α of the motor $\hat{a}$, and the coordinates $B_{\alpha\beta}$ of $\hat{B}$ functions of the coordinates b_α of the motor $\hat{b}$. Therefore, we also write:

$$\hat{A} \equiv \hat{a}^\times, \quad \hat{B} \equiv \hat{b}^\times, \tag{27}$$

and call $\hat{A}$ and $\hat{B}$ the cross motor tensors of the motor $\hat{a}, \hat{b}$, respectively. With $\underline{\hat{N}}$ and $\underline{\delta}$ we write for (26):

$$\hat{c} = \hat{a} \times \hat{b} = \underline{a}^T \underline{\hat{n}} \times \underline{\hat{n}}^T \underline{b} = \underline{a}^T \underline{\hat{N}}\underline{b} = \hat{A} \circ \hat{b} = \underline{\hat{n}}^T \underline{A}\hat{n} \circ \underline{\hat{n}}^T \underline{b} = \underline{\hat{n}}^T \underline{A}\delta \underline{b}$$

or

$$\hat{c} = \hat{a} \times \hat{b} = \underline{a}^T \underline{\hat{n}} \times \underline{\hat{n}}^T \underline{b} = \underline{a}^T \underline{\hat{N}}\underline{b} = \hat{a} \circ \hat{B} = \underline{a}^T \underline{\hat{n}} \circ \underline{\hat{n}}^T \underline{B}\hat{n} = \underline{a}^T \underline{\delta B}\hat{n}$$

and find

$$(\underline{a}^T \underline{\hat{N}} - \underline{\hat{n}}^T \underline{A}\delta)\underline{b} = 0, \quad \underline{a}^T (\underline{\hat{N}}\underline{b} - \underline{\delta B}\hat{n}) = 0,$$

from which

$$\underline{a}^T \underline{\hat{N}} = \underline{\hat{n}}\underline{A}\delta \quad \text{and} \quad \underline{\hat{N}}\underline{b} = \underline{\delta B}\hat{n}$$

follow.

With $\underline{\hat{n}} \circ \underline{\hat{n}}^T = \underline{\delta}$ and the 6×6 identity matrix $\underline{\delta\delta} = \underline{I}$ we obtain:

$$\underline{A} \equiv \underline{a}^\times = \underline{\delta\hat{n}} \circ (\underline{a}^T \underline{\hat{N}})\underline{\delta} \quad \text{and} \quad \underline{B} \equiv \underline{b}^\times = \underline{\delta}(\underline{\hat{N}}\underline{b}) \circ \underline{\hat{n}}^T \underline{\delta}, \tag{28}$$

or written in components:

$$
\underline{a}^\times = \begin{vmatrix}
0 & 0 & 0 & 0 & -a_3 & a_2 \\
0 & 0 & 0 & a_3 & 0 & -a_1 \\
0 & 0 & 0 & -a_2 & a_1 & 0 \\
0 & -a_3 & a_2 & 0 & -a_6 & a_5 \\
a_3 & 0 & -a_1 & a_6 & 0 & -a_4 \\
-a_2 & a_1 & 0 & -a_5 & a_4 & 0
\end{vmatrix}, \quad
\underline{b}^\times = \begin{vmatrix}
0 & 0 & 0 & 0 & -b_3 & b_2 \\
0 & 0 & 0 & b_3 & 0 & -b_1 \\
0 & 0 & 0 & -b_2 & b_1 & 0 \\
0 & -b_3 & b_2 & 0 & -b_6 & b_5 \\
b_3 & 0 & -b_1 & b_6 & 0 & -b_4 \\
-b_2 & b_1 & 0 & -b_5 & b_4 & 0
\end{vmatrix}. \quad (29)
$$

As these two matrices are skew, they are formally identical. With $\underline{\hat{N}}^T = -\underline{\hat{N}}$, $\underline{\delta}^T = \underline{\delta}$ and $\underline{A}^T = -\underline{A}$ it follows from (28) that:

$$
\underline{A}^T = -(\underline{\delta}(\underline{\hat{N}}a) \circ \underline{\hat{n}}^T \underline{\delta}) \Rightarrow \underline{A} = \underline{\delta}(\underline{\hat{N}}a) \circ \underline{\hat{n}}^T \underline{\delta},
$$

which is formally identical to $\qquad \underline{B} = \underline{\delta}(\underline{\hat{N}}b) \circ \underline{\hat{n}}^T \underline{\delta}.$

With the cross motor tensor it is easy to find the elements of a motor tensor which arises from a motorial multiplication of a motor by a motor tensor. For example:

$$
\hat{C} = \hat{A} \times \hat{b} = \hat{A} \circ \hat{b}^\times = \underline{\hat{n}}^T \underline{A}\hat{n} \circ \underline{\hat{n}}^T \underline{b}^\times \underline{\hat{n}} = \underline{\hat{n}}^T \underline{A}\underline{\delta}\underline{b}^\times \underline{\hat{n}} = \underline{\hat{n}}^T \underline{C}\underline{\hat{n}}
$$

yields $\qquad\qquad\qquad \hat{C} = \hat{A} \times \hat{b} \Rightarrow \underline{C} = \underline{A}\underline{\delta}\underline{b}^\times \qquad\qquad\qquad (30)$

Similarly, the motorial premultiplication of a motor by a tensor motor leads to:

$$
\hat{D} = \hat{b} \times \hat{A} = \hat{b}^\times \circ \hat{A} = \underline{\hat{n}}^T \underline{b}^\times \underline{\hat{n}} \circ \underline{\hat{n}}^T \underline{A}\hat{n} = \underline{\hat{n}}^T \underline{b}^\times \underline{\delta}\underline{A}\hat{n} = \underline{\hat{n}}^T \underline{D}\hat{n}
$$

or $\qquad\qquad\qquad\qquad \hat{D} = \hat{b} \times \hat{A} \Rightarrow \underline{D} = \underline{b}^\times \underline{\delta}\underline{A} \; . \qquad\qquad\qquad (31)$

5. The Motor Space as a Lie Algebra

A Lie Algebra [17] is a linear space L over a field F on which a product $[\,,\,]$, the Lie bracket (commutator) is defined with the following four properties:

$$
X, Y \in L \Rightarrow [X, Y] \in L,
$$
$$
\alpha, \beta \in F, \quad X, Y, Z \in L \Rightarrow [X, \alpha Y + \beta Z] = \alpha[X, Y] + \beta[X, Z],
$$
$$
[X, Y] = -[Y, X],
$$
$$
[X, [Y, Z]] + [Y, [Z, X]] + [Z, [X, Y]] = 0. \qquad (32)
$$

It will be shown that, with the motorial product as the commutator, the space of motors is a real Lie Algebra. The first three requirements are evidently fulfilled for the motorial product of motors, and the last, the Jacobi identity for motors, can easily be proved via the Jacobi identity for three vectors. For $\mathbf{a}, \mathbf{b}, \mathbf{c}$ it is found that

$$
\mathbf{a} \times (\mathbf{b} \times \mathbf{c}) + \mathbf{b} \times (\mathbf{c} \times \mathbf{a}) + \mathbf{c} \times (\mathbf{a} \times \mathbf{b}) =
$$

$$((a \circ c)b - (a \circ b)c) + ((b \circ a)c - (b \circ c)a) + ((c \circ b)a - (c \circ a)b) = 0. \tag{33}$$

The double motorial product $\hat{a} \times (\hat{b} \times \hat{c})$ and the equation (4) leads to a motor whose representation in point A is

$$\{a \times (b \times c), \quad a \times (b \times c_A) + a \times (b_A \times c) + a_A \times (b \times c)\},$$

the vector parts of the motor $\hat{a} \times (\hat{b} \times \hat{c}) + \hat{b} \times (\hat{c} \times \hat{a}) + \hat{c} \times (\hat{a} \times \hat{b})$ related to point A will then be:

$$a \times (b \times c) + b \times (c \times a) + c \times (a \times b),$$

and

$$a \times (b \times c_A) + a \times (b_A \times c) + a_A \times (b \times c) +$$
$$b \times (c \times a_A) + b \times (c_A \times a) + b_A \times (c \times a) +$$
$$c \times (a \times b_A) + c \times (a_A \times b) + c_A \times (a \times b).$$

Equation (33) shows that both vectors are zero. This gives the Jacobi identity for motors:

$$\hat{a} \times (\hat{b} \times \hat{c}) + \hat{b} \times (\hat{c} \times \hat{a}) + \hat{c} \times (\hat{a} \times \hat{b}) = 0. \tag{34}$$

It follows that the 6 dimensional space of motors with the motorial product as the commutator, is a real Lie Algebra.

6. Motor Equation of the Motion of a Rigid Body

The distribution of the velocities within a rigid body is given by the angular velocity vector ω together with the velocity v_A of one of its points A. The pair of vectors ω and v_A represents the speed motor $\hat{s}$ of the rigid body related to point A. The velocity of any mass point of the body, whose position gives the vector r leading from the point A to this mass point can be calculated by:

$$v = v_A + \omega \times r \tag{35}$$

The momentum p of the body and its angular momentum L_A about point A, are defined by

$$p = \int v \, dm, \quad L_A = \int r \times v \, dm \tag{36}$$

respectively. The pair of vectors p and L_A represent a motor $\hat{p}$, the momentum motor of the body, because the transformation rule (2) for L is

$$L_B = \int (r + r_{AB}) \times v \, dm = L_A + r_{AB} \times p. \tag{37}$$

From (35) one obtains for the pair of vectors of this momentum motor $\hat{p}$:

$$p = \int v \, dm = v_A m + \omega \times r_C m \tag{38}$$

and

$$L_A = \int r \times v \, dm = r_C \times v_A m + J_A \circ \omega, \tag{39}$$

where m denotes the mass of the body, r_C the position vector of its center of mass C leading from point A to C, and J_A the inertia tensor of the body about A:

$$\mathbf{r}_C = \int \mathbf{r}\,dm/m \quad \text{and} \quad \mathbf{J}_A = \int ((\mathbf{r}\circ\mathbf{r})\mathbf{I} - \mathbf{r}\mathbf{r})dm. \tag{40}$$

The components $p_1, p_2, p_3, L_{A1}, L_{A2}, L_{A3}$ of the vectors $\mathbf{p}$, $\mathbf{L}_A$ in the cartesian coordinate system $(A\,|\,\mathbf{n}_1, \mathbf{n}_2, \mathbf{n}_3)$ are given by:

$$p_1 = m(v_{A1} - \omega_2 r_{C3} + \omega_3 r_{C2}), \quad p_2 = m(v_{A2} - \omega_3 r_{C1} + \omega_1 r_{C3}), \quad p_3 = m(v_{A3} - \omega_1 r_{C2} + \omega_2 r_{C1}),$$
$$L_{A1} = m(v_{A2} r_{C3} - v_{A3} r_{C2}) + J_{A11}\omega_1 + J_{A12}\omega_2 + J_{A13}\omega_3,$$
$$L_{A2} = m(v_{A3} r_{C1} - v_{A1} r_{C3}) + J_{A12}\omega_1 + J_{A22}\omega_2 + J_{A23}\omega_3,$$
$$L_{A3} = m(v_{A1} r_{C2} - v_{A2} r_{C1}) + J_{A13}\omega_1 + J_{A23}\omega_2 + J_{A33}\omega_3. \tag{41}$$

Identifying the components p_1, p_2, p_3 with the first three, and L_{A1}, L_{A2}, L_{A3} with the last three components of the motor $\hat{\mathbf{p}}$ we can, according to (15),(16), write the motor $\hat{\mathbf{p}}$ as a linear combination of the unit motors $\hat{\mathbf{n}}_\alpha$ defined in (14):

$$\hat{\mathbf{p}} = p_1\hat{\mathbf{n}}_1 + p_2\hat{\mathbf{n}}_2 + p_3\hat{\mathbf{n}}_3 + p_4\hat{\mathbf{n}}_4 + p_5\hat{\mathbf{n}}_5 + p_6\hat{\mathbf{n}}_6 = p_\alpha\hat{\mathbf{n}}_\alpha = \underline{p}^\mathsf{T}\underline{\hat{\mathbf{n}}} = \underline{\hat{\mathbf{n}}}^\mathsf{T}\underline{p}. \tag{42}$$

With the matrix $\underline{T}$:

$$\underline{T} = \underline{T}^\mathsf{T} = \|\,T_{\alpha\beta}\,\| = \begin{Vmatrix} m & 0 & 0 & 0 & -mr_{C3} & mr_{C2} \\ 0 & m & 0 & mr_{C3} & 0 & -mr_{C1} \\ 0 & 0 & m & -mr_{C2} & mr_{C1} & 0 \\ 0 & mr_{C3} & -mr_{C2} & J_{A11} & J_{A12} & J_{A13} \\ -mr_{C3} & 0 & mr_{C1} & J_{A12} & J_{A22} & J_{A23} \\ mr_{C2} & -mr_{C1} & 0 & J_{A12} & J_{A23} & J_{A33} \end{Vmatrix}, \tag{43}$$

and the matrix $\underline{s}$, which contains the element s_α of the speed motor $\hat{\mathbf{s}}$

$$\underline{s}^\mathsf{T} = \begin{vmatrix} \omega_1 & \omega_2 & \omega_3 & v_{A1} & v_{A2} & v_{A3} \end{vmatrix}, \tag{44}$$

we write instead of (42):
$$\hat{\mathbf{p}} = \underline{\hat{\mathbf{n}}}^\mathsf{T}\underline{T}\delta\underline{s}, \tag{45}$$

and with $\underline{\delta} = \underline{\hat{\mathbf{n}}}\circ\underline{\hat{\mathbf{n}}}^\mathsf{T}$ as well as the inertia motor tensor $\hat{\mathbf{T}}$:

$$\hat{\mathbf{T}} = \underline{\hat{\mathbf{n}}}^\mathsf{T}\underline{T}\underline{\hat{\mathbf{n}}} = T_{\alpha\beta}\hat{\mathbf{n}}_\alpha\hat{\mathbf{n}}_\beta, \tag{46}$$

finally obtain for $\hat{\mathbf{p}}$:
$$\hat{\mathbf{p}} = \hat{\mathbf{T}}\circ\hat{\mathbf{s}}. \tag{47}$$

Therefore, the momentum motor of the rigid body is found by scalar postmultiplication of the speed motor by the second order inertia motor tensor. With $\omega_\alpha = 0$ and $J_{A\alpha\beta} = 0$ formula (47) yields for the first vector part of the momentum motor, the momentum vector of the mass point mechanics: $\mathbf{p} = m\mathbf{v}_A = m\mathbf{n}_i v_{Ai}$. The differential equations which rule the motion of a rigid body are:

$$\frac{d\mathbf{p}}{dt} \equiv \dot{\mathbf{p}} = \mathbf{F} \text{ and } \frac{d\mathbf{L}_A}{dt} + \mathbf{v}_A \times \mathbf{p} \equiv \dot{\mathbf{L}}_A + \mathbf{v}_A \times \mathbf{p} = \mathbf{M}_A. \qquad (48)$$

If the coordinate system $(A \mid \mathbf{n}_1, \mathbf{n}_2, \mathbf{n}_3)$ is fixed to the rigid body, the absolute time rates of $\mathbf{p}$ and $\mathbf{L}_A$ are:

$$\frac{d\mathbf{p}}{dt} = \frac{d}{dt}(p_\alpha \mathbf{n}_\alpha) = \dot{p}_\alpha \mathbf{n}_\alpha + p_\alpha \dot{\mathbf{n}}_\alpha = \mathbf{p}^\circ + \omega \times \mathbf{p},$$

and
$$\frac{d\mathbf{L}_A}{dt} = \frac{d}{dt}(L_{A\alpha} \mathbf{n}_\alpha) = \dot{L}_{A\alpha} \mathbf{n}_\alpha + L_{A\alpha} \dot{\mathbf{n}}_\alpha = \mathbf{L}_A^\circ + \omega \times \mathbf{L}_A.$$

Therewith we obtain:
$$\mathbf{p}^\circ + \omega \times \mathbf{p} = \mathbf{F} \qquad (49)$$

$$\mathbf{L}_A^\circ + \omega \times \mathbf{L}_A + \mathbf{v}_A \times \mathbf{p} = \mathbf{M}_A. \qquad (50)$$

In these formulas the operator $(\ )^\circ$ stands for the Jaumann time derivative which measures the time rate in relation to the coordinate system $(A \mid \mathbf{n}_1, \mathbf{n}_2, \mathbf{n}_3)$. As this system is fixed to the body the coordinates r_{C1}, r_{C2}, r_{C3} of $\mathbf{r}_C$, and the moments of inertia $J_{A11}, J_{A12}, J_{A13}, J_{A22}, J_{A23}, J_{A33}$ in (43) are constant. Like $\mathbf{p}$ and $\mathbf{L}_A$, $\mathbf{p}^\circ$ and $\mathbf{L}_A^\circ$ are vector parts of a corresponding motor. With (4), the force motor $\hat{\mathbf{f}}$, the momentum motor $\hat{\mathbf{p}}$ and the speed motor $\hat{\mathbf{s}}$ we can replace (49),(50) by the following motor differential equation:

$$\hat{\mathbf{p}}^\circ + \hat{\mathbf{s}} \times \hat{\mathbf{p}} = \hat{\mathbf{f}}, \qquad (51)$$

or :
$$\dot{\hat{\mathbf{p}}} = \hat{\mathbf{f}}. \qquad (52)$$

There is no other formula containing the equations of the rigid body motion which has a comparable information density. With the dual vectors $\hat{\mathbf{p}} = \mathbf{p} + \varepsilon \mathbf{L}_A$, $\hat{\mathbf{s}} = \mathbf{w} + \varepsilon \mathbf{v}_A$ and $\hat{\mathbf{f}} = \mathbf{F} + \varepsilon \mathbf{M}_A$ A. T. Yang's formula (8) in [16] is

$$\hat{\mathbf{p}}^\circ + \hat{\mathbf{s}} \times \hat{\mathbf{p}} = \hat{\mathbf{f}}, \qquad (53)$$

which is formally identical to (51). Instead of a formula similar to (52), for the absolute time rate of the dualvector $\dot{\hat{\mathbf{p}}}$ one obtains

$$\dot{\hat{\mathbf{p}}} = \hat{\mathbf{f}} - \varepsilon \mathbf{v}_A \times \mathbf{p}. \qquad (54)$$

The linear motorial relation (47) between the speed motor $\hat{\mathbf{s}}$ and the momentum motor $\hat{\mathbf{p}}$ together with formula (52) and the equation $\hat{\mathbf{T}}^\circ = 0$ yields:

$$\hat{\mathbf{T}} \circ \hat{\mathbf{s}}^\circ + \hat{\mathbf{s}} \times \hat{\mathbf{T}} \circ \hat{\mathbf{s}} = \hat{\mathbf{f}}. \qquad (55)$$

This equation can also be written as $\hat{\mathbf{T}} \circ \hat{\mathbf{s}}^\circ + \hat{\mathbf{s}}^\times \circ \hat{\mathbf{T}} \circ \hat{\mathbf{s}} = \hat{\mathbf{f}}$, from which the matrix form is immediately obtained in the form:

$$\underline{T} \delta \dot{s} + \underline{s}^\times \delta \underline{T} \delta s = \underline{f}. \qquad (56)$$

To convert a motor equation into a matrix equation, there is a simple rule to follow: first the motorial product should be written as a scalar product with the aid of the motor cross tensor (29), and then the motor and the motor tensors are replaced by their corresponding matrices and the scalar product operator is to be exchanged with the $\underline{\delta}$ matrix (24).

References

[1] v. Mises, R., " Motorrechnung, ein neues Hilfsmittel in der Mechanik", *Zeitschrift für Angewandte Mathematik und Mechanik,* Band 4, Heft 2, S. 155-181, 1924.
    v. Mises, R.," Anwendungen der Motorrechnung", *Zeitschrift für Angewandte Mathematik und Mechanik,* Band 4, Heft 3, S. 193-213, 1924.

[2] v. Mises, R.," Motor Calculus, a New Theoretical Device in Mechanics", *Forthcoming English Translation of* [1], (J. E. Baker, K. Wohlhart).

[3] Clifford, W. K.," Preliminary Sketch of Biquaternions", *Proc. Mathematic Society,* vol.4, pp. 381-395, 1873.

[4] Study, E.," Geometrie der Dynamen", *B.G. Teubner Verlag,* Leipzig, 1903.

[5] Winkelmann, M., and Grammel, R., " Kinetik der starren Körper", Kap.8, Handbuch der Physik, Hg. Geiger, H., and Scheel, K., *J. Springer Verlag,* Berlin, 1927.

[6] Magnus, K.," Über die Anwendungen der allgemeinen Bewegungs-gleichungen starrer Körper in bewegten Bezugssystemen", *Zeitschrift für Angewandte Math. und Mechanik,* Band 23, Heft 6, S. 336-356, 1942.

[7] Raher, W. and Selig, F.," Die Verwendung der Motorsymbolik in der theoretischen Mechanik", *Sitzungsberichte der Österr. Akad.der Wiss.,* Math. Naturw. Klasse, Abt. II, Bd.163, Heft 5 bis 7, 1954.

[8] Birkhoff, G., et al., " Selected Papers of Richard von Mises", Vol.1, *American Mathematical Society,* Providence. Rhode Island, 1963.

[9] Sugimoto, K. and Duffy, J.," Application of Linear Algebra to Screw Systems", *Mechanism and Machine Theory,* Vol. 17, No.1, pp. 73-83, 1982.

[10] Parkin, I.A., " A Third Conformation with the Screw Systems; Finite Twist Displacement of a Directed Line and a Point", *Mech. Mach. Theory,* Vol. 27, No. 22, pp. 177-188, 1992.

[11] Keler, M.,"Analyse und Synthese der Raumkurbelgetriebe mittels Raumliniengeometrie und dualer Größen", Diss. München, 1958.

[12] Yang, A. T., and Freudenstein, F.," Application of Dual-Number Quaternions Algebra to the Analysis of Spatial Mechanisms".Trans. ASME, *Journal of Engineering for Industry,* Series E, Vol.86, pp. 300-308, June,1964.

[13] Dimentberg, F. M., " Wintowoje Istschislenie", Izdat." *Nauka*", Moskow, USSR, 1965. English Translation: AD680993 (Clearing House for Federal and Scientific Technical Information).

[14] Kislitzin, S.G.," Wintowoje Affinory i nekatorie ich priloschenija k woprossam kinematiki twordawa tela" (Screw Affinors and Some of Their Applications to Problems in the Kinematics of the Solid Body), Utsch. Sap. Len. Gos. Ped. Inst. Im. A.I. Herzena (Scientific Annals of the A.I. Herzen Leningrad State Pedagogical Institute), Vol. 10, 1938.

[15] Shoham, M. and Brodsky, V. " Analysis of Mechanisms by the Dual Inertia Operator", Computational Kinematics, ed. by J. Angeles et. al. *Kluwer Acdemic Publishers,* Netherlands, 1993.

[16] Yang, A.T.," Inertia Force Analysis of Spatial Mechanisms", Trans. ASME, *Journal of Engineering for Industry,* Series B, Vol.93, No.1, pp. 27-32, Feb. 1971.

[17] Euler, N. and Steeb, W.H." Continuous Symmetries, Lie Algebras and Differential equations", *B.I. Wissenschaftsverlag,* Mannheim, 1992.

THE INVERTED SLIDER-CRANK USED FOR THE DESIGN
OF AN APPROXIMATED STRAIGHT-LINE MECHANISM

Evert A Dijksman (author)   &   Anton T J M Smals (software-support)

Precision Engineering & Automation
Faculty of Mechanical Engineering
Eindhoven University of Technology
Eindhoven, The Netherlands

1. Abstract   The design of a centric inverted slider-crank producing a symmetrical curve touching a common tangent at three pairs (PP-PP-PP) of instantaneously near coupler-point positions, requires, apart from a scale-factor only *one* design-parameter ($\phi_0 = \sphericalangle K_1B_0K_3$) The mechanism may be seen as a *degenerate* λ-mechanism, the λ-mechanism being a 4-bar linkage ($A_0A-K-BB_0$) for which $\overline{AB} = \overline{B_0B} = \overline{KB}$ Then, if B goes to infinity, the λ-mechanism turns into an inverted slider-crank Approximations of the straight-line by the coupler curve of these mechanisms are usually better than those obtained by the λ-mechanisms with the same parameter ϕ_0 A graph showing the double relative deviation ($\Delta h/L$) of the straight-line plotted against the dimensionless length (L/h_{mn}) of the straight-line demonstrates this result In practice, the graph may be used to determine the singular design-parameter (ϕ_0) from the desired length of the straight-line In the special case for which $\phi_0 = 0$, a well-known result is recovered, where the tracing-point happens to be Ball's point (Bl_2) with excess 2 (figure **4**) (The curve then touches the common tangent at 6 infinitesimally near coupler point positions)

2. Introduction

Symmetrical curves produced by inverted slider-cranks may be approximated by a straight-line (See figure 1) This can be done by catching a stretch of the curve within a rectangular box of length L and width Δh (See figure **5**) The middle of L then joins the symmetry-axis intersecting the rectangle and the curve at two infinitesimally near positions ($K_3 = K_4$) of the lower tangent This tangent coincides with a common tangent for two other pairs of infinitesimally near positions of the tracing point Thus,

103

104

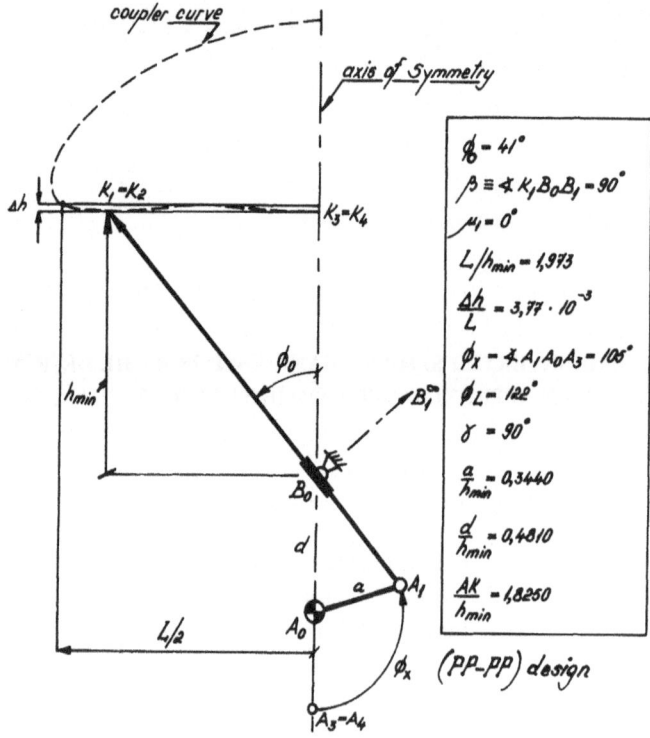

coupler curve

axis of Symmetry

$K_1 = K_2$

$K_3 = K_4$

Δh

ϕ_0

h_{min}

B_1

B_0

d

a A_1

$L/2$

A_0

ϕ_x $(PP\text{-}PP)$ design

$A_3 = A_4$

$\phi_0 = 41°$

$\beta \equiv \angle K_1 B_0 B_1 = 90°$

$\mu_1 = 0°$

$L/h_{min} = 1.973$

$\dfrac{\Delta h}{L} = 3.77 \cdot 10^{-3}$

$\phi_x = \angle A_1 A_0 A_3 = 105°$

$\phi_L = 122°$

$\gamma = 90°$

$\dfrac{a}{h_{min}} = 0.3440$

$\dfrac{d}{h_{min}} = 0.4810$

$\dfrac{AK}{h_{min}} = 1.8250$

Figure 1   *Inverted slider-crank used as a straight-line mechanism*
(Degenerate crank-and-rocker mechanism of type λ)

the 6 tracing- or coupler point positions $K_1 = K_2$, $K_3 = K_4$ and $K_5 = K_6$ all join the lower (common) tangent of the curve for which $\overline{K_1 K_3} = \overline{K_4 K_6}$. The upper tangent of the rectangle touches the curve at two symmetrical but separate positions of K and intersects the curve at two other locations, leading to the length of the straight-line. (Note that both the lower - and the upper tangent intersect the curve at 6 intersections, being the maximum number possible for a 6th order curve.) We conclude that the design of the inverted slider-crank may be based on 3 pairs of infinitesimally near *accuracy* points K_i on the *lower* tangent, representing an application of *multiply separated position* (MSP)-theory, here being a mixture of separate - and instantaneously near positions. The symmetry of the curve is obtained by considering only *centric* inverted slider-cranks, that is to say by obliterating the two possible eccentricities in the mechanism. That leaves only 3 lengths, namely $a = \overline{A_0 A}$, $d = \overline{A_0 B_0}$ and $l = \overline{AK}$. Of these, one is needed to have a common lower tangent. Thus, apart from a scaling-factor, only *one* design degree of freedom is left. For this, one takes the angle $\phi_0 = \angle K_1 B_0 K_3$.

105

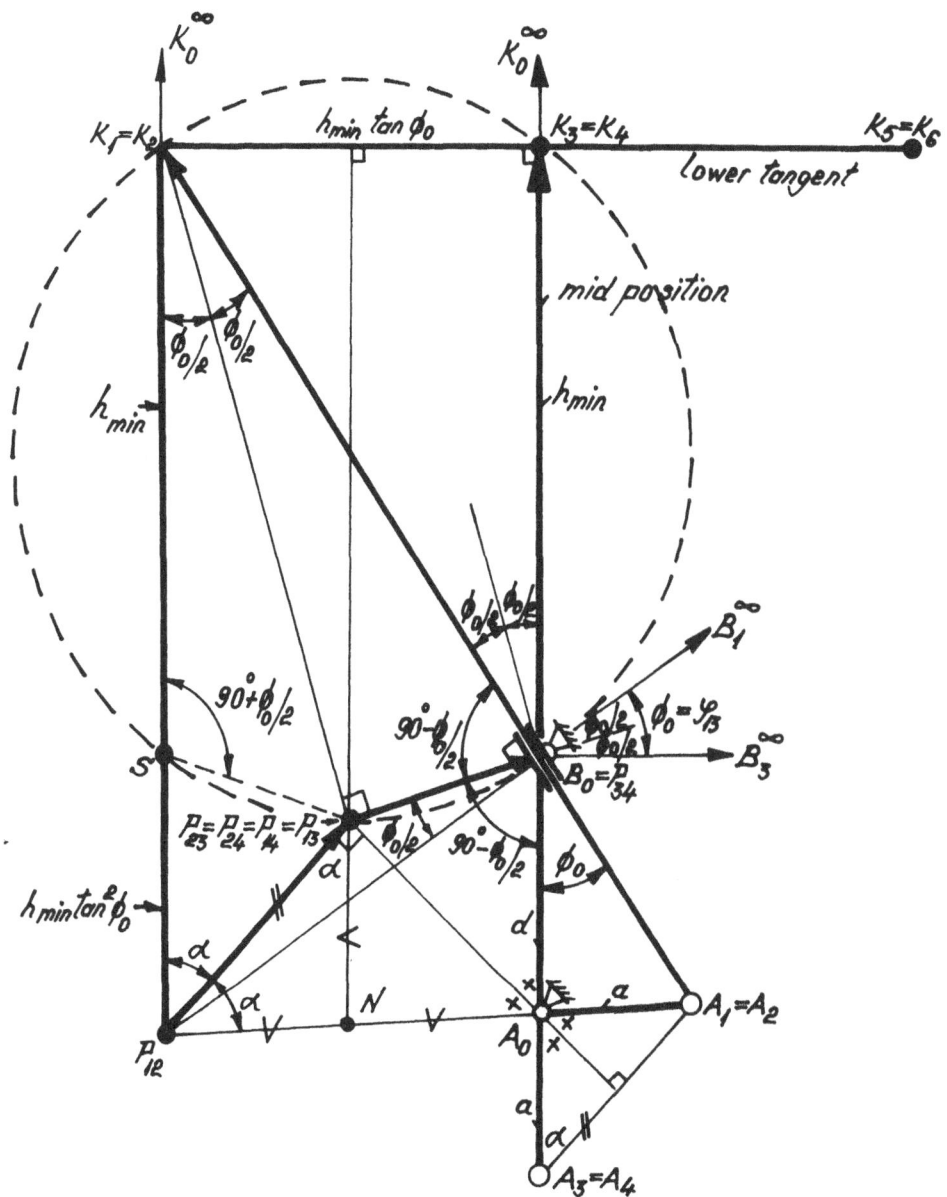

Figure 2   *Geometrical design of an inverted slider-crank*

106

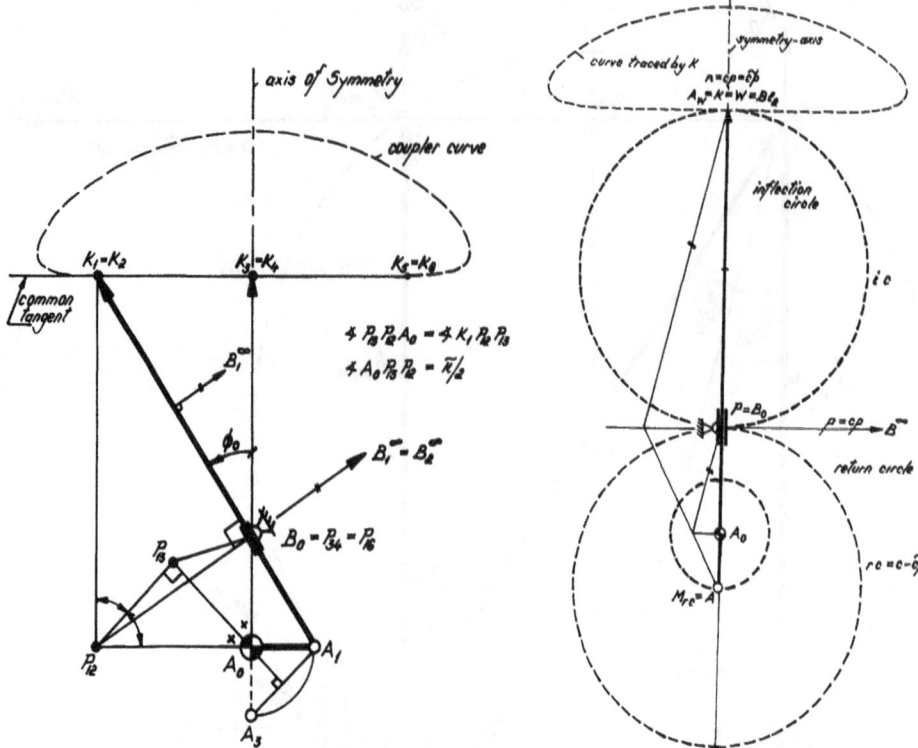

Figure 3 *Simplified design for an inverted slider-crank very neatly approximating a straight-line*

Figure 4 *Inverted slider-crank having a tracing point being Ball's Undulation point with excess 2*

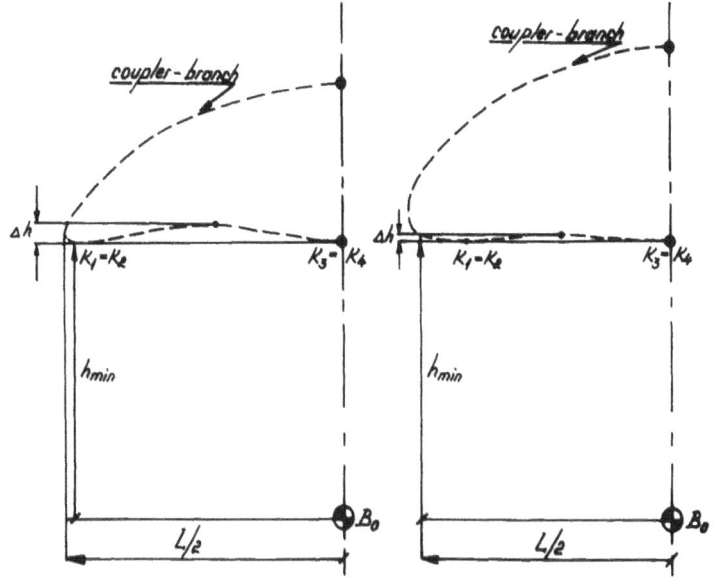

Figure 5   *Determination of the length L of the straight-line*

3. Geometrical design

λ-type 4-bar linkages for which $\overline{KB} = \overline{B_0B} = \overline{AB}$, producing symmetrical 4-bar coupler curves, turn into centric inverted slider-crank mechanisms when the joint B goes to infinity. (ref.[5]). Thus, if $(A_0A-K-BB_0)$ represents the 4-bar and $B \Rightarrow B^*$, then the circle about B joining the points A, B_0 and K , merges into the slider KB_0 containing the crank-joint A. (See figure 2). Indeed, by taking B to infinity, the inverted slider-crank, while being centric, will have a coupler point joining the slider. Thus, the centric inverted slider crank, i.e. the one without eccentricities, represents a limit-mechanism for the λ-type four-bar linkage having A_0A as a revolving crank. Naturally, the symmetrical-axis of the curve, produced by such a limit-mechanism, has to join the centers (A_0 and B_0) of the frame. Because of the symmetry of the curve and of the mechanism, not three but only two pairs of coincident positions are needed for its design. They correspond to the pairs ($K_1 = K_2$) and ($K_3 = K_4$). The two times two positions define 6 *rotation poles* P_{12} , $P_{32} = P_{24} = P_{41} = P_{13}$ and P_{34}. Of them, four are all coinciding at one and the same location. The remaining poles, namely P_{12} and P_{34} , become (instantaneous) velocity poles. The pole P_{12} for instance, may then be found at the intersection of the normals to the slider at B_0 and to the curve at $K_1 = K_2$. Similarly, one so finds the velocity pole P_{34} at the frame-center B_0.

The condition for the common (lower) tangent to touch the curve at the same height, h_{mm}, needs *one* design degree of freedom. Thus, if point S is defined as the intersection of the path-normal $P_{12}K_1$ with the diameter-circle about $\overline{B_0K_1}$, the condition is represented by the equation $\overline{SK_1} = (h_{mm}) = \overline{B_0K_3}$. Hence, the mechanism is fully determined by the (scaling) length of h_{mm} and by the independent parameter $\phi_0 = \sphericalangle K_1B_0K_3$. The rotation pole P_{13}, being the coincident pole of four of them, joins the intersection of the midnormal of $\overline{K_1K_3}$ with the angle bisector of the $\sphericalangle B_1 \ddot{}B_0B_3 \ddot{}$. Hence,

$\sphericalangle P_{13}B_0K_1 = 180^0 - (\sphericalangle K_1B_0B_1 \ddot{} + \frac{1}{2} \phi_0) = 90^0 - \frac{1}{2} \phi_0$, while the equality $\overline{SP_{13}} = \overline{P_{13}B_0}$ leads to $\sphericalangle B_0K_1P_{13} = \sphericalangle P_{13}K_1S = \frac{1}{2} \phi_0$, finally yielding to $\sphericalangle K_1P_{13}B_0 = 90^0$.

Thus, the rotation pole P_{13} joins the diameter-circle about $\overline{K_1B_0}$ containing the points K_1 , K_3 , B_0 , P_{13} and S. Further, the points P_{12}, A_0 and A_1 all join the path-normal of the crank-joint A_1 , while P_{13} meets the midnormal of $\overline{A_1A_3}$. The rotation-pole P_{13} then joins the angle-bisectors through the vertices K_1, B_0 and A_0 of the quadrilateral $K_1P_{12}A_0B_0$. Thus, P_{13} represents the center of a circle being inscribed in the quadrilateral $K_1P_{12}A_0B_0$, whereas the line $P_{12}P_{13}$ coincides with the 4th angle-bisector of that quadrilateral. Hence, $\sphericalangle K_1P_{12}P_{13} = (\alpha) = \sphericalangle P_{13}P_{12}A_0$, (See figure 3). The midnormal $\overline{K_1K_3}$ intersects the path-normal $P_{12}A_0$ of A_1 at a point N for which $\overline{P_{13}N} = \overline{P_{12}N} = \overline{NA_0}$ giving $\sphericalangle A_0P_{13}P_{12} = 90^0$. This determines the exact location of the crank-center A_0 at the axis of symmetry from the already known locations of the poles P_{12} and P_{13} .

4. Computational Design

Application of the Law of Sines at $\Delta P_{12}P_{13}K_1$ leads to a relation between the angle α and the design-parameter ϕ_0:

$$\frac{\sin \alpha}{\sin (\alpha + \frac{1}{2}\phi_0)} = \cos \phi_0 \cdot \cos (\frac{1}{2}\phi_0)$$

or
$$\cot \alpha = \frac{4}{\sin 2\phi_0} - \cot \frac{1}{2}\phi_0 = 2 \tan \phi_0 - \tan \frac{1}{2}\phi_0 \qquad (1)$$

The Law of Sines applied at $\Delta A_0A_1B_0$ leads to the formula

$$\frac{d}{a} = \frac{\sin (\phi_0 + 2\alpha)}{\sin \phi_0} \qquad (2)$$

in which the angle α may be established through equation (1). Further,

$$\overline{A_0B_0} + \overline{B_0K_3} = \overline{P_{12}K_1} - \overline{K_1K_3} \cdot \tan(90^0 - 2\alpha)$$

or

$$d + h_{mn} = (h_{mn}/cos^2\phi_0) - h_{mn}.\tan\phi_0.\cotan 2\alpha$$

leading to the expression

$$d/h_{mn} = \frac{1}{2}.\tan\phi_0.\{\tan(\frac{1}{2}\phi_0) + \tan\alpha\} \tag{3}$$

Division of equation (3) by equation (2) then leads to

$$a/h_{mn} = \frac{1}{2} + \frac{1}{2}.\tan\phi_0.\{\tan(\frac{1}{2}\phi_0) - \tan\alpha\} \tag{4}$$

Clearly,

$$\overline{A_3A_0} + \overline{A_0B_0} + \overline{B_0K_3} = \overline{A_1K_1} = l$$

whence,

$$a + d + h_{mn} = l$$

and

$$l/h_{mn} = 1 + (a/h_{mn}) + (d/h_{mn}) \tag{5}$$

Substituting the obtained expressions (3) and (4) into the expression for l/h_{mn} yields

$$\frac{l}{h_{min}} = \frac{3}{2} + \tan\phi_0.\tan\frac{1}{2}\phi_0 = \frac{3}{2} - \frac{2}{1-\cotan^2(\frac{1}{2}\phi_0)} \tag{6}$$

So, the dimensions of the inverted slider-crank mechanism, producing the straight-line, may all be established as a function of the scale-value $h_{mn} = \overline{B_0K_3}$ and of the singular design-parameter ϕ_0, to be taken from the graph in which the relative width $\Delta h/h_{mn}$ of the rectangle has been plotted against the relative length L/h_{mn} of the straight-line being approximated. The straight-line to be approximated runs horizontal at the height of $(h_{mn}+\frac{1}{2}\Delta h)$ intersecting the curve at 6 *separate* coupler point positions.

(Since first the locations of these 6 points were unknown, they have not been used as accuracy points for the design, though older programs based on a probing Chebyshev-spacing have done so, ref.[1] and [3].) The method developed here represents a so-called (PP-PP-PP) approach to a shifted straight-line (i.e. shifted by a distance of $\frac{1}{2}\Delta h$) to be produced by the inverted slider-crank. Thus, the maximum deviation of the curve from the straight-line just equals $\frac{1}{2}\Delta h$, in so far it runs within the rectangular box of length L.

The crank-angle needed to reach position 1 from the mid-position 3 of the mechanism, satisfies the equation

$$\sphericalangle A_1A_0A_3 = 180^0 - 2\alpha \tag{7}$$

in which the angle α is to be calculated through equation (1)

In the limit-case $\phi_0 = 0$, the two design-positions merge into a singular position, being the inverse of the elliptic position, sometimes named the *cardioidal position*, (ref.[4]). Then, a *boundary - mechanism* appears with the dimensions $d/h_{min} = \frac{1}{3}$; $a/h_{min} = 1/6$ and $l/h_{min} = 3/2$ \tag{8}

110

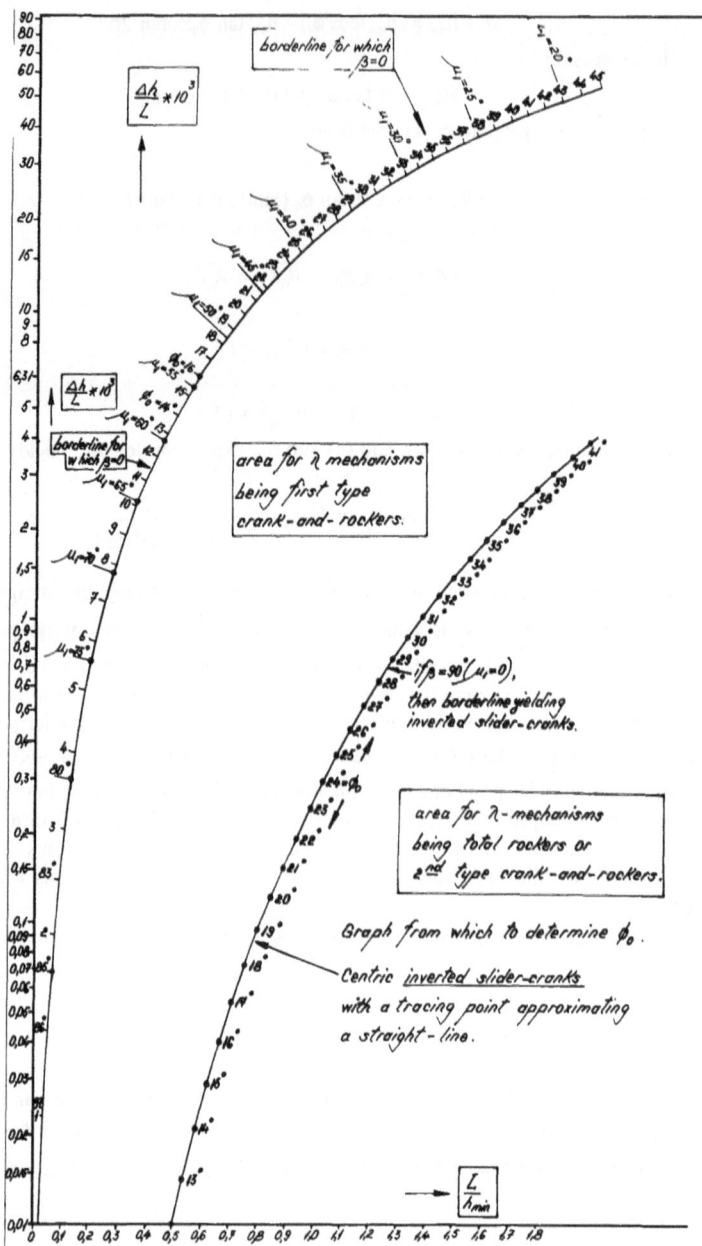

Figure 6 — *Graph from which to determine the design-parameter ϕ_0*

Whence, $d/a = 2$ and $l/a = 9$, (See figure 4). Thus, when $\phi_0 = 0$, all 6 accuracy points coincide (i.e. $K_1 = K_2 = K_3 = K_4 = K_5 = K_6$), yielding a 6-point contact between lower tangent and coupler curve. The coupler point then represents Ball's point with excess 2.

Larger values for ϕ_0 however, always lead to longer stretches L of the straight-line that has to be approached Of course, larger deviations of $\frac{1}{2}\Delta h$ have then to be taken into account. The designer may choose a suitable value for ϕ_0 by observing the graph of figure 6 showing $10^3 \Delta h/L$ as a function of L/h_{min}. The length L of the straight-line has been calculated through the coupler curve equation and depends on whether the curve does have, or does not have, a vertical tangent before leaving the area lying under the upper tangent, (See figure 5). If it does, the length L will be the horizontal distance between the two vertical tangents. Otherwise, the length L represents the distance between the outermost intersections of the curve with the upper tangent.

5. Conclusions

A simplified design, as demonstrated in figure 3, provides the mechanical engineer with an easy design for a straight-line mechanism using the kinematic chain of an inverted slider-crank. The singular design parameter ϕ_0, needed for such a design, may be taken from a graph (figure 6) in which twice the relative deviation $(\frac{\Delta h}{L})$ from the straight-line has been plotted against the dimensionless length $(\frac{L}{h_{min}})$ of that line. Accurate dimensions of the mechanism may be obtained through the equations 1 to 6 derived from the figure. Note that the *accuracy-line* will run at a distance $(h_{min} + \frac{1}{2} \Delta h)$ from the frame-center B_0.

6. Literature

1   VDI/AWF-Handbuch Getriebetechnik, VDI-Richtlinien, Nr 2137, (August 1959) *Ebene Kurbelgetriebe, Konstruktion zentrischer Kurbelschleifen mit Geradführung*

2   Gierse,F J ,Gunzel,D ,Schaeffer,T **Fortschrittberichte VDI**, *Untersuchung und Klassifizierung von Maßsynthese-Verfahren für die Bestimmung der Kinematik ebener Führungsgetriebe* Reihe 1 **Konstruktionstechnik/Maschinen - elemente nr.218,** VDI-Verlag (1993), S 103-S 105 (zentrische Kurbelschleife)

3   Karelin,v S *On the synthesis of the inverted slider-crank mechanisms for approximate straight-line motion,* Mechanism and Machine Theory, vol **21**,(1986), Nr 1, pp 13-18

4   Dijksman,E A **Motion Geometry of Mechanisms**, pp 112-113 (The *cardioidal position* of the moving plane), Cambridge University Press, London, New York, Melbourne (1976)

5   Dijksman,E A & Smals,A T J M *Het benaderen van een recht lijnstuk met symmetrische koppelkrommen*, (in dutch) De Constructeur, (1988), nr 8, pp 34-44, (Figures 6 and 17)

6   Dijksman,E A & Smals,A T J M *The translation-position of λ-formed 4-bar linkages used for the design of mechanisms approximating a straight-line*, Anales de Ingeniería Mecánica, Revista de la Asociación Española de Ingeniería Mecánica, año 10, volumen **1**, noviembre 1994, pp 79-86 Actas del xi Congreso Nacional de Ingeniería Mecánica

ON THE CHOICE OF INDEPENDENT LOOPS IN MECHANISM KINEMATICS

Pietro Fanghella and Carlo Galletti
Istituto di Meccanica Applicata alle Macchine
Università di Genova
Via Opera Pia 15-A, 16145 Genova, ITALY

Abstract. This paper addresses the issue as to whether the choice of a set of independent loops in a mechanism may reduce the chain, i.e., to ascertain if it is
possible to find closure equations in hierarchical form. The selection of the
shortest loops and the search for sinks are considered as possible strategies.
Theoretical considerations and application examples demonstrate that these
choices are useful in analyzing many technical mechanisms, but cannot in general ensure a chain reduction.

1. Introduction

The advantages of relative kinematics over absolute kinematics in modelling
multibody systems are well known [1]. In order to be applied to closed-loop
mechanisms, the relative-kinematics approach requires that a complete set of
independent loops be recognized, in order to obtain a corresponding set of closure equations. Despite the fact that any complete set of independent loops can
be used to obtain closure equations, a proper choice of the loops provides the
"best" equations, i.e., the minimum number of equations with the minimum
coupling among them. It has been shown that, for many technical mechanisms,
such equations for each loop contain as unknowns only the pairs of the loop itself, and that the equations can be solved in closed form. So, in these cases, it is
possible to reduce a mechanism to a hierarchy of single-loop chains and to obtain a system of equations that can be written and solved in triangular form. This
topic has been investigated for about 80 years, from the early work by Assur on
planar mechanisms up to recent contributions in several fields of kinematics
(references are given in [2] and [3]). More recently, Kramer [4] used symbolic

113

J.-P. Merlet and B. Ravani (eds.), Computational Kinematics, 113–122.

geometry, including explicit reasoning; Chieng and Hoeltzel [5] adopted symbolic pattern searching to solve planar mechanisms.

Fanghella and Galletti [6] and Kecskeméthy [7] presented two systematic approaches to the search for a complete set of independent loops. Their methods establish automatically, for any user-defined spatial multiloop mechanism, if a triangular system of equations can be found and how to obtain closed-form solutions. In both approaches, the search for loops is based on a recursive process aimed at finding a hierarchy of single-loop kinematically determined chains (SLDCs): the loops at the top of the hierarchy are the root loops; the loops at the bottom are the leaf loops. An SLDC is a single-loop chain in which:

a) independent values can be assigned to a certain number (>0) of pair variables (driving pairs);

b) the relative positions of all links depend only on the geometry of the chain and on the values of its driving pairs.

A particular type of single-loop determined chain (i.e., an SLDC_0) can be obtained by removing all driving pairs from any SLDC. The basic differences between these approaches lie in the strategies used for searching candidate loops and for performing the recursive process.

The first approach [6] requires that a correct number of drivers and their locations be assigned as problem data. All chains that contain at least one driver are analyzed in order to find if one or more of them are SLDCs; if so, every determined chain forms an independent loop of the required set (a root loop). Then, a single fictitious deformable body replaces each SLDC, and a further analysis of loops is made to find other determined chains. This process continues until no more SLDCs are found. At this stage, if all mechanism pairs and bodies are present in the set of loops, the mechanism can be analyzed by means of a hierarchy of subsystems of equations (one subsystem for each SLDC), and the corresponding set of independent loops has been generated.

The second approach [7] requires that a candidate set of loops be chosen in advance, but no driver assignments are needed. Then, it is verified if, in this set, there exists at least a loop which form an SLDC when all its pairs shared with other loops are regarded as driving pairs (more driving pairs are introduced, if required). If such a loop is found, this is called a sink loop: it is removed from the loop set and a further search for another sink is performed. If, according to this procedure, all given loops become sinks, the mechanism can be analyzed by a hierarchy of subsystems of equations and the corresponding set of loops is the candidate one. Kecskeméthy suggested forming the candidate set of independent loops by choosing the shortest loops.

In short, the first approach requires that driver locations be assigned in advance, performs a forward (from roots to leaves) search for chains, and generates a correct set of independent loops. The second approach does not require that drivers be located but imposes that a set of independent loops be selected in advance; then this set is checked for consistency with the uncoupling strategy through a backward (from leaves to roots) analysis.

It is known that a given mechanism can be reduced or not, depending on the choice of driver locations. Therefore, the first approach can indicate if a mechanism can be reduced by means of a given choice of driver locations. On the contrary, the second approach can find where to locate drivers in order to have hierarchically coupled equations. The latter possibility may appear a minor advantage if one wants to perform a kinematic analysis of a mechanism and its real motors, but is of major importance in dynamic analysis, where free coordinates can be chosen in such a way as to obtain uncoupled equations.

In the final discussion, at the Workshop on Computational Kinematics in Dagstuhl Castle, Germany (1993), the participants addressed the problem of establishing if the choice of the shortest loops can always provide correct results when one uses the second approach. Further work on this matter was suggested. The aim of this paper is to contribute to such a discussion by examples and theoretical considerations. The implications of the choice of the shortest loops are considered. The tools we use are concepts derived from definitions of Assur's groups, and applications of displacement groups to kinematics .

2. Independent Loops and Assembling Mechanisms

In this and the following sections, two issues are considered:

a) whether the choice of a set of shortest loops in a mechanism is always the correct choice to detect if the mechanism contains sinks;

b) whether the search for sinks can always allow one to find the leaves of the hierarchy to which a mechanism can be reduced.

Several examples given in the paper concern planar mechanisms, but most of the related discussions can also be applied to spatial mechanisms and to mechanisms where different chains of links have displacements in different subgroups of the spatial displacement group. In the 6-bar mechanism in Fig. 1, three loops (I, II, and III) are recognized; any couple of such loops provides a complete independent set. If loops I and II are regarded as a complete set, we see that:

i) the loops share two pairs (C and D);

ii) if values are assigned to the shared pairs, the chain A, B, C, D corresponding to loop I is overconstrained;

iii) if values are assigned to the shared pairs, the chain G, F, E, D, C corresponding to loop II is kinematically determined.

Therefore, a sink corresponding to loop II is found.

116

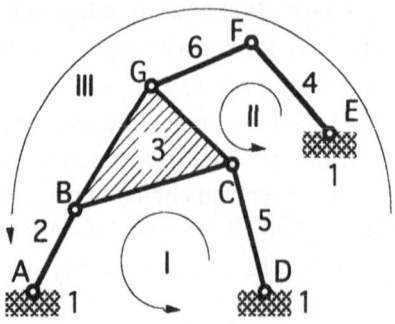

Figure 1 Loops of a mechanism.

The remaining loop I is a 1-d.o.f. chain and the mechanism can be reduced to the sequence of loops I and II. As a consequence, closure equations for loop I are independent of the variables of the pairs G, F, E, and such equations can be solved for three unknowns of the pair variables of loop I when a value is given to the variable of the fourth pair. Equations for loop II depend on the variables of the shared pairs (which are provided by the previous solution) and furnish the values of the other pairs G, F, E.

Now, concerning loops II and III, we observe that:

i) the loops share three pairs (G, F, and E);

ii) if values are assigned to the shared pairs, the chain A, B, G, F, E corresponding to loop III is overconstrained;

iii) if values are assigned to the shared pairs, the chain G, F, E, D, C corresponding to loop II is overconstrained.

Therefore, no sinks can be recognized for such a set of loops.

Because of topological symmetry, the choice of the set of loops I and III does not allow sink recognition.

From this example, we deduce that a generic choice of independent loops does not ensure that sinks will be recognized, even when they exist, and that the corresponding mechanism cannot be reduced to single-loop chains. Reference [7] suggests that the set of loops be formed by using loops of minimum length. The example confirms that, in this case, this is not only a convenient choice but also the only one. The key to the problem lies in the fact that, in the mechanism in Fig.1, the existence of a sink can be recognized only if the sequence G, F, E is contained in only one loop; if so, this loop is a sink.

In order to provide general observations, we consider the general planar SLDC_0, which is a 3-pair closed chain with at least one revolute. To obtain an SLDC, we can expand a link of the SLDC_0 by inserting one or more pairs that allow a relative displacement between the link parts (Fig. 2). References and a deeper discussion on this subject can be found in [2].

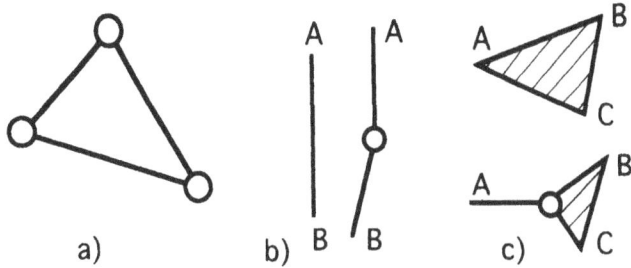

Figure 2   Three-pair chain and link expansions.

It is possible to build multiloop kinematically determined chains with any number of links by using the following recursive rule:

• a kinematically determined chain is obtained by replacing any link of a 3-pair chain with a kinematically determined chain.

The chains obtained in this way have loops with three or more pairs (in any case, at least one loop with only three pairs exists); therefore, a mechanism is obtained by expanding at least one link of each remaining 3-pair chain.

Figure 3 shows an example of this procedure. In the figure, link 7 of chain II is replaced by chain I, using links 3 and 1 to get chain connections. A 3-pair chain of links A, B, D still exists; then one of its links, for instance, link 3, is expanded, and the same mechanism as in Fig. 1 is obtained. Because all mechanisms obtained in this way are assembled by using only 3-pair chains, they can always be reduced and the corresponding hierarchy of equations derived. A "natural" set of independent loops coincides with the two original 3-pair chains: A, B, D and G, F, E, (i.e., after link substitutions and expansions), with the loops A, B, C, D and G, F, E, D, C.

Note that a kinematic analysis scheme with a hierarchical set of equations follows the solution sequence from chain I to chain II, as a link of chain II is replaced by chain I. Then, the loop A, B, C, D is a root and the loop G, F, E, D, C is a leaf of the reduction process. If only one loop in the set of independent loops contains the pairs E, F and G of chain II, this can be recognized as a sink, as stated above; otherwise, this recognition is impossible.

Therefore, we can state the following 3-pair chain rule: in a given planar mechanism, a sink can be recognized iff at least one serial 3-pair chain is contained in only one of the chosen independent loops. In the mechanism shown in Fig. 3-c, the choice of the loops A, B, C, D and G, F, E, D, C, or of the loops A, B, C, D and A, B, G, F, E, allows the pairs G, F and E forming a 3-pair chain to be considered in only one loop, thus confirming the possibility of using both sets of loops. As previously stated, the set of loops A, B, G, F, E and G, F, E, D, C does not allow sink recognition.

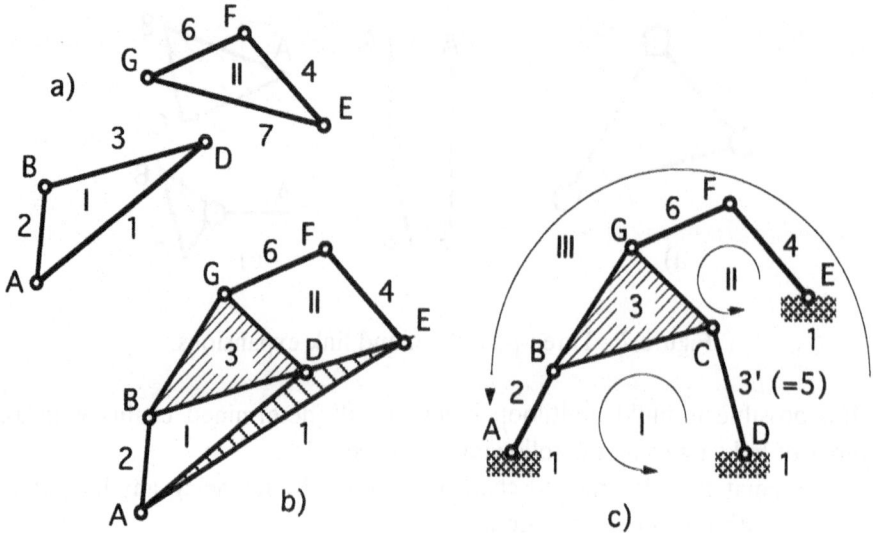

Figure 3 Generation of a mechanism from 3-pair chains.

3. Complex Mechanism Assemblies

We are now able to build a more complex, reducible, kinematic chain and to determine a set of independent loops based on the 3-pair chain rule. Starting from the chain of pairs A, B, C in Fig. 4-a, we replace each link with a 3-pair chain (D, E, F; G, H, L; M, N, P), and obtain the kinematically determined chain in Fig. 4-b. Then, we replace a link of a further 3-pair chain (Q, R, S) with the chain in Fig. 4-b, making connections to links 2 and 5. In this way, we obtain the chain in Fig. 4-c. A mobile chain can now be created by expanding links 3, 4 and 6 of the remaining 3-pair loops. The 3-d.o.f. mechanism is shown in Fig. 4-d.

By construction, this is a reducible mechanism, and the corresponding set of 5 independent loops is given by the hierarchy of 3-pair chains, after link substitutions and expansions:

roots (3 loops): D, E, F, F'; G, H, L, L'; M, N, P, P'
intermediate (1 loop): A, F, F', B, P', P, C, L', L
leaf (1 loop): C, L', L, A, Q, R, S (or F, F', B, P', P, S, R, Q).

Note that any leaf can be a sink, as the pairs Q, R, and S belong only to the leaf loop . A kinematic analysis of the relative displacements can be made by following the loop sequence.

However, the following set of 5 shortest independent loops is found from among the 23 loops of the chain:
D, E, F, F'; G, H, L, L'; M, N, P, P'; C, L', L, A, Q, R, S; F, F', B, P', P, S, R, Q. Unfortunately, the pairs Q, R, S are included twice in this set of loops; therefore, no

sinks can be recognized, even though the mechanism can be reduced to a single-loop hierarchy.

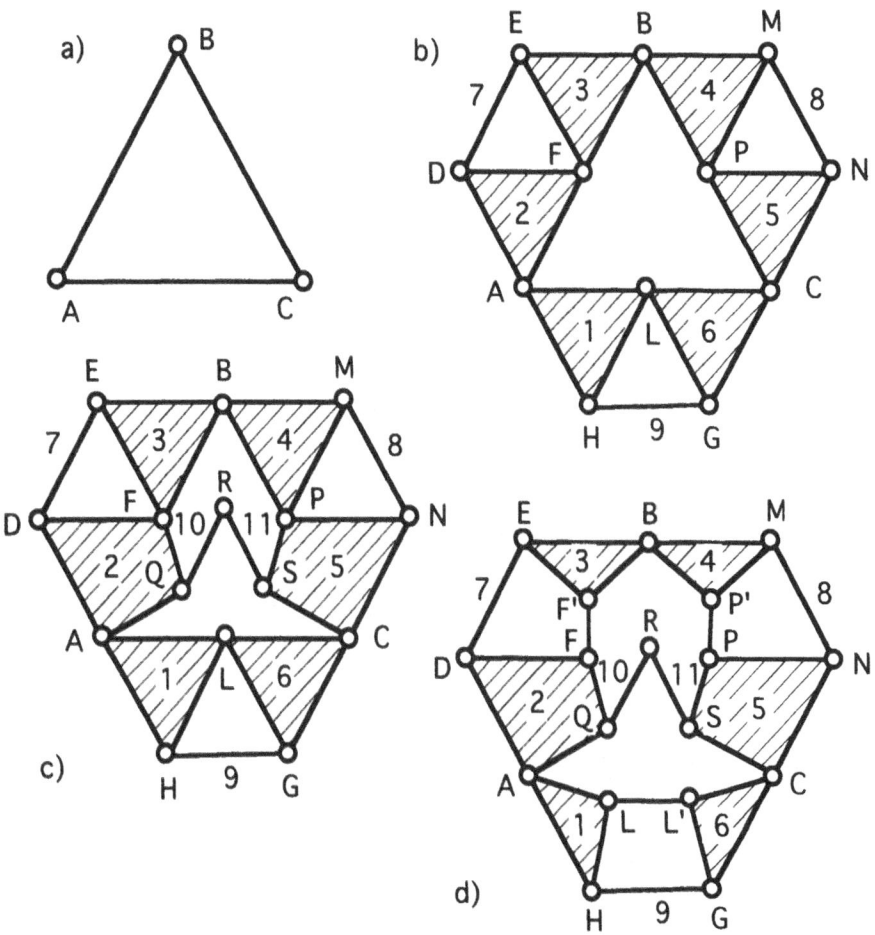

Figure 4   Generation of a multiloop chain from 3-pair chains.

This demonstrates that the 3-pair rule does not coincide with the choice of the shortest loops, and that, in general, this choice does not allow one to recognize sinks and to reduce a mechanism.

4. Passive Mobility Problems

Let us reconsider the mechanism in Fig. 1. As stated in Section 2, the choice of loops II and III does not allow the recognition of sinks; moreover, these loops are not the shortest. In order to find a new kind of exception to the shortest-loop strategy, it is interesting to explore whether it is possible to "force" such loops to be the shortest, without changing the mechanism structure. To this end, let us consider each pair A, B, C, D replaced by a sequence of two coaxial pairs A+A', B+B', and so on (Fig. 5-a). This sequence of pairs corresponds to the technical construction shown in Fig. 5-b. In this case, loops II and III are shorter than loop I; the sum of their pair connectivities are 7 and 8 respectively. Therefore, the shortest-loop strategy does not provide loops for reducing this mechanism .

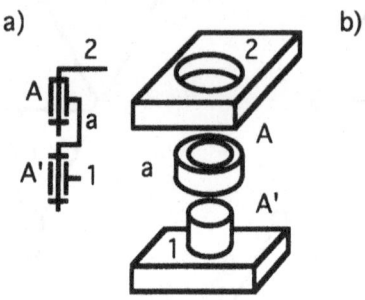

Figure 5   Passive mobility in a revolute pair

This problem depends on the presence of passive mobilities inside the constraints A, B, C, D; such mobilities could be easily detected and suppressed by an automatic mobility analysis [8].

5.   Spatial Mechanism with Intersecting Groups

A more complex situation occurs in spatial mechanisms, where, in several cases of technical interest, it is possible to find chains that do not fulfil the standard mobility rules, and that are still movable. Fig. 6-a shows a single-loop, 1-d.o.f. chain with links connected by revolute and prismatic pairs. The axis of revolute A is parallel to C; the axis of revolute D is parallel to F. The chain A, B, C realizes a translating gimbal constraint with connectivity = 3 (i.e., RP3,3) between links 1 and 4, and the rotation axis parallel to A; the chain D, E, F realizes another translating gimbal constraint with connectivity = 3 (i.e., RP3,3) between links 1 and 4, and the rotation axis parallel to D. The positive mobility of the chain is due to the non-empty intersection of the two gimbal groups (i.e., RP3), and the resulting displacement between links 1 and 4 is a 1-d.o.f. spatial translation (i.e., P3,1) [8]. We can use this mobile chain to replace link 0 of the unique SLDC_0 existing in group P3 (Fig. 6-b). As in Section 3 for planar chains, a multiloop mechanism with 1-d.o.f. is obtained (Fig. 6-c) that can be reduced to the sequence of the following chains: root: A, B, C, D, E, F; leaf: A, B, C, G, H, L (or D, E, F, L, H, G).

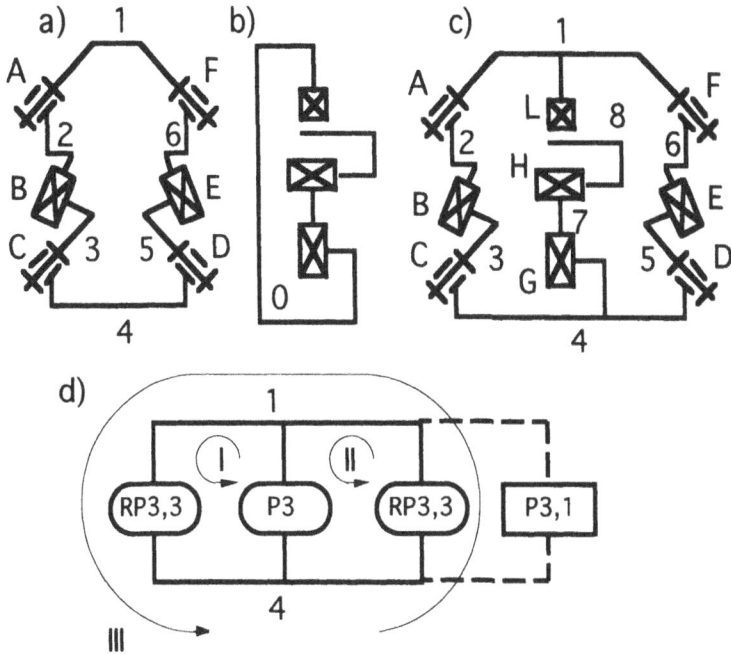

Figure 6   Spatial kinematic chain with group intersections.

Given the value of one pair variable of the root chain, we can solve the systems of closure equations of both loops sequentially.

For the complete chain (Figs. 6-c and 6-d), three loops of the same length exist: I): A, B, C, G, H, L; II): D, E, F, L, H, G; III): A, B, C, D, E, F. All three loops have the same length (6 pairs with 1-d.o.f.).

Mobility analysis of loops I and II reveals that both have 2-d.o.f.; mobility analysis of loop III reveals that it has 1-d.o.f.

Let us now discuss the three alternatives for the choice of a couple of independent loops.

Loops I and II have three common pairs (G, H, L); neither of them can be used as a sink because it is impossible to assign independent values to the variables of these pairs, for both loops have 2-d.o.f..

Loops I and III share three pairs (A, B, C); neither of them can be used as a sink because it is impossible to assign independent values to the variables of these pairs, for chain I has 2-d.o.f. and chain III has 1-d.o.f. A symmetric situation occurs for the couple of loops II and III. Therefore, even though the multiloop chain considered can be reduced to two single-loop chains (a root and a leaf), a simple analysis of single loops does not allow one to recognize the reduction by performing a search for sinks.

122

6. Conclusions

By theoretical discussion and with the aid of a few examples, we have explored the possibility of reducing a mechanism. We have shown that, in some cases, the shortest-loop strategy fails to find sinks, whereas, in other cases, a reduction is possible, although no sinks exist. However, the given examples do not seem to have technical applications. Therefore, we conclude that the shortest-loop and sink strategies, even though they do not always succeed, are useful tools for obtaining reduced kinematic models of real mechanisms.

References

1. Hiller, M. et al., 1992, "Modelling and Simulation of Mobile Robots and Large Manipulators," *Proc. 3rd Int. Workshop Advances in Robot Kinematics*, Parenti-Castelli and Lenarcic, Eds., Felloni, Ferrara (I), pp. 27-36.
2. Galletti, C., 1986, "A Note on Modular Approaches to Planar Linkage Kinematic Analysis," *Mech. Mach. Theory*, pp. 385-391.
3. Fanghella, P. and Galletti, C., 1990 "Kinematics of Robot Mechanisms with Actuating Closed Loops," *Int. J. Robotic Research*, pp. 19-24.
4. Kramer, G., 1992 *Solving Geometric Constraint Systems*, MIT Press.
5. Chieng, W. H. and Hoeltzel, D. A., 1990 "Computer-Aided Kinematic Analysis of Planar Mechanisms Based on Symbolic Pattern Matching of Independent Loops," *ASME J. Mechanical Design*, pp. 337-346.
6. Fanghella, P. and Galletti, C., 1992 "Hierarchical Generation of Independent Loops for Symbolic Kinematics of Robot. Mechanisms," *Proc. 3rd Int. Workshop Advances in Robot Kinematics*, Parenti-Castelli and Lenarcic, Eds., Felloni, Ferrara (I), pp. 96-103.
7. Kecskeméthy, A., 1993, "On Closed Form Solutions of Multiple-Loop Mechanisms," *Computational Kinematics*, J. Angeles et al., Eds., Kluwer Academic Publisher, pp. 263-274.
8. Fanghella, P. and Galletti, C., 1994, "Mobility Analysis of Single-Loop Kinematic Chains: an Algorithmic Approach Based on Displacement Groups," *Mech. Mach. Theory*, pp. 1187-1204.

Acknowledgement
This work was supported by italian MURST.

SYMBOLIC AND NUMERIC COMPUTATION FOR THE REAL-TIME SIMULATION OF A CAR BEHAVIOR

C. GARNIER

Simulog
1 rue James Joule
78286 Guyancourt Cedex

B. MOURRAIN

INRIA
Safir Project, UR de Sophia-Antipolis
2004 route des lucioles
BP-93 06902 Valbonne

AND

P. RIDEAU

Aérospatiale
100 Bd du Midi, BP 99
06322 Cannes-la-Bocca Cedex

Abstract. French car manufacturers have started a project named SARA. The goal of the project is to build a real-time interactive simulator including realistic car dynamic behavior. For that purpose, a specific approach has to be used for suspension links and vehicle modeling in order to combine computational efficiency and correct behavior modeling. A benchmark has been defined to test some software on these specifications.

This paper first describes the project and suspension specific problems. Then the approaches mixing symbolic and numerical computation, used by Simulog and Inria to meet the benchmark, are described. An implementation in **James**, a multi-body software based on **Maple**, is presented. **James** is used to build and generate the whole vehicle dynamic simulator. Simulation results and computational efficiency analysis conclude this paper.

J.-P. Merlet and B. Ravani (eds.), Computational Kinematics, 123–132.
© 1995 *Kluwer Academic Publishers.*

1. SARA Project

1.1. PROJECT DESCRIPTION

The SARA project involves three of the principal French car manufacturers: PSA, Renault, and the INRETS (French National Institute for Research and Security in Transport) and other partners. The principal goal of SARA is to build an interactive real-time simulator which can be used on the one hand like an experimental tool for vehicle suspension design, and on the another hand as a tool to study any kind of virtual vehicle/driver/environment system.

For that purpose, the simulator is built from hardware and software components which must act in a real-time environment. The hardware consist of a moving cabin with large motions in order to be able to create sufficient accelerations (1g in longitudinal and lateral motions). The cabin is driven by a real-time simulator which integrates the equations of motion.

One challenge is that the dynamical model must be flexible and modular in order to take into account actual and future vehicles. Two main constraints are the ability to execute in a real-time environment and to simulate complex behavior. This paper describes how **James** deals with such aspects.

1.2. BENCHMARK DESCRIPTION

The goal of the benchmark is to build a dynamic model of a Peugeot 605 vehicle with all the classical features: suspension links with elasto-kinematic effects, motor and brakes, rear and driving wheels, tires etc. and to measure the CPU time needed for one 10 ms cycle. The different steps of the benchmark introduce all these features.

The first step is dedicated to the study of the rear suspension of the vehicle. This suspension is a "double wishbone" and is a closed-loop mechanism (4 bodies and 6 pin, ball and gimbal joints). Motion is described by a set of algebro-differential equations which cannot be solved in time. One solution is to describe each suspension as a general slider whose main degree of freedom is wheel vertical displacement with induced lateral and rotational motions. Induced motions are precomputed and are implemented using look-up tables during simulation. Other steps are dedicated to:

- the introduction of elasto-kinematic effects: modification of suspension characteristics with forces and torques acting on the vehicle,
- the building of the slider joint in the multi-body software,
- complete vehicle model assembly: introduction of models for brakes, motor, tires, etc
- the driving maneuver simulations and a performance assessment.

This paper focuses on the suspension modeling, and the building and use of the new slider joint.

2. Kinematic Analysis of a rear suspension

A representation of the suspension: a *double wishbone* is presented in figure 1. Modeling is done using:

- 4 bodies: the spindle (triangle E-F-H), the upper (A-B-E) and lower (C-D-F) wishbone and the tie rod (bar H-L),
- 6 joints: two rotational (pin) joints between ground and upper and lower wishbone, one universal (gimbal) between ground and tie rod and three ball (spherical) joints between spindle and other bodies.

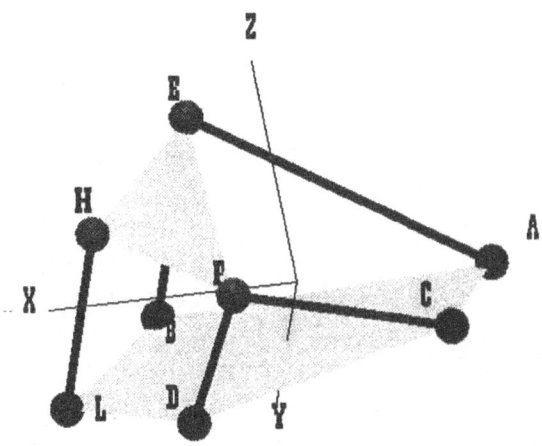

Figure 1. Suspension model

Points A, B, L, D, C are located on the ground, so the mechanism has one degree of freedom (dof) and we choose the spindle vertical displacement as the main dof. The goal is to compute other displacements (translational or rotational) as functions of such a dof for the numerical values given in table 1.

Two approaches are used. The first one uses the **James** software to generate symbolically the mechanism equations which are then solved numerically. The second approach uses the **Maple** computer algebra system to construct a series expansion (with respect to the main dof) of the induced motions.

2.1. JAMES: SYMBOLIC MODELING AND NUMERICAL SOLVING

James is a multi-body software developed by Aerospatiale and Simulog. It splits the analysis of a mechanical system into two steps:

- symbolic step: modeling of the system using bodies, joints, forces, computation of equations of motion, and generation of a dedicated simulator. All these manipulations are done symbolically in **Maple**.
- numeric step: use of the simulator with appropriate data: mass, inertia, etc

For the suspension model, **James** computes the equations of a tree-like underlying model (some joints are "forgotten") and it adds constraint equations for the neglected joints. The numerical code of the equations is generated by the code generation module. These equations are solved using a Newton algorithm for different values of the wheel vertical displacement. These numbers make up the input of a spline function which will model the behavior of the suspension.

2.2. A SERIES EXPANSION APPROACH

The purpose of this section is to present an approach based on series expansions for a precise description of the motion of the rear suspension around its initial position. This method uses symbolic computations and no help from numerical techniques is required. All these computations are carried out in **Maple** using some procedures which have been specifically developed. This approach though natural from a mathematical point of view seems unusual in the kinematic analysis of mechanisms. We refer here to the general presentation [Erdman] which divides computational kinematics in 3 broad themes (matrix computations and iteration methods, continuation methods, algebraic manipulations and elimination). Examples of the algebraic/geometric approach can be found in [Bottema *et al.*]. The problem that we consider here is a special case of a parallel robot (see [Merlet]) where we forget one leg.

One characteristic of the motion we want to study is that it remains in a neighborhood of the initial position (.1 meter for z the vertical displacement of the spindle). So we use a classical tool in differential geometry, which is the development in series. An advantage of the series is that they give local information without heavy computation. Moreover, if a change of parameters is needed, we can compute the series of the new function by simple algebraic manipulations via the implicit function theorem. More precisely, what we are going to use is the following: if a parameter X has a series in t of the form $X = \sum_{i=1}^{N} a_i t^i + O(t^{N+1})$ then we can obtain the development $t = \sum_{i=1}^{N} b_i X^i + O(X^{N+1})$ by linear operations on the coefficients a_i. This allows us to choose "good" parameters for the initial computation of the suspension and when everything is known with respect to these parameters, to invert the series and express the quantities in terms of the required parameters. Let $E(t)$ be any position of point E. It is located

A = [-1.2170573, 0.47000,-0.21300]	E = [-1.5007573, 0.69696,-0.15141]
B = [-1.5192473, 0.39716,-0.16548]	F = [-1.5200973, 0.72984,-0.36806]
C = [-1.1350573, 0.35700,-0.35400]	H = [-1.6513273, 0.71159,-0.24225]
D = [-1.5200573, 0.31400,-0.35400]	L = [-1.6470573, 0.31435,-0.24930]

TABLE 1. Numerical values of point locations

on a circle and can be obtained as the intersection of the two spheres $S(A, \|AE\|)$, $S(B, \|BE\|)$ (where $\| \|$ is the Euclidean norm).

As E(t) is located on a circle (radius R), we use parametric coordinates of the form $x = R\frac{1-t^2}{1+t^2}, y = R\frac{2t}{1+t^2}$ using $t = \text{tg}(\frac{\theta}{2})$ where θ is the polar angle. By a linear transformation, we can map this circle (with adequate radius $R = \frac{\|AB \wedge AE\|}{\|AB\|}$) onto the circle described by $E(t)$. This is the computation that has been implemented in **Maple**. Using numerical values defined in table 2.2, this yields

$$E(t) = \left[\quad -\frac{1.50075+1.3742001t^2-0.0971287t}{1+t^2}, \frac{0.69696+0.1367790t^2-0.0326306t}{1+t^2}, \right.$$
$$\left. -\frac{0.15141+0.2052665t^2-0.5676452t}{1+t^2} \right]$$

Let $F(t)$ be the position of F. It is also on a circle defined by the intersection of the two spheres $S(C, \|CF\|)$, $S(D, \|DF\|)$. We also express that $\|E(t)F(t)\|$ is constant by locating $F(t)$ on the sphere $S(E(t), \|EF\|)$. Now $F(t)$ is located at the intersection of the three spheres $S(E(t), \|EF\|)$, $S(C, \|CF\|)$, $S(D, \|DF\|)$.

This yields a priori two points and we choose the one which at $t = 0$ is in the initial position (if these two points coincide at $t = 0$, this would imply that the initial position is critical which is not the case). The same computations are applied to describe the location of $H(t)$.

At this step, we have a description of the complete motion of the rear suspension with respect to t. We can compute the altitude z of the center with respect to t. This yields for $z(t)$:

$$z = 0.589063\,t+0.006536\,t^2-0.545698\,t^3+0.052204\,t^4+0.503558\,t^5+O\left(t^6\right)$$

As what we need is the description of the movement with respect to z, we can invert this development and we obtain:

$$t = 1.697610\,z - 0.031977\,z^2 + 4.533362\,z^3 - 1.162952\,z^4 + 24.36332\,z^5 + O\left(z^6\right)$$

The values of t can now be back-substituted in the previous computation to use z as the main parameter. A-posteriori computation of errors by substituting results in equations show that, for instance, the error in the distance between $E(t)$ and A is low. The accuracy of the computation is at least five digits for $|z| \leq 0.1$ (this can also be verified directly on the graph of this error for $|z| \leq 0.1$). The same results hold for the computation of the other quantities.

This approach has also been applied when the position of L depends on a parameter p (rack position) giving the direction of the front wheels. In this case, the series depends on two parameters p, z but the computations can be handled in exactly the same way.

2.3. DISCUSSION

Figure 2 shows the lateral displacement of the spindle with respect to the spindle vertical motion. The two approaches used give the same results. The approach using a series expansion is very interesting for many reasons.

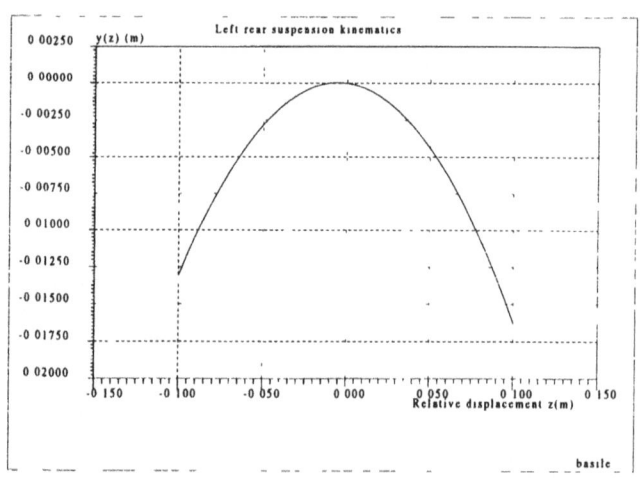

Figure 2. Lateral displacement

First, numerical approaches are very CPU intensive especially for front suspensions where two parameters, vertical displacement and steering rack position have to be considered. One may also compute the formal series expansion without any numerical quantities (position and orientation of link points). The series coefficients, in this case, will be functions of symbolic parameters and will be generic for one kind of suspension.

3. Use in Real-Time Simulation

3.1. BUILDING OF A GENERIC LINK

The **James** software allows us to use a generic link. The aim was twofold:
- the possibility for the user to introduce "exotic" links easily,
- the computation of the equations of motion for a generic system (a system in which all the links are generic) and optimization of these equations.

Let us consider a link j between two bodies I and J. The link joins the point ij of body I to the point ji of body J. The motion of each body is defined by the translation of its reference point (translation D_I (D_J) of reference point R_I (R_J) for body I (J)) and its rotation vector (Ω_I (Ω_J) for body I (J)).

The definition of the link is done by expressing D_J and Ω_J as functions of D_I and Ω_I:

$$D_J = D_I + r_{ij} + U_j(X_j, Rg_j) - r_{ji} \tag{1}$$

$$\Omega_J = \Omega_I + V_j \omega_j \tag{2}$$

The degrees of freedom of the link are the vectors X_j for the translation and Rg_j for the rotation. Their derivatives are:

$$\frac{d}{dt} X_j = \dot{X}_j$$

$$\frac{d}{dt} Rg_j = Mg_j(Rg_j)\, \omega_j$$

The term $Mg_j(Rg_j)$ expresses that, for free bodies or some joints like spherical ones, angular velocity parameters are quasi-coordinates and are not derivatives of angular parameters (Euler angles for example).

The other notations are:
- r_{ij} (r_{ji}): vector describing location of point ij (ji) in I (J),
- $U_j(X_j, Rg_j)$: translation of the link. It can depend on Rg_j (rack and pinion joint),
- V_j: direction of rotation.

Then we define $\dot{D}_J$, translational velocity of body J, as a function of $\dot{D}_I$, translational velocity of body I. This is done by differentiating equation 1. First we differentiate $U_j(X_j, Rg_j)$:

$$\frac{d}{dt} U_j(X_j, Rg_j) = \tilde{\Omega}_I U_j(X_j, Rg_j) + dU_j \dot{X}_j + dW_j Mg_j \omega_j \tag{3}$$

Vectors expressed in the frame attached to body I are noted with a superscript ^{0}. The new vectors introduced in equation 3 have the following definition:

$$dU_j^0 = \frac{\partial}{\partial X_j} U_j^0(X_j, Rg_j)$$

$$dW_j^0 = \frac{\partial}{\partial Rg_j} U_j^0(X_j, Rg_j)$$

Finally we get for $\dot{D}_J$:

$$\dot{D}_J = \dot{D}_I - \widetilde{R_I R_J}\Omega_I + dU_j \dot{X}_j + (dW_j Mg_j + \widetilde{r_{ji}}V_j)\omega_j \qquad (4)$$

$R_I R_j$ is the vector between the reference points of bodies I and J ($R_I R_J = r_{ij} + U_j(X_j, Rg_j) - r_{ji}$).

Using equations 4 and 2, the generic link can be finally written:

$$\begin{pmatrix} \dot{D}_J \\ \Omega_J \end{pmatrix} = \begin{pmatrix} I_3 & -\widetilde{R_I R_J} \\ 0 & I_3 \end{pmatrix} \begin{pmatrix} \dot{D}_I \\ \Omega_I \end{pmatrix} + \begin{pmatrix} dU_j & dW_j Mg_j + \widetilde{r_{ji}}V_j \\ 0 & V_j \end{pmatrix} \begin{pmatrix} \dot{X}_j \\ \omega_j \end{pmatrix}$$

All the links used in **James** are based on this generic link. The next section will show how it has been used to simulate a car suspension behavior.

3.2. INSTANTIATION FOR A CAR SUSPENSION

James gives to users an interface to build new joints. Classical joints (rotational, translational, universal...) which are provided with the software are built using the same kind of scheme and can be used as examples.

A new joint is defined by a **Maple** table where the indices are predefined names (U, OM ...) and the values are specializations of these names. For example, to describe a general slider where the main dof is the vertical displacement:

$$X_i = \begin{pmatrix} f_i(z_i) \\ g_i(z_i) \\ z_i \end{pmatrix}.$$

The same mechanism is used for the first and second derivatives. So all quantities are defined using symbolic expressions.

Thanks to symbolic computation, one can check very quickly, by defining an elementary model (such as one body connected to ground using one joint) that there is no trouble with the definition of the joint. The fact that in a symbolic computation software like **Maple**, one manipulates and visualizes formulae instead of numbers is a very interesting device.

3.3. BUILDING THE WHOLE MODEL

As the new joint is defined, the whole model of the car is built using:

- five bodies: the central one plus four spindles,
- four Sara joints between the main body and spindles,
- many forces to introduce tire effects (forces and torques), brakes, motor torque, spring/damper suspensions...

The wheels' self rotations are introduced as supplementary dof.

The equations are computed in the central body frame. That reduces the number of elementary operations as some projections are avoided. The equations can also be linearized with respect to some degrees of freedom, thanks to symbolic computation, resulting in a constant mass matrix automatically detected during code generation. More than 2000 lines of C code are generated for the non-linear model. Specific code is added to define specific features such as motors or brakes characteristics, tire behavior ...

3.4. RESULTS

At this time, the non-linear simulator has been used to perform some tests of usual behavior such as longitudinal maneuvers (braking or acceleration), lateral or coupled maneuvers (left cornering or throttle off in curve).

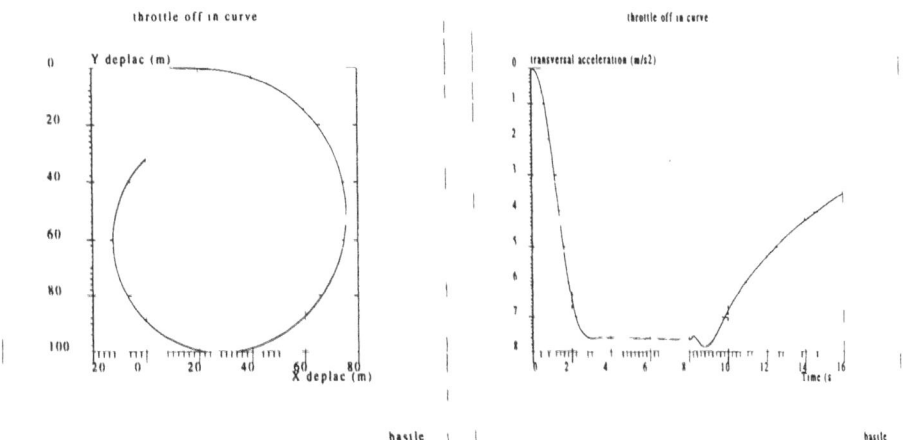

Figure 3. Vehicle trajectory   *Figure 4.* Transversal acceleration

The diagrams show the dynamic of the vehicle during the "throttle off in curve" maneuver. At time=8s, the accelerator pedal is released and the vehicle slows down, so the radius of curvature of the vehicle trajectory is modified.

Computations are carried out on two workstations: an Onyx Silicon Graphics with a R4400 Mips processor, and a DEC Alpha 3000/600. With both stations, computation time for one cycle (using the Runge-Kutta 4 integration algorithm) is less than cycle time (.01 s as cycle frequency is

100 Hz), so real-time can be achieved. A ratio of more than 2 is obtained without any manipulation or optimization of the generated code or any simplification of the model. Moreover, the simulator runs faster than a hand-written one.

4. Conclusion

This paper has shown that new computational techniques can be successfully used to perform some tasks usually performed only with very powerful computers or by hand: analysis of mechanisms, generation of code for real-time simulation. A first advantage is the time saved for model generation and coding. Another important advantage is the new level of confidence one can have in the code as no human intervention is needed to produce it. An significant amount of validation time is also saved.

The combination of both symbolic and numerical computations opens new application fields and allows one to think about Automatic Model Generation. Nevertheless this cannot be achieved only by one tool: symbolic or numeric, the two domains are required. So one of the most important questions is to understand and define the boundary between these two domains and how to go from one to the other.

References

Bottema, O. and Roth, B. (1979) Theoretical Kinematics. Dover Pub. Inc

Erdman, A.G. (1993) MODERN KINEMATICS, Development in the Last Forty Years. *Wiley Series in Design Engineering*

Garnier, C. and Rideau, P. (1989) James: a New Tool for Dynamic Modeling. In ESA SP-289, *Proceedings of International Conference on Spacecraft Structure and Mechanical Testing*, Noordwick, Holland

Merlet, J.-P., (1994) Parallel manipulators: State of the Art and Perspective. *Advanced Robotics* **Vol. 8. 6**

Rideau P., (1993) Computer Algebra and Mechanics: James Software. In A.M. Cohen, editor, *Computer Algebra and Industry*, John Wiley & Sons

ON MINIMUM JOINT TORQUE CONFIGURATIONS
OF MULTIPLE-LINK MANIPULATOR

J. LENARČIČ
J. Stefan Institute, University of Ljubljana
Jamova 39, Ljubljana, Slovenia

Abstract. For a given position of the end effector, the optimum configuration of a planar multiple-link hyper-redundant manipulator is calculated between two arbitrary points. The optimality criterion is defined by a linear combination of joint torques. Analytical and time-consuming numerical procedures are used to show that the optimum solution can effectively be approximated by a set of straight sections of the mechanism.

1. Introduction

Designs of multiple-link hyper-redundant robots date back to the earliest days of robotics. They were reported by Hirose and Umetani (1876), Morecki *et al.* (1989), Chirikjian and Burdick (1993), and others. However, the implementation in industry is still far away because of problems in their realisation, as well as in their mathematical treatment and control (Lenarčič, 1993). The objective of recent investigations in this area is to develop numerical algorithms that do not depend on the number of degrees of freedom (Chirikjian and Burdick, 1991,1994). The idea of the present work is proposed in Lenarčič (1993). It is discovered that the contigurations of a mulitple-link manipulator achieve typical forms with respect to different optimality criteria. For example, in Lenarčič (1994) we indentiy two typical configurations relative to a linear criterion of joint torgues, one associated to a force applied to the end effector and one to the gravity of the links. The purpose of the present work is to extend these results to a more general case where the end points of the treated part of the mechanism are arbitrarily placed in the workarea. These findings can be utilised to derive optimum configurations of multiple-link mechanisms in a space filled of obstacles.

J.-P. Merlet and B. Ravani (eds.), Computational Kinematics, 133–142.

2. Problem statement

The n-R planar manipulator moves in the plane $\vec{x}$ and $\vec{y}$. It possesses identical links (they do not collie) and unlimited joints. The positions of joints are

$$x_i = x_{i-1} + d\cos(Q_i), y_i = y_{i-1} + d\sin(Q_i) \qquad (1)$$

where $Q_i, i = 1, ..., n$, are the angles between the links and $\vec{x}$, and d are the link lengths. The manipulator carries a load whose gravity force F is applied to the end effector at x_n. When the gravity of the links is neglected, the joint torques correspond to $\tau_i = F(x_n - x_i), i = 1, ..., n$. The objective of the manipulator's motion is to minimise the linear criterion

$$f = \frac{1}{n}\sum_i |\tau_i| = \frac{F}{n}\sum_i |x_n - x_i| = \frac{Fd}{D}\sum_i |x_n - x_i| \qquad (2)$$

subject to the desired position of the end effector specified by x_n and y_n. Here, D is the total link length of the manipulator.

The optimum configuration of the manipulator, correspondent to the cost function (2), is sought as a series of straight sections. A "positive" straight section is the sub chain that possesses equal orientation whose joints are positioned on the left side of the force axis and whose joint torques are taken as positive. The cost function f is the sum of $f_k, k = 1, ..., K$ - the cost functions of the included straight sections. By limiting the number of joints n to infinity and the link lengths d to zero we get

$$f_k = D_k(x_n - x_k + \tfrac{1}{2}D_k c_k), f_k = -D_k(x_n - x_k + \tfrac{1}{2}D_k c_k) \qquad (3)$$

for a positive and a negative section, respectively. Here, D_k is the length of the related section, x_k is the x component of the end point of the section, and c_k is the cosine of the orientation (inclination) of the section. It is also assumed that the total length of the manipulator $D = 1$, as well as the applied force $F = 1$. It is important that the cost function (2) depends only on the number of straight sections included in the given configuration.

3. End points on a horizontal line

Here, the problem is to find the optimum configuration of the mechanism that connects a point P and a vertical line L as given in Fig. 1 (left). We divide the mechanism into a series of small straight sections, each characterised by a constant increment ε_x. It is evident that the cost function depends on the horizontal position of the sections (as well as their length and inclination), but it doesn't change if the sections are shifted vertically. Consider two straight sections with the same x and increment ε_x, one in-

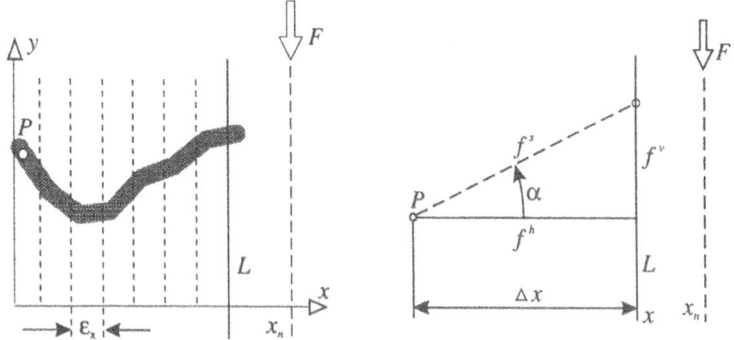

Figure 1. Connecting point P and vertical line L

clined for $\alpha, 0 \leq \alpha \leq 90$, its cost function is f, and the other for α', its values are between $\alpha \leq \alpha' < 90$, its cost function is f'. The ratio

$$R = \frac{f}{f'} = \frac{\dfrac{\varepsilon_x}{\cos(\alpha)}(x_n - x + \frac{1}{2}\varepsilon_x)}{\dfrac{\varepsilon_x}{\cos(\alpha')}(x_n - x + \frac{1}{2}\varepsilon_x)} = \frac{\cos(\alpha')}{\cos(\alpha)} \tag{4}$$

is always smaller than one and it doesn't depend on x and y. It is certain that a less expensive section contains a smaller α and that the best is when $\alpha = 0$. The optimum configuration that connects P and L is composed of a series of small horizontal sections. They form a horizontal straight section between P and L, denoted by the cost function f^h (Fig. 1, right)

$$f^h = \Delta x (x_n - x + \frac{1}{2}\Delta x) \tag{5}$$

It is useful to study the ratio between f^s and $f^h + f^v$ that is relative to the combination of straight sections termed "internal triangle" as shown in Fig. 1 (right). The aim in to determine the best configuration with the end points P and P' and choosing between f^s and $f^h + f^v$ (note that $f^s + f^v$ is always greater than f^h). It follows from (3)

$$f^s = \frac{\Delta x}{\cos(\alpha)}(x_n - x + \frac{1}{2}\Delta x) = \frac{f^h}{\cos(\alpha)} \tag{6}$$

$$f^v = \Delta x \tan(\alpha)(x_n - x) \tag{7}$$

This gives

$$R = \frac{f^h + f^v}{f^s} = \frac{x_n - x}{x_n - x + \frac{1}{2}\Delta x}\sin(\alpha) + \cos(\alpha) \tag{8}$$

It can be expressed in the following form (if $x \neq x_n$)

$$R = R_0 \sin(\alpha + \alpha_0) \tag{9}$$

$$R_0 = \sqrt{1 + (\frac{x_n - x}{x_n - x + \frac{1}{2}\Delta x})^2}, \alpha_0 = \arctan(\frac{x_n - x + \frac{1}{2}\Delta x}{x_n - x}) \tag{10}$$

R_0, α_0 depend on values $\alpha, x_n, x, \Delta x$, and are limited by $1 \leq R_0 \leq \sqrt{2}$ and $45 \leq \alpha_0 \leq 90$.

When $x = x_n$, it follows that $R_0 = 1$ and $R < 1$ for any $0 \leq \alpha < 90$. This means that $f^h + f^v$ is less expensive than f^s for every point P' located on $x = x_n$. This result is eqivalent to what we reported in Lenarčič (1994). However, when $x < x_n$, it follows that $R_0 > 1$ and that for some values α the cost function f^s is less expensive than $f^h + f^v$. This is when $0 < \alpha < 189 - 2\alpha_0$.

4. End points on a vertical line

Here, the goal is to find the configuration of the mechanism that connects a point P and a horizontal line L and minimises the given cost function of joint torques (Fig. 2, left). If we cut the region between P and L into a series of infinitesimal horizontal slices of a height ε_y, the optimum solution for each slice will depend on those of other slices. In general, the optimum configuration must be obtained by using time-consuming optimisation procedures.

Suppose that P is connected with L by a straight section as presented in (Fig. 2, right). The angle α of the slant section f^s can change between 0 and

$$\alpha_{max} = \arctan(\frac{x_n - x}{\Delta y}) \tag{11}$$

The cost functions of the vertical, slant, and the horizontal section are

$$f^v = \Delta y(x_n - x) \tag{12}$$

$$f^s = \frac{\Delta y}{\cos(\alpha)}(x_n - x - \frac{1}{2}\Delta y \tan(\alpha)) \tag{13}$$

$$f^h = \Delta y \tan(\alpha)(x_n - x - \frac{1}{2}\Delta y \tan(\alpha)) \tag{14}$$

The combination of these is termed "external triangle". If

$$R = \frac{f^v}{f^s} = \frac{(x_n - x)\cos(\alpha)}{x_n - x - \frac{1}{2}\Delta y \tan(\alpha)} \tag{15}$$

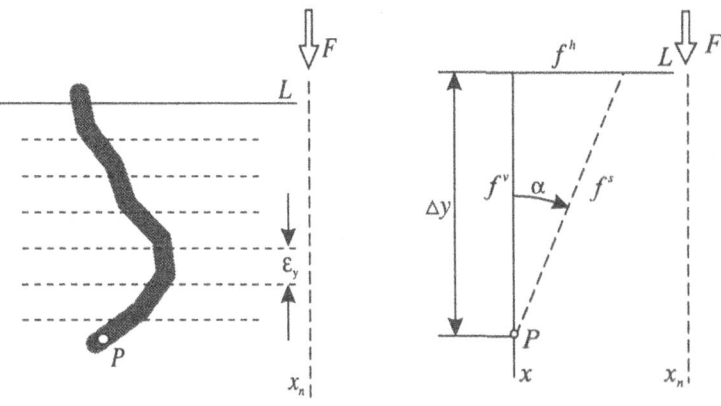

Figure 2. Connecting point P and horizontal line L

it is possible to verify that f^s is smaller than f^v if the values of α satisfy

$$\frac{\tan(\alpha)}{1 - \cos(\alpha)} > \frac{2(x_n - x)}{\Delta y} \tag{16}$$

Hence, there exists an optimum α that minimises the cost function of a straight section connecting P and L. It depends on the values of Δy, as well as $x_n - x$. It is interesting that the the optimum $\alpha > 0$ if $\Delta y > 0$. It means that f^v is never the optimum solution for a mechanism connecting P and L as presented in Fig. 2 (right) if P doesn't lie on L. It follows that $f^v + f^h$ is always more expensive than f^s, but also that $f^s + f^h$ isn't always more expensive than f^v. If $\Delta y >> x_n - x \Rightarrow f^s + f^h < f^v$.

A numerical scheme based on a comparison between all possible combinations of straight sections is extremely time-consuming and can only serve to find a very approximate solutions. If the region between P and L is divided into a net of points as shown in Fig. 3 (left), we can form the mechanism as a combination of straight sections between the selected points, starting in point $(1,1)$ and ending in $(1,m), (2,m), ..$ or (m,m). A simplification is that the inclination of each straight section must be between 0 and 90. Therefore, the solution lies entirely inside the rectangle of $m \times m$ points, and the relation between the end point of a section (i,j) and the end point of the following section (k,l) is $i \leq k \leq m$ and $j \leq l \leq m$.

Experiments made with this algorithm at $m = 7$ help to identify some basic characteristics of the optimum configuration. The optimum solution

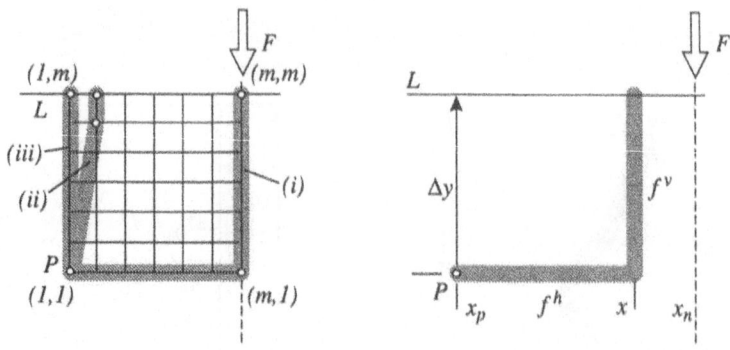

Figure 3. Region of possible solutions between point P and horizontal line L

depends on the ratio between Δy and $x_n - x$ as follows

$$\text{(i)} \quad \frac{\Delta y}{x_n - x} > 0.53 \Rightarrow (1,1) - (7,1) - (7,7) \tag{17}$$

$$\text{(ii)} \quad 0.43 \leq \frac{\Delta y}{x_n - x} \leq 0.53 \Rightarrow (1,1) - (2,6) - 2,7) \tag{18}$$

$$\text{(iii)} \quad \frac{\Delta y}{x_n - x} > 0.43 \Rightarrow (1,1) - (1,7) \tag{19}$$

We can speculate that the optimum configuration is composed of one horizontal and one vertical straight section as shown in Fig. 3 (right)

$$f^h = (x - x_p)(x_n - \tfrac{1}{2}x - \tfrac{1}{2}x_p), f^v = \Delta y(x_n - x) \tag{20}$$

We can show that $f^h + f^v$ is a quadratic function of x between x_p, x_n

$$f^h + f^v = -\tfrac{1}{2}x^2 + (x_n - \Delta y)x + C_0 \tag{21}$$

whose maximum is at $x_0 = x_n - \Delta y$. The smallest value of $f^h + f^v$, denoted by S, is achieved at x_s, where $x_s = x_p$ or x_n

$$x_0 < (x_p + x_n) \Rightarrow x_s = x_n, S = f^h(x_n) \tag{22}$$

$$x_0 = (x_p + x_n) \Rightarrow x_s = x_p \vee x_n, S = f^h(x_n) = f^v(x_p) \tag{23}$$

$$x_0 < (x_p + x_n) \Rightarrow x_s = x_p, S = f^v(x_p) \tag{24}$$

A first approximation of the most convenient configuration connecting P and a horizontal line L is either the vertical section represented by (iii), with the cost function specified in (19), or a combination of the horizontal and the vertical section represented by (i), with the cost function specified in (17).

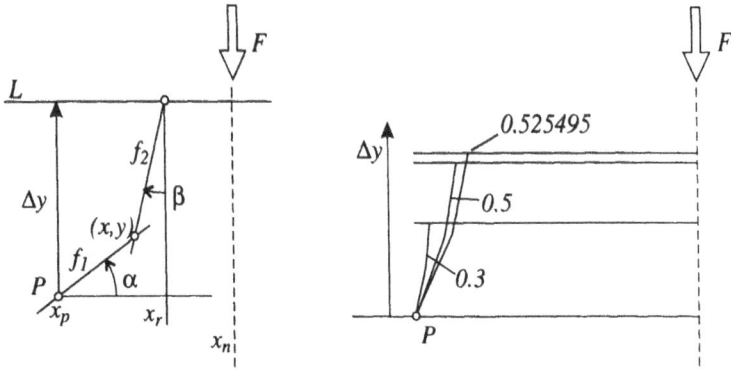

Figure 4. Optimum solution composed of two inclined linear sections

A generalisation is to search for the optimum solution with two straight sections which may be inclined for angles α, β at x_r (Fig. 4, left). It is assumed that $y_p = 0$. The intersection between the lines

$$y = \tan(\alpha)(x - x_p), y = \Delta y + \tan(90 - \beta)(x - x_r) \qquad (25)$$

is established by

$$\hat{x} = \frac{\Delta y + \tan(\alpha)x_p - \tan(90 - \beta)x_r}{\tan(\alpha) - \tan(90 - \beta)} \qquad (26)$$

$$\alpha < \alpha_{max} \wedge \beta < \beta_{max} \Rightarrow \hat{y} = \tan(\alpha)(\hat{x} - x_p) \qquad (27)$$

$$\alpha = \alpha_{max} \wedge \beta = \beta_{max} \Rightarrow \hat{x} = \frac{x_p + x_r}{2}, \hat{y} = \frac{\Delta y}{2} \qquad (28)$$

For all combinations of α, β, x_r we must compute the cost functions

$$f_1 = \sqrt{(\hat{x} - x_p)^2 + (\hat{y} - y_p)^2}(x_n - \tfrac{1}{2}\hat{x} - \tfrac{1}{2}x_p) \qquad (29)$$

$$f_2 = \sqrt{(x_r - \hat{x})^2 + (y_r - \hat{y})^2}(x_n - \tfrac{1}{2}x_r - \tfrac{1}{2}\hat{x}) \qquad (30)$$

The minimum value of $f_1 + f_2$ depends on the ratio between Δy and $x_n - x_p$

$$\frac{\Delta y}{x_n - x_p} \approx 0 \Rightarrow \alpha \approx 0, \beta \approx 90, \frac{\Delta y}{x_n - x_r} \approx 0 \qquad (31)$$

$$\frac{\Delta y}{x_n - x_p} = 0.3 \Rightarrow \alpha = 76.51, \beta = 4.92, \frac{\Delta y}{x_n - x_r} = 0.316 \qquad (32)$$

$$\frac{\Delta y}{x_n - x_p} = 0.5 \Rightarrow \alpha = 65.97, \beta = 8.35, \frac{\Delta y}{x_n - x_r} = 0.588 \qquad (33)$$

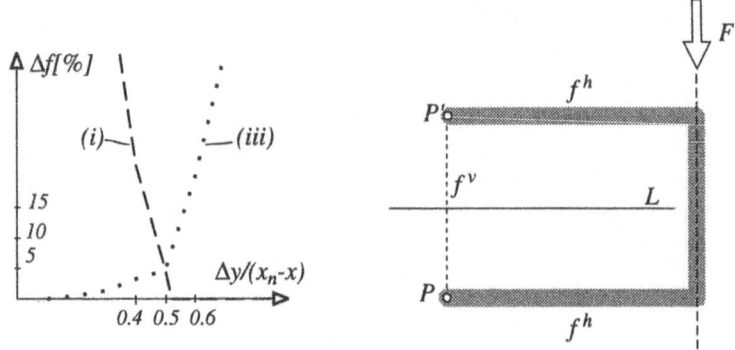

Figure 5. Comparison between the optimum and the suboptimum solution

$$\frac{\Delta y}{x_n - x_p} = 0.525495 \Rightarrow \alpha = 63.785, \beta = 9.025, \frac{\Delta y}{x_n - x_r} = 0.0.634 \quad (34)$$

$$\frac{\Delta y}{x_n - x_p} > 0.525495 \Rightarrow \alpha = 0, \beta = 0, x_r = x_n \quad (35)$$

The associated configurations are shown in Fig. 4 (right). Obviously, $\alpha = 0, \beta = 0, x_r = x_n$ is the optimum if the ratio between Δy and $x_n - x_p$ is greater than 0.525495. This is equivalent to solution (i) in (17). The solution with $\alpha \neq 0, \beta \neq 0, x_r \neq x_n$ collapses into $\alpha = 0, \beta = 0, x_r = x_n$ after an increment of the ratio between Δy and $x_n - x_p$ that is smaller than 0.000005. The vertical connection between P and L is equivalent to (iii) in (19) and is never the optimum if $\Delta y \neq 0$.

If we choose (i) or (iii), we will make an error smaller than five percents (in our computations smaller than 4.53 percents - Fig. 5, left). Hence, in a practical implementation, we suggest to utilise (i) or (iii) for the most convenient configuration with the initial point at P and the final point located on a horizontal line L. It implies the same conclusions when we search for the optimum configuration between two points P and P' that lie on a vertical line as presented in Fig. 5 (right). This situation can be combined of two symmetrical parts divided by L. The most convenient configuration is obtained at $x = x_s$ based on

$$x_n - x_p < \Delta y \Rightarrow x_s = x_n, S = 2f^h(x_n) \quad (36)$$

$$x_n - x_p = \Delta y \Rightarrow x_s = x_p \lor x_n, S = 2f^h/x_n) = f^v(x_p) \quad (37)$$

$$x_n - x_p > \Delta y \Rightarrow x_s = x_p, S = f^v(x_p) \quad (38)$$

Clearly, S is an approximate solution whose value is in the worst case, when $x_n - x_p \approx \Delta y$, less than 5 percents more costly than the theoretical optimum.

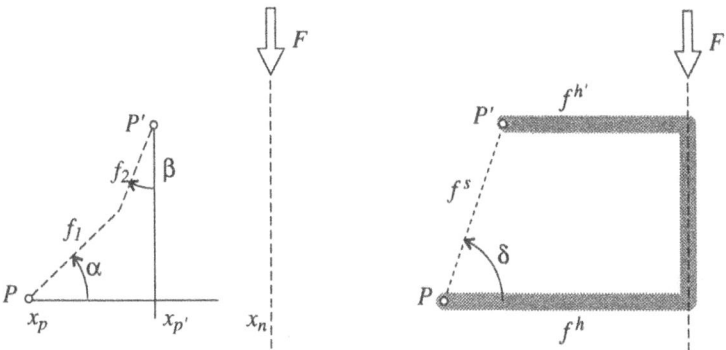

Figure 6. Connecting points P and P' on a slant line

5. End points on a slant line

Here, the goal is to find the optimum configuration connecting points P and P' positioned on a slant line as shown in Fig. 6 (left). To search for the optimum configuration, we use the approach described in the previous section. The difference is that $x_r = x_{p'}$ is fixed. We compare the cheapest combination of two slant straight sections characterised by $f_1 + f_2$ and the slant section f^s or the combination of two horizontal and one vertical section characterised by $f^f + f^{h'}$ (note that the cost function of the vertical section coincident with the force axis is zero) as illustrated in Fig. 6 (right).

The experiments show that the best combination of $f_1 + f_2$ depends on the angle δ and the distance between P' and the force axis. It is evident that the error between f^s or $f^h + f^{h'}$ and the best $f_1 + f_2$ diminishes when $\delta = 0$ (the end points are on a horizontal line.) For a given δ, the maximum error is when $f^s = f^h + f^{h'}$, but the worst case is when $\delta = 90$. Since

$$f^s = \sqrt{(x_{p'} - x_p)^2 + \Delta y^2}(x_n - \tfrac{1}{2}x_{p'} - \tfrac{1}{2}x_p) \tag{39}$$

$$f^h + f^{h'} = (x_n - x_{p'})^2 + (x_{p'} - x_p)(x_n - \tfrac{1}{2}x_{p'} - \tfrac{1}{2}x_p) \tag{40}$$

and by introducing $(x_{p'} \neq x_p)$

$$r_x = \frac{x_n - x_{p'}}{x_{p'} - x_p}, r_y = \frac{\Delta y}{x_{p'} - x_p} \tag{41}$$

we find the following condition

$$\frac{r_x(r_x + 1)}{r_x + \tfrac{1}{2}} \geq r_y \Rightarrow f^h + f^{h'} \geq f^s \tag{42}$$

that is a generalisation of (22-24). We propose to utilise f^s or $f^h + f^{h'}$ selected by (42) as the approximation of the multiple link mechanism connecting two arbitrary points.

6. Conclusions

Optimum configurations of a planar multiple-link n-R manipulator are investigated with respect to a criterion defined as absolute joint torques. The objective is to find the optimum form of the mechanism's configuration passing through two arbitrary points. The approach is based on dividing the mechanism into a series of straight sections. This minimises the complexity of the numerical procedures and enables to find very effective approximations for the optimum solutions. It is shown that we can select between two simple configurations, one is a single straight section between the end points and the other contains two horizontal sections that connect the end points with the force axis and a vertical straight section between the first two.

References

Hirose, S.,Umetani, Y. (1976) Kinematic control of active cord mechanism with tactile sensors, *Proc. 2nd Int. CISM IFToMM*, pp. 241-252

Morecki, A., Busko, Z., Fraczek, J., Kaminski, D. (1989) Modelling and synthesis of elastic manipulators, *Proc. 20th ISIR*, Tokyo, Japan, pp. 545-552

Chirikjian, G.S., Burdick, J.W.(1993) Design, implementation, and experiments with a thirty degree-of-freedom hyper-redundant robot, *Proc. IEEE Conf. Robotics Automat.*, Atlanta.

Lenarčič, J. (1993) Computational considerations on kinematic inversion of multi-link redundant manipulators, *Computational Kinematics (J. Angles, G. Hommel, P. Kovacs, Eds.)*, Kluwer Academic Publishers, Dordrecht, pp. 75-84

Chirikjian, G.S., Burdick, J.W. (1991) Kinematics of hyper-redundant manipulators, *Advances in Robot Kinematics (S. Stifter, J. Lenarčič, Eds.)*, Springer-Verlag, Vienna, pp. 392-399

Chirikjian, G.S., Burdick, J.W. (1994) A modal approach to hyper-redundant manipulator kinematics, *IEEE Trans. Robotics Automat.*,**Vol. 10, No. 3**, pp. 343-354

Lenarčič, J. (1993) Optimum configurations of planar n-R hyper-redundant mechanism, *Proc. ICAR'93*, Tokyo, Japan.

Lenarčič, J. (1994) Minimum joint torque configurations of planar multiple-link manipulator, *Advances in Robot Kinematics and Computational Geometry (J. Lenarčič, B. Ravani, Eds.)*, Kluwer Academic Publichers, Dordrecht, pp. 281-288

A DETERMINATION OF SINGULAR CONFIGURATIONS OF SERIAL NON-REDUNDANT MANIPULATORS, AND THEIR ESCAPEMENT FROM SINGULARITIES USING LIE PRODUCTS

JOSÉ M. RICO AND JAIME GALLARDO A.
Instituto Tecnológico de Celaya, 38000 Celaya, Gto. México

AND

JOSEPH DUFFY
CIMAR, University of Florida, Gainesville, Fl 32611

Abstract. Given a serial non-redundant manipulator in a singular configuration, this paper describes a procedure that identifies the screws which are involved in the singular configuration. Then, the paper shows that the Lie products of the involved screws provide information to decide whether a motion out of the singular configuration is possible and, if so, which screw or screws must be actuated for the manipulator to move out of the singular configuration. An example of the application of the method is provided.

1. Introduction.

A manipulator is in a singular configuration when the end-effector is unable to move with an arbitrary velocity state. Singular configurations of serial manipulators have been analyzed using the Jacobian matrix. By setting the determinant of the Jacobian matrix to zero, Waldron and his coworkers (Waldron *et al.*, 1985; Wang and Waldron, 1987) found important results regarding the singular configurations of serial manipulators. Another approach to the analysis of singular configurations of serial manipulators consists in a differential analysis of the displacement function of the manipulator, see (Litvin, 1980; Litvin, 1986). In contrast, the escapement of a serial manipulator from singular configurations has received much less attention. As far as the authors are aware, the only comprehensive study is that of Hunt (Hunt, 1986). There, Hunt employed the matrix of cofactors of

143

J.-P. Merlet and B. Ravani (eds.), Computational Kinematics, 143–152.

144

the Jacobian matrix to identify the screws associated with a singular configuration. Then, Hunt applied the theory of screw systems to determine which kinematic pair(s) is (are) necessary to actuate for the manipulator to escape from the singular configuration. However, the matrix of cofactors of the Jacobian matrix may fail to identify the screws involved in some singular configurations when the loss of freedoms is greater than one.

In this paper, the screws responsible for a singular configuration will be isolated by looking at the dimension of the subspaces generated by different subsets of the set of screws of a non-redundant serial manipulator. Then, it will be shown that the Lie products of the isolated screws provide information to decide whether a escapement of the manipulator from a singular configuration is possible. If the answer is in the affirmative, the Lie products also indicate which kinematic pair(s) is (are) required to actuate to execute the escape. The method illustrated in this paper does not attempt to sidestep screw theory. Rather, it is an application of the motor product of the classical screw theory— which is identified with the Lie product of modern differential geometry— to the study of singular configurations of serial non-redundant manipulators.

2. Fundamentals.

In the spatial case, a non-redundant serial manipulator has six kinematic pairs, which are usually revolute or prismatic. Then, the velocity state of the end-effector; i.e. its first order instantaneous motion, is given by

$$\vec{V} \equiv \begin{bmatrix} \vec{w} \\ \vec{v}_0 \end{bmatrix} = w_1 \$_1 + w_2 \$_2 + w_3 \$_3 + w_4 \$_4 + w_5 \$_5 + w_6 \$_6, \qquad (1)$$

where

$\vec{w}$ = Angular velocity of the end-effector with respect to the base link.

$\vec{v}_0$ = The velocity of a point fixed in the end-effector that, in the instant considered, coincides with the origin O of the chosen coordinate system, OXYZ.

w_i = Joint rate velocity of the i-th kinematic pair.

$\$_i$ = Infinitesimal screw that represents the i-th kinematic pair with respect to the coordinate system OXYZ.

Moreover, $\vec{V}, \$_1, \$_2, \$_3, \$_4, \$_5$ and $\$_6$ can also be regarded as elements of the classical screw algebra $\mathcal{S}$. The elements of this real six-dimensional algebra can be described as ordered pairs of three-dimensional vectors, with the following three operations, $\forall (\vec{w}_1, \vec{v}_{01}), (\vec{w}_2, \vec{v}_{02}) \in \mathcal{S}$ and $\forall \lambda \in \Re$

1. Addition
$$(\vec{w}_1, \vec{v}_{01}) + (\vec{w}_2, \vec{v}_{02}) = (\vec{w}_1 + \vec{w}_2, \vec{v}_{01} + \vec{v}_{02}), \qquad (2)$$

2. Scalar Multiplication
$$\lambda(\vec{w}_1, \vec{v}_{01}) = (\lambda \vec{w}_1, \lambda \vec{v}_{01}), \qquad (3)$$

3. Motor Product or Dual Vector Product
$$[(\vec{w}_1, \vec{v}_{01}) (\vec{w}_2, \vec{v}_{02})] = (\vec{w}_1 \times \vec{w}_2, \vec{w}_1 \times \vec{v}_{02} - \vec{w}_2 \times \vec{v}_{01}), \qquad (4)$$

where the addition, scalar multiplication and cross product in the right hand side of equations follow the ordinary rules of three-dimensional vector algebra. In particular, the motor product is distributive, anticommutative, non-associative, and it satisfies the Jacobi identity. Thus, S is a non-associative algebra that satisfies the Jacobi identity.

Within this framework, it is possible to propose a definition of a singular configuration of a non-redundant serial manipulator.

Definition 1. Let $\$_1, \$_2, \$_3, \$_4, \$_5$, and $\$_6$ be the screws which represent the six kinematic pairs of a non-redundant serial manipulator, and $[\$_1, \$_2, \$_3, \$_4, \$_5, \$_6]$ the vector subspace of S generated by the screws. The non-redundant serial manipulator is in a singular configuration if

$$dim[\$_1, \$_2, \$_3, \$_4, \$_5, \$_6] < 6 = dim\, S \qquad (5)$$

Further, the difference, $6 - dim[\$_1, \$_2, \$_3, \$_4, \$_5, \$_6]$, is called the loss of freedom. From a physical point of view, if a non-redundant serial manipulator is in a singular configuration, there are elements $\$ \in S$ such that $\$ \notin [\$_1, \$_2, \$_3, \$_4, \$_5, \$_6]$, and the end-effector **cannot** have a velocity state of the form $\vec{V} = \vec{V}^* + \lambda\$$, where $\vec{V}^* \in [\$_1, \$_2, \$_3, \$_4, \$_5, \$_6]$, and $\lambda \in \Re$.

Proposition 1. The following statements are equivalent

(1) A non-redundant serial manipulator is in a singular configuration.
(2) The set $\{\$_1, \$_2, \$_3, \$_4, \$_5, \$_6\}$ is linearly dependent.
(3) The matrix $J = [\$_1, \$_2, \$_3, \$_4, \$_5, \$_6]$ is singular; i.e. $|J| = 0$.

Unfortunately, proposition 1 does not provide information about which subset (or subsets) of $\{\$_1, \$_2, \$_3, \$_4, \$_5, \$_6\}$ is (are) the one(s) linearly dependent. Thus, it is evident that the study of singular configurations of manipulators is intertwined with the theory of subspaces of the screw algebra, S. In classical kinematics, those subspaces are called screw systems.

3. Lie Product: Definition and Analysis.

In Section 2, the motor product or dual vector product of classical screw algebra S was introduced. Further, in differential geometry, there exists the general notion of the Lie product, which is closely related to the derivation of vector fields. Here, an attempt will be made to relate the notion of the Lie product, as studied in differential geometry, within the framework of spatial kinematics. The results obtained in this Section are not new. Nevertheless, they provide the theoretical foundation for their application in Section 4.

Definition 2 (Lie product, adapted from Hausner and Schwartz (Hausner and Schwartz, 1968)). Let $\$_j, \$_k \in e(3)$, the Lie algebra of the Euclidean group, $E(3)$, and let $\$_j$ be defined as the tangent vector of an Euclidean motion $m_j(t)$ at $t = 0$. Then

$$[\$_j\ \$_k] \equiv \frac{d}{dt}\{m_j(t)\$_k[m_j(t)]^{-1}\}|_{t=0}. \qquad (6)$$

Definition 3 (Motor product). Let $\$_j, \$_k \in \mathcal{S}$; i.e.

$$\$_j = (\vec{w}_j; \vec{v}_{0j}) \text{ and } \$_k = (\vec{w}_k; \vec{v}_{0k}),$$

where $\vec{w}$ represents the angular velocity of the rigid body, and $\vec{v}_0$ represents the velocity of the point of the rigid body that, in the instant considered, coincides with an arbitrary point O fixed in the reference frame. Then, the motor product of $\$_j$ and $\$_k$ is given by

$$[\$_j \ \$_k] \equiv (\vec{w}_j \times \vec{w}_k; \vec{w}_j \times \vec{v}_{0k} - \vec{w}_k \times \vec{v}_{0j}), \tag{7}$$

where $\times$ stands for the usual three-dimensional vector product.

Obviously, the first task is to prove that both definitions are equivalent. For that purpose, the homogeneous representation of an Euclidean motion, and the isomorphism between $e(3)$, the Lie algebra of the Euclidean group $E(3)$, and classical screw algebra, $\mathcal{S}$, will be employed.

Assume that, given an arbitrary coordinate system OXYZ, the Euclidean motion $m_j(t)$ is represented by

$$m_j(t) = \begin{bmatrix} R_{\theta,\hat{u}} & \vec{t} \\ 0 & 1 \end{bmatrix}, \tag{8}$$

where, $R_{\theta,\hat{u}}$ represents the rotation matrix associated with a rotation of θ degrees around the axis given by the unit vector $\hat{u}$, and $\vec{t} \in \Re^3$ represents the displacement of the origin O, of the coordinate system OXYZ. It should be noted that both are functions of time. Further, assume that $\$_k$ is represented, in the same coordinate system OXYZ, by

$$\$_k = \begin{bmatrix} M_k & \vec{v}_{0k} \\ 0 & 0 \end{bmatrix}, \tag{9}$$

where M_k is the skewsymmetric matrix

$$M_k = \begin{bmatrix} 0 & -\omega_{kz} & \omega_{ky} \\ \omega_{kz} & 0 & -\omega_{kx} \\ -\omega_{ky} & \omega_{kx} & 0 \end{bmatrix}. \tag{10}$$

It should be noted that the angular velocity vector associated to the k-th pair is given by $\vec{\omega}_k = (\omega_{kx}, \omega_{ky}, \omega_{kz})$. Further $\vec{v}_{0k} \in \Re^3$.

Carrying out the computations, it follows that

$$\frac{d}{dt}\{m_j(t)\$_k[m_j(t)]^{-1}\}|_{t=0} = \begin{bmatrix} M_j M_k - M_k M_j & M_j \vec{v}_{0k} - M_k \vec{v}_{0j} \\ 0 & 0 \end{bmatrix}, \tag{11}$$

where

$$\frac{d(\vec{t})}{dt}|_{t=0} = \vec{v}_{0j}, \tag{12}$$

and

$$M_j = \begin{bmatrix} 0 & -\omega_{jz} & \omega_{jy} \\ \omega_{jz} & 0 & -\omega_{jx} \\ -\omega_{jy} & \omega_{jx} & 0 \end{bmatrix}.$$

It should be noted that the angular velocity vector associated with the j-th pair is given by $\vec{\omega}_j = (\omega_{jx}, \omega_{jy}, \omega_{jz})$. Finally, using the isomorphism between the Lie algebra $e(3)$, and classical screw algebra S (Karger and Novák, 1985) (Murray et al., 1994), it follows that

$$\frac{d}{dt}\{m_j(t)\$_k[m_j(t)]^{-1}\}|_{t=0} = (\vec{\omega}_j \times \vec{\omega}_k; \vec{\omega}_j \times \vec{v}_{0k} - \vec{\omega}_k \times \vec{v}_{0j}). \tag{13}$$

Therefore, the following proposition has been proved.

Proposition 2. The motor product or dual vector product of classical screw algebra S, and the Lie product of the Lie algebra of the Euclidean group, $e(3)$, are algebraically equivalent.[1]

Proposition 2 is only part of a far more reaching result which states that $e(3)$ and S are isomorphic as algebras. However, in this paper, only the equivalence between the Lie product of $e(3)$, and the motor product of S is required.

4. A Theory of Singular Screw Systems.

This Section develops a theory of singular subspaces of the Lie algebra of the Euclidean group, and provides with necessary conditions for a non-redundant serial manipulator to escape from a singular configuration.

Definition 4. Let $S = \{\$_1, \$_2, \$_3, \$_4, \$_5, \$_6\}$ be the **ordered** set of the screws which represent the kinematic pairs of a non-redundant serial manipulator, with respect to a fixed but arbitrary coordinate system OXYZ. **Then S is called the ordered screw set of the manipulator.** Here, $\$_1$ represents the kinematic pair that connects the open serial chain to the ground link, and $\$_6$ represents the kinematic pair that connects the end-effector to the open serial chain.

In the following development, it will be necessary to consider several different classes of subsets of the ordered screw set of a manipulator.

Definition 5. Let $S = \{\$_1, \$_2, \$_3, \$_4, \$_5, \$_6\}$ be the ordered screw set of a non-redundant serial manipulator. A subset $S_o \subset S$ is said to be an **ordered subset of** S if $\$_j, \$_k \in S_o$ with $j < k$ implies that $\$_i \in S_o$ for all $j < i < k$. Unless explicitly stated, a subset $S_n \subset S$ is a **regular subset** of S.

Definition 6. Let $S_o = \{\$_j, \$_{j+1}, \ldots, \$_k\}$, with $1 \leq j < k \leq 6$ be an ordered subset of the screw set S, of a non-redundant serial manipulator in

---

[1]It is important, however, to note that while the definition of the motor product or dual vector product **does not** indicate any restriction; i.e. it is possible to multiple two arbitrary screws; the definition of the Lie product indicates that the product is **physically meaningful only** when the motion around the j-th screw produces a change on the k-th screw; further analysis is still required in this issue.

a singular configuration. The subset S_o is said to be a **singular ordered subset** of S, denoted S_{os}, if

$$dim[\$_j, \$_{j+1}, \ldots, \$_k] = k - j. \tag{14}$$

Further, the subset S_o is said to be a **minimal singular ordered subset of** S, denoted S_{oms}, if in addition to (14), also satisfies the property that no **ordered proper subset** of S_o satisfies (14).

Definition 7. Let $S_n \subset S$ be a subset of S. The subset S_n is said to be a **singular subset** of S, denoted S_{ns}, if[2]

$$dim[S_n] = \text{Cardinality } S_n - 1. \tag{15}$$

Further, the subset S_n, is said to be a **minimal singular subset** of S, denoted S_{nms} if in addition to (15), also satisfies the property that no **proper subset** of S_n also satisfies (15)[3].

It should be noted that definitions 4-7 are independent of the coordinate system employed in the representation of the ordered screw set S. This result follows from the well known fact that screw transformations induced by a change of coordinate system, or a change of the unit of length, are non-singular. Further, non-singular transformations preserve the dimension of the subspaces and the ordering of the screw set.

It is interesting to note that every singular ordered subset, S_{os}, and every minimal singular ordered subset, S_{oms}, contain a singular subset, S_{ns}, and a minimal singular subset, S_{nms}, respectively. Further, a minimal singular ordered subset, S_{oms}, and its corresponding minimal singular subset, S_{nms}, identify the screws responsible, at least partially, for the singular configuration of the serial non-redundant manipulator.

Example 1. Consider the example proposed by Hunt (Hunt, 1986) (page 174). It involves the spatial manipulator shown in Figure 1, where the screws representing the kinematic pairs are given, with respect to the coordinate system shown in the figure, by

$$\$_1 = (1, 0, 0; 0, 0, -1), \$_2 = (0, 1, 0; 0, 0, 0),$$

$$\$_3 = (0, 1, 0; 0, 0, 5), \$_4 = (1, 0, 0; 0, 0, 0),$$

$$\$_5 = (0, \frac{1}{\sqrt{2}}, \frac{1}{\sqrt{2}}; 0, \frac{-10}{\sqrt{2}}, \frac{10}{\sqrt{2}}), \$_6 = (0, \frac{1}{\sqrt{2}}, \frac{-1}{\sqrt{2}}; 0, \frac{10}{\sqrt{2}}, \frac{10}{\sqrt{2}}).$$

The ordered screw set of the manipulator is $S = \{\$_1, \$_2, \$_3, \$_4, \$_5, \$_6\}$, and it is straightforward to show that $dim[S] = 4$. Thus, the manipulator is in a singular configuration. Moreover, it is straightforward to show that the subset $\{\$_1, \$_2, \$_3, \$_4\}$ is singular; i.e. $dim[\$_1, \$_2, \$_3, \$_4] = 3$. Further, removing anyone of the screws yields linearly independent subsets. Thus

---

[2] The cardinality of a set is the number of elements in the set.

[3] In a recent article Lipkin and Pohl (Lipkin and Pohl, 1991) defined dependent sets and minimally dependent sets, they are equivalent respectively to singular subsets and minimal singular subsets

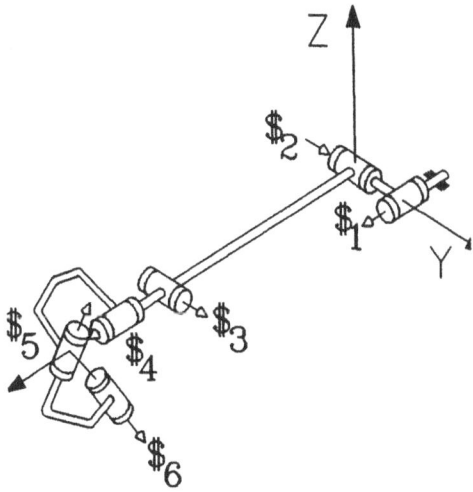

Figure 1. **Examples of Singular Subsets.**

$S^1_{oms} = \{\$_1, \$_2, \$_3, \$_4\}$, is a minimal singular ordered subset of S, and its corresponding minimal singular subset is $S^1_{nms} = S^1_{oms}$. Similarly, it is also easy to determine that the subset $\{\$_2, \$_3, \$_4, \$_5, \$_6\}$ is also singular; i.e. $dim[\$_2, \$_3, \$_4, \$_5, \$_6] = 4$. In addition, removing either $\$_2$ or $\$_6$ yields linearly independent subsets. Thus $S^2_{oms} = \{\$_2, \$_3, \$_4, \$_5, \$_6\}$ is another minimal singular ordered subset of S. Finally, since $dim[\$_2, \$_3, \$_5, \$_6] = 3$, and removing any screw from the set yields linearly independent subsets[4], the corresponding minimal singular subsets is $S^2_{nms} = \{\$_2, \$_3, \$_5, \$_6\}$.

The following proposition is a generalization of a previous result reported by Waldron (Waldron *et al.*, 1985) which indicates that the first and last kinematic pairs of a serial manipulator cannot guide the manipulator into, or permit the manipulator to move out of, a singular configuration.

Proposition 3. Let $S_{oms} = \{\$_j, \$_{j+1}, \ldots, \$_k\}$ be an ordered subset of a serial manipulator. Then, the dimension of S_{oms} is not affected by displacements around $\$_1, \$_2, \ldots, \$_j$, and $\$_k, \$_{k+1}, \ldots, \$_6$.

Proof: Rotations around $\$_i$, where $i = 1, 2, \ldots, j$ induce a non-singular transformation over the space generated by S_{oms}. Rotations around $\$_i$, where $i = k, k+1, \ldots, 6$, induce the identity mapping over the space generated by S_{oms}; in both cases its dimension remains unaffected. Then, it is obvious that the displacements around $\$_1, \$_2, \ldots, \$_j$, and $\$_k, \$_{k+1}, \ldots, \$_6$

---

[4]An ordered singular subset is minimal if the subset does not have a proper ordered subset which is also singular. However, a well known result in linear algebra states that all the subsets of a linearly independent set are linearly independent; i.e. they are not singular. Therefore, the result follows. This result is also valid for singular subsets

cannot lead the manipulator into, or let escape the manipulator out from, the singular configuration associated with S_{oms}.

Proposition 4. Let S be the screw set of an open serial chain at time $t = 0$, and $\$_j(0), \$_k(0), \in S$,with $j < k$. Then, the screw $\$_k(\Delta t)$, due to the motion $m_j(t)$ around $\$_j$, is approximated by

$$\$_k(\Delta t) \approx \$_k(0) + [\$_j(0) \ \$_k(0)](\Delta t) = \$_k + [\$_j \ \$_k](\Delta t).$$

Proposition 5. Let S be the ordered screw set of a non-redundant serial manipulator, $S_{oms} = \{\$_j, \$_{j+1}, \ldots, \$_f, \ldots, \$_h, \ldots, \$_k\}$ a minimal singular ordered subset, and S_{nms} its corresponding minimal singular subset. Assume that $\$_h \in S_{nms}$, and $[\$_f \ \$_h] \notin [S_{oms}]$. Then, S'_n, the subset obtained from S_{nms} by substituting $\$_h$ by $\$_h(\Delta t) = \$_h + [\$_f \ \$_h](\Delta t)$ with $\Delta t \neq 0$, is no longer singular. Further, S'_o, the subset obtained from S_{oms} by making the same substitution; i.e.

$$S'_o = \{\$_j, \$_{j+1}, \ldots, \$_f, \ldots, \$_h + [\$_f \ \$_h](\Delta t), \ldots, \$_k\}$$
is no longer singular.

Proposition 6. Let $S_{oms} = \{\$_j, \$_{j+1}, \ldots, \$_f, \ldots, \$_h, \ldots, \$_k\}$ be a minimal singular ordered subset of a serial non-redundant manipulator in a singular configuration. Let $S_{nms} \subset S_{oms}$ be its corresponding minimal singular subset. Then, a necessary condition for the manipulator to escape from the partial singular configuration associated with S_{oms} is the existence of a pair of screws $\$_f, \$_h \in S_{oms}$ with $f < h$, and $\$_h \in S_{nms}$ such that $[\$_f \ \$_h] \notin [S_{oms}]$.

Proof: Let $S_{oms} = \{\$_j, \$_{j+1}, \ldots, \$_f, \ldots, \$_h, \ldots, \$_k\}$ be a minimal singular ordered subset of a serial non-redundant manipulator in a singular configuration. Then a necessary condition for the manipulator to escape from the singular configuration associated with S_{oms} is that there exists a $\$_h \in S_{oms}$ such that

$$dim[S'_o] = dim[\{S_{oms}/\{\$_h\}\} \cup \{\$_h + [\$_f \ \$_h](\Delta t)\}] = k - j.$$
Then, by proposition 5, the result follows.

Using proposition 6, it is not difficult to prove the following result.

Proposition 7. Let $S_{oms} = \{\$_j, \$_{j+1}, \ldots, \$_k\}$ be a minimal singular ordered subset of a serial non-redundant manipulator in a singular configuration. If $[S_{oms}]$ is a subalgebra of the Lie algebra of the Euclidean group, $e(3)$, then the manipulator is in a **permanent singular configuration**. [5]

As a corollary of proposition 7, it is possible to state the following rule.

Kinematic Design Rule for Serial Non-Redundant Manipulators. Let $S = \{\$_1, \$_2, \$_3, \$_4, \$_5, \$_6\}$ be the ordered screw set of a serial non-redundant manipulator; then, in order to avoid permanent singular configurations of the manipulator, no minimal singular ordered subset of S

---

[5] A permanent singular configuration is a configuration that it is impossible to escape from, by displacing the kinematic pairs of the manipulator.

should generate a subalgebra of $e(3)$; i.e. no ordered subset of n $(2 \leq n \leq 6)$ adjacent screws must belong to a subalgebra of $e(3)$ of dimension $n - 1$.[6]

It is important to note that proposition 6 provides only necessary conditions for the manipulator to escape from the partial singular configuration associated with a minimal singular ordered subset S_{oms}. There are two reasons why it is more difficult to provide necessary and sufficient conditions for the manipulator to escape from a global, as opposed to, a partial singular configuration of a manipulator.

1. The existence of more than one minimal singular ordered subsets, and their mutual interactions.

2. The necessary conditions introduced so far involve the motion of only one screw associated with a partial singular configuration. It must be evident that any necessary and sufficient condition for the global singular configuration must take into account the motion of all the remaining screws[7].

However, it is still possible to provide necessary conditions for the escapement of a serial non-redundant manipulator from a singular configuration with more than one minimal singular ordered subset.

Proposition 8. Let $S = \{\$_1, \$_2, \$_3, \$_4, \$_5, \$_6\}$ be the ordered screw set of a serial non-redundant manipulator in a singular configuration; and let S_{oms}^1 and S_{oms}^2 be minimal singular ordered subsets[8] of S. Then necessary conditions for the manipulator to escape from the global singular configuration are

1. Each of the minimal singular ordered subsets of S, satisfies the necessary conditions indicated in proposition 6; i.e. for $i = 1, 2$ there exists $\$_f^i, \$_h^i \in S_{oms}^i$, with $\$_h^i \in S_{nms}^i$ such that $[\$_f^i \ \$_h^i] \notin [S_{oms}^i]$.

2. The dimension of the space generated by the set S' obtained by substituting $\$_h^i$ for $\$_h^i + [\$_f^i \ \$_h^i]\Delta t$ for i=1,2, satisfies that $dim[S'] = 6$.

Of course, it is possible to find simpler conditions for particular cases of singular configurations of manipulators. However, a complete analysis of all these possibilities is, in the authors' opinion, beyond the scope of the present work.

Example 2. Consider again the manipulator shown in Figure 1. As indicated in Section 4. $S = \{\$_1, \$_2, \$_3, \$_4, \$_5, \$_6\}$, and $dim[S] = 4$. Further, the first minimal singular ordered subset is $S_{oms}^1 = \{\$_1, \$_2, \$_3, \$_4\}$, with

---

[6] A list of the subalgebras of $e(3)$ can be found in Hunt (Hunt, 1986).

[7] The authors conjecture that if the ordered subset of a serial non-redundant manipulator in a singular configuration satisfy the necessary conditions given by proposition 8, it would be very unlikely that the manipulator remain in the singular configuration.

[8] In a properly designed serial non-redundant manipulator, it is very unlikely that a global singular configuration have more than a pair of minimal singular ordered subsets. In any case, this proposition can be modified accordingly.

$S^1_{nms} = S^1_{oms}$. Therefore, according with proposition 5, it is necessary to consider the following Lie products

$$[\$_2 \ \$_3] = (0,0,0;5,0,0), [\$_2 \ \$_4] = (0,0,-1;0,0,0), [\$_3 \ \$_4] = (0,0,-1;0,5,0).$$

It is easily shown that none of them belong to $[S^1_{oms}]$. Therefore, actuating either the second or the third revolute, the manipulator can escape from the partial singular configuration associated with $[S^1_{oms}]$.

Moreover, a second minimal singular ordered subset is $S^2_{oms} = \{\$_2, \$_3, \$_4, \$_5, \$_6\}$ with $S^2_{nms} = \{\$_2, \$_3, \$_5, \$_6\}$. This singular configuration is easily solved, the conclusion is that both $[\$_3 \ \$_5]$, and $[\$_3 \ \$_6]$ satisfy the necessary conditions; i.e. they do not belong to $[S^2_{oms}]$. Since

$$dim[\$_1, \$_2, \$_3, \$_4 + [\$_3 \ \$_4]\Delta t, \$_5, \$_6 + [\$_3 \ \$_6]\Delta t] = 6, \forall \Delta t \neq 0.$$

The conclusion is that actuating the third revolute, alone, the manipulator can move out of the partial singular configuration associated with both $[S^1_{oms}]$ and $[S^2_{oms}]$.

5. Acknowledgements.

The first author thanks Professor Peter Crouch, Center for Systems Science at Arizona State University, whose continuous questions about the role of the Lie product in spatial kinematics led to the present work. The first author also thanks the Mechanical and Aerospace Engineering Department, the Center for Integrated Manufacturing, and the Center for Systems Science at Arizona State University, and in particular to Prof. Joseph Davidson, for the facilities given to him during his sabbatical stay. The authors also thank the University of Florida Center of Excellence Fund for its support.

References

Waldron, K. J., Wang, S. L., and Bolin, S. J. (1985), A Study of the Jacobian Matrix of Serial Manipulators. *Journal of Mechanisms, Transmissions, and Automation in Design*, Vol. 107, pp. 230-238.

Wang, S. L. and Waldron, K. J. (1987), A Study of the Singular Configurations of Serial Manipulators. *Journal of Mechanisms, Transmissions, and Automation in Design*, Vol. 109, pp. 14-20.

Litvin, F. L. (1980), Application of theorem of implicit function system existence for analysis and synthesis of linkages. *Mechanism and Machine Theory*, Vol. 15, pp. 115-125.

Litvin, F. L., Zhang, Y., Parenti Castelli, V., and Innocenti C. (1980), Singularities, Configurations, and Displacement Functions for Manipulators. *The International Journal of Robotic Research*, Vol. 5, pp. 52-65.

Hunt, K. H. (1986), Special configurations of robot-arms via screw theory, Part 1. The Jacobian and its Matrix of Cofactors. *Robotica*, Vol. 4, pp. 171-179.

Hausner, M. and Schwartz, J. T. (1968), *Lie Groups, Lie Algebras*, New York: Gordon and Breach, pp. 50.

Karger, A. and Novák, J. (1985) *Space kinematics and Lie Groups*, New York: Gordon and Breach, pp. 197-206.

Murray, R. M., Zexiang, L., and Sastry, S. S. (1994) *A Mathematical Introduction to Robotic Manipulation*, Boca Raton, Florida: CRC Press, pp. 175-176.

Lipkin, H. and Pohl, E. (1991) Enumeration of Singular Configurations for Robotic Manipulators. *Journal of Mechanical Design*, Vol. 113, pp. 272-279.

A GENERAL CRITERION FOR THE IDENTIFICATION OF NONSINGULAR POSTURE CHANGING 3-DOF MANIPULATORS

J. EL OMRI AND P. WENGER
Laboratoire d'Automatique de Nantes,
C.N.R.S. Unit 823
Ecole Centrale de Nantes, University of Nantes
1, rue la Noe 44072 NANTES. e-mail: wenger@ec-nantes.fr

Abstract. This paper provides a new necessary and sufficient condition for a 3-DOF serial manipulator to be type-2, i.e. nonsingular posture changing. This condition is based on the existence in the workspace of cusps. In addition to this new condition, the most recent results in the kinematics of serial manipulators are reviewed in the first section of this work. Several illustrative examples are given in the end of the paper.

1. INTRODUCTION

It is now established that robot manipulators do not necessarily have to run into a singularity when changing posture [1, 2, 3]. In fact, only a limited category of robot manipulators must meet a singularity when going from one posture to another. The class of such manipulators is called type-1. Type-1 manipulators include most of usual industrial manipulators. Type-2 manipulators are the nonsingular posture changing manipulators. Since type-2 manipulators may have more than one inverse kinematic solution in one singularity-free domain of the jointspace, a new partition of the jointspace into uniqueness domains has been defined, using the concept of characteristic surfaces [3]. A relationship between genericity [4], solvability [5] and nonsingular posture changing ability was suggested in [6]. However, no generic, stringent scheme has yet been proposed for classifying type-1 and type-2 architectures. Roughly speaking, type-2 manipulators include all general manipulators, while manipulators with simplifying kinematic conditions are more likely to be type-1. More stringently, two separate

153

J.-P. Merlet and B. Ravani (eds.), Computational Kinematics, 153–162.

conditions for a 3-DOF manipulator to be type-1 (one is sufficient and the other one is necessary) were provided in [3]. Using the aforementioned two conditions, a rather heavy numerical identification procedure was given, and a list of type-1 3-DOF was provided in [7]. This list was completed with some 6-DOF type-1 manipulators in [8].

In this paper, a novel fundamental necessary and sufficient condition is stated for a 3-DOF to be type-2. It is proved that a given 3-DOF is type-2 if and only if there exists at least one cusp in the workspace, i.e. point for which three equal inverse kinematic solutions exist. It is shown that this condition is very straightforward to use, by just examining the geometry of the internal boundary surfaces of the workspace.Section 2 of this paper is devoted to a review on the most recent results in robot kinematics. Finally, illustrative application examples are given in section 4 for several type-1 and type-2 3-DOF manipulators.

2. RECENT RESULTS IN THE KINEMATICS OF SERIAL MANIPULATORS

2.1. TYPE-1 AND TYPE-2 MANIPULATORS

Type-1 manipulators are those manipulators which must pass through a singularity when changing posture. Typical examples of such manipulators are 3-R manipulators with two parallel or intersecting joint axes. Type- 2 manipulators can change posture without running into a singularity.

2.2. SINGULARITIES

Robotic singularities analysis is a topic of high interest since the singularities play a central role in the kinematic behavior of robot manipulators. Also, this is a very difficult issue, mainly because many particular cases occur. Singularities arise when the Jacobian matrix becomes rank deficient. In a positional (resp. rotational) singularity, a direction exists along which (resp. around which) the end- effector (EE) cannot assume a non zero velocity. In a general singularity, a full screw motion of the EE with non zero velocity cannot be executed (this singularity arises only for 6-DOF manipulators). Since the singularities are always independent of q_1, the singularities of 3-DOF architectures can be depicted in the plane (q_2, q_3).

The singularities can be classified into general singularities and special singularities. General singularities have a finite number of inverse kinematic solutions, and special singularities have an infinity of inverse solutions. For 3-R positional manipulators, general singularities occurs when the EE lies on a line that intersects all the joint axes, and special singularities arise when the EE meets joint axis 1 or 2. It was shown in [2] that if $\alpha_2 \neq \pm 90°$ and $d_3^2 tan^2(\alpha_2) = a_3^2 - a_2^2$ (using the standart DH parameters), the EE can meet joint axis 2 for one value of $\theta_3 = atan2[d_3 tan(\alpha_2/a_3), -a_2/a_3]$.

In this case, one singular circle is swept out in the workspace, and one singular surface is generated in the jointspace. If $\alpha_2 = \pm 90°$, it was shown in [3] that under the conditions $d_3 = 0$ and $a_2 \leq a_3$, the EE can also meet joint axis 2 for $\theta_3 = \pm arcos(-a_2/a_3)$, yielding two singular surfaces in the jointspace, and two singular circles in the workspace.

2.3. MAXIMAL SINGULARITY-FREE DOMAINS OF THE JOINTSPACE

The maximal singularity-free domains of the jointspace were first introduced in [10] and referred to as the aspects. They were called regions in [1] and c-sheets in [2] and [11]. Note that, unlike the aspects, the definitions of c-sheets and regions assume unlimited joints. The word "region" is more often referred to as a subset of the workspace. In order to avoid confusion, the word "c-sheet" will be used in this paper. The c-sheets are intrinsic to the robot architecture. The aspects, instead, may differ in shape and multiplicity according to the joint limits of the robot. When no joint limits exist, aspects and c-sheets coincide. The c-sheets are the maximal uniqueness domains of the kinematic map for type-1 manipulators only.

2.4. SINGULAR SURFACES AND BOUNDARY SURFACES FOR 3-DOF MANIPULATORS

The singularities are commonly derived by equating to zero the determinant of the Jacobian matrix :

$$det(J) = 0 \tag{1}$$

Burdick provided an alternative and more compact way for representing the general singularities of 3-R geometries [2]. However, since only the general singularities can be obtained, any special singularity must be found out by analysing the occurence of particular cases. Also, generalization to 3-DOF with one or more prismatic joints is not straighforward. Recently, a linear combination of trigonometric functions which are linear in q_2 and q_3 was derived in [12] for general RRR architectures. This form can be easily extended to the case of prismatic joints.

The singularities generally form surfaces in the joint space, called singularity surfaces, and surfaces in the workspace, called boundary surfaces. Mapping the boundary surfaces can be achieved by using eq.(1) along with the forward kinematic map, or they can be generated in a more straighforward way by searching for points where the polynomial of the inverse kinematics has repeated roots.

The boundary surfaces appear within the workspace or on the external workspace boundary. When crossing a singularity in the joint space which gives rise to a boundary surface in the workspace, the EE reflects back on the corresponding boundary surface (should it be an internal or an external boundary) [1, 12, 13]. This property, however, does not hold anymore for the special singularities occurring when the EE meets axis 2. In this case, instead, the EE simply crosses the corresponding singular circle in the workspace.

The singularity surfaces in the workspace separate regions with different numbers of inverse kinematic solutions, usually called postures [2], [14]. This number is even for unlimited revolute joints. The number of postures is related to the number of real roots of the polynomial in one of the joint variables, which appear as pairs. It is well known that the maximum number of postures is 16 for a general 6-R manipulator, and 4 for a general 3-R robot. When going from one region to a region with a lower degree of accessibilty, one pair of solutions coincide on the boundary. This pair becomes a pair of complex roots when arriving in the region with lower accessibility.

2.5. CHARACTERISTIC SURFACES AND PSEUDO-SINGULAR MANIFOLDS

The characteristic surfaces (CS) have been defined in [3] for any serial non-redundant manipulator with or without joint limits. Let Q denote the robot jointspace. Let A_i be an aspect in Q, that is, a connected component of $Q \dot{-} S$ ($\dot{-}$ means the difference between sets) where S is the set of singularities. The CS of A_i, denoted $Sc(A_i)$, are defined as the preimage in A_i of the boundary A_i^* of A_i (f is forward kinematic map, and $f^{-1}()$ means the preimage of a set) :

$$Sc(A_i) = f^{-1}(f(A_i^*)) \cap A_i \qquad (2)$$

For a given aspect, the CS are the set of configurations which place the EE on a boundary surface, which is the image of a boundary of the aspect at hand. Note that the boundaries of an aspect include singularity surfaces as well as joint limits boundaries (if there are any). Thus, the CS may be generated by both singularities and joint limits. The CS are independent of joint 1 only when unlimited (unlike the singularities which are, in any case, independent of joint 1). Representation of singularities and CS of 3-DOF geometries can be made, thus, in plane (q_2, q_3) only when joint 1 is unlimited.

The CS induce a partition on each aspect into open connected sets called basic components. They were introduced with the aim of defining the new maximal uniqueness domains of the kinematic map for type-2 manipulators. Note that, by definition, CS exist only for type-2 manipulators. A type-1 manipulator has no CS.

The pseudo-singular manifolds were defined in [11] only for 3-DOF geometries with unlimited joints, as the preimage in Q of the boundary surfaces. Unlike the CS, the pseudo-singular manifolds are not related to an aspect, and may arise for both type-1 and type-2 manipulators. They were introduced for mapping back in the jointspace, the regions with different number of accessible postures in the workspace.

2.6. BASIC COMPONENTS AND PATCHES

The basic components were defined according to the CS.

Let A_i be an aspect. The basic components of A_i are the connected components of the set $A_i - Sc(A_i)$. The existence of basic components was already intuitively established in [1]. The patches were defined similarly in [11] according to the pseudo-singular manifolds. It is worth noting that the basic components divide an aspect only when multiple inverse kinematics solutions exist in it. The patches, instead, may divide an aspect even when only one inverse solution exist in it. An illustrative example can be found in section 4.

3. A NECESSARY AND SUFFICIENT CONDITION

We first recall here the conditions stated in [3].

Sufficient Condition: *Any 3-DOF manipulator with no more than two inverse kinematic solutions is type-1.*

Necessary condition: *For a 3-DOF manipulator with four solutions to be type-1, there must exist at least four c-sheets. This condition is not sufficient.*

Degree of accessibility on the boundary surfaces (BS):

The number of inverse kinematic solutions is given by the number of roots of the polynomial in one of the joint variables. This polynomial is a quartic for general 3-DOF manipulators. For a 3-DOF manipulator with unlimited joints, there may exist 0, 2 or 4 solutions out of the BS. On the BS, it has been shown in [11] and [15] that there are 1, 2, 3, or an infinity of distinct solutions according to the following cases :

case 1: One solution appears on a BS when two equal roots exists on it. Such BS are those separating regions with two and 0 solutions (like the outer workspace boundary or an internal boundary generating a void). One solution may appear when there are four equal roots. This occurs only for special 3-DOF manipulators, i.e. manipulators whose inverse kinematics can be solved with polynomials of lower degree.

case 2: Three solutions exist on a BS when there are two equal roots and two distinct roots. They are located on the internal boundary surfaces separating two and four solutions regions.

case 3: Two solutions exist at the the intersection between two BS. In this case, there are two pairs of equal roots. Such points are referred to as "nodes".

<u>case 4:</u> Two solutions may also exist at the connection between two BS with three solutions. At those points, there are three equal roots and one single root. It was shown [11] that such points define cusps on the BS.

<u>case 5:</u> An infinity of solutions may exist at points where two BS intersect. This arises when one special singularity surface intersects one general singularity surface in the jointspace.

Theorem 1: *If one or more cusps exist in the workspace of a 3-DOF manipulator, this manipulator is type-2* (i.e. nonsingular posture changing)

Proof: Let X be a cusp. There exist three equal solutions and one distinct solution at X. At X, two BS which bound a 4-solution region meet. Let X' belong to the 4-solution region. As X' goes to X, three solutions among the four admissible go closer to each other, and merge when arriving at X. If each of these three solutions were separated by singular surfaces, we would have one point with four equal roots at the preimage of X, which is not the case by definition of X. Thus, at least two solutions are not separated by a singularity surface, which proves that a nonsingular change of posture is possible. Consequently the manipulator is type-2.

Theorem 2: *a type-2 manipulator has one or more cusps in its workspace.*

Proof: The basic components of a type-2 manipulator are domains of the aspects where only one solution exists. The inverse mapping of the 4-solution region of a type-2 manipulator leads to four distinct basic components. We know that when leaving the 4-solution regions from any initial posture, at least one BS can be crossed and one BS cannot. This means that the basic components are bounded by at least one singularity surface and one characteristic surface. Thus, there are points in the jointspace where one singularity surface meets one characteristic surface. Since there are two equal solutions on the singularity surface, and one solution on the characteristic surface, there are three equal solutions at the intersection point. Thus there is a cusp in the workspace.

Consequence: *The nonsingular change of posture is done by encompassing one cusp in the workspace. Also, two CS which meet at a singularity are crossed successively in the jointspace.*

Now, the necessary and sufficient condition can be stated:

Theorem 3: *A 3-DOF manipulator is type-2 if and only if there is at least one cusp in its workspace.*

Consequently, type-2 3-DOF manipulators can be identified by simply tracing a cross section of the boundary surfaces. This can be done quickly using one of the methods presented in section 2, and substituing x, y, z by $\rho = \sqrt{x^2 + y^2}$, 0 and z, respectively. A cusp appears only inside the workspace, and can be easily recognised as a "turning back point", where two BS meet tangentially (see figure 1). A type-2 manipulator may have several 4-solution regions in its workspace. In those regions for which at

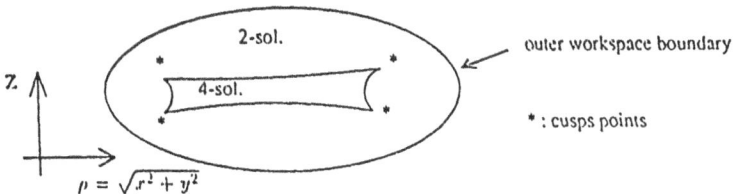

Figure 1.   Typical cusps in the workspace

least one corner point is a cusp, a nonsingular change of posture is possible. In any other region, the change of posture is necessarily singular.

Equations of the cusps can be obtained from the polynomial in $tan(q_1/2)$. Several attempts have been made to search for existence conditions of cusps using the aforementioned equations. However, no general relation involving only the DH parameters has been found yet. This is the subject of current research work from the authors.

4.   ILLUSTRATIVE EXAMPLES

A huge number of examples has been treated which confirm the results stated above. Some of them are analysed in this section. The BS are traced in a cross section of the workspace (in the plane $\rho = \sqrt{x^2 + y^2}, z$) and the singularity and CS are drawn in the plane of the last two joint variables. The singularity surfaces have been drawn in bold lines to distinguish them from the charcteristic surfaces. For more information, the number of accessible solutions inside the regions has been indicated.

4.1.   TYPE-2 MANIPULATORS WITH TWO C-SHEETS

These manipulators are common 3-DOF type-2 manipulators. Figure (2)

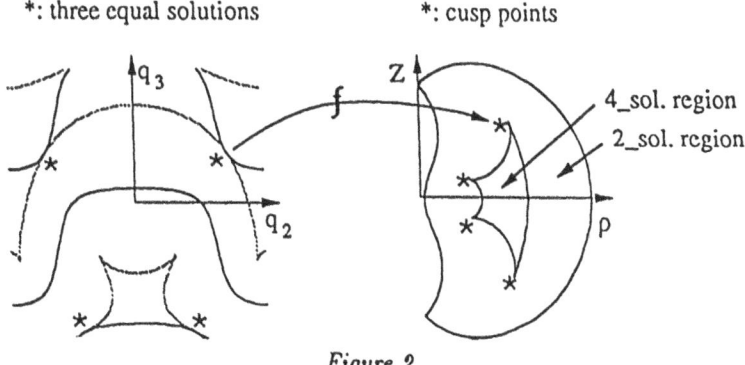

Figure 2.

depicts a 3-R manipulator with DH parameters $\alpha_1 = -90°, \alpha_2 = 90°, a_1 = 1, a_2 = 2, a_3 = 1.5, d_1 = 0, d_2 = 1, d_3 = 0$. There are two distinct singularity

160

surfaces for this robot, and four cusps in the workspace, which are the corner points of the 4-solution region.

Another example is a 3-R robot such that $\alpha_1 = 45°, \alpha_2 = 80°, a_1 = 1.5, a_2 = 1, a_3 = 2, d_1 = 0, d_2 = 1, d_3 = 1$ (figure 3). It is worth noting that this manipulator has one single singularity surface (this can be more easily seen after identification of the opposite sides of the jointspace). There is one node and four cusps (the four other corners of the 4-solution region).

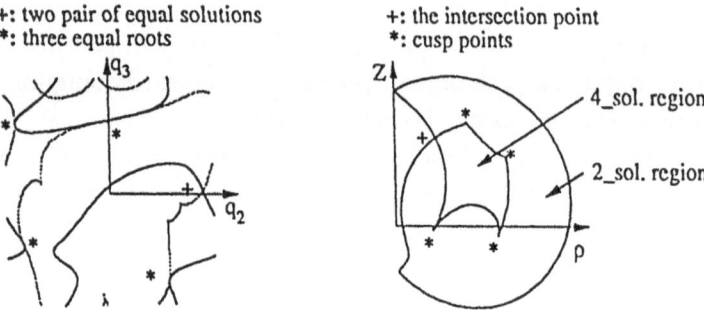

Figure 3.

4.2. TYPE-2 MANIPULATORS WITH FOUR C-SHEETS

Most manipulators with four sheets are type-1. However, some are type-2 as is the following 3R robot analysed in figure 4 (DH parameter : $\alpha_1 = -90°, \alpha_2 = 90°, a_1 = 1, a_2 = 2, a_3 = 2.6, d_1 = 0, d_2 = 1, d_3 = 0$). The two additional singularities are special singularities which map onto two isolated points in the cross section of the workspace. We note that there are only two cusps, and there is one node. Also, there are two distinct 4-solution regions. In one of them, the nonsingular change of posture can occur. In the other one, the manipulator must run into a singularity for changing posture. In figure 5, the pseudo-singular manifolds have been traced, instead of the CS, in order to illustrate the difference already discussed in section 2.

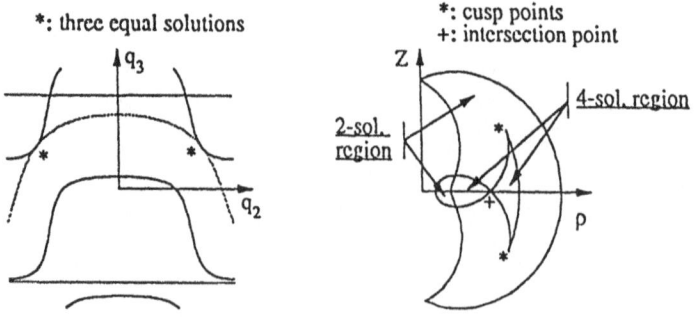

Figure 4. Boundary surfaces and CS

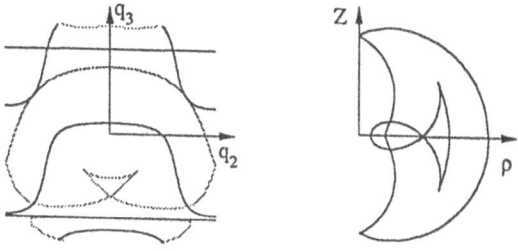

Figure 5. Pseudo-singular manifolds

4.3. TYPE-1 MANIPULATORS WITH FOUR C-SHEETS

Many type-1 3-DOF manipulators have four c-sheets. Figure 6 shows a 3-R robot which should be compared to the type-2 robot presented just above. The DH parameters are $\alpha_1 = -90°, \alpha_2 = 90°, a_1 = 1, a_2 = 0.5$ (instead of 2 above), $a_3 = 3$ (instead of 2.6), $d_1 = 0, d_2 = 1, d_3 = 0$). Unlike the manipulator given above, this one is type-1 and no cusp appears in the workspace (note, the points appearing on the external workspace boundary are not cusps, but nodes located on the Z_1 axis).

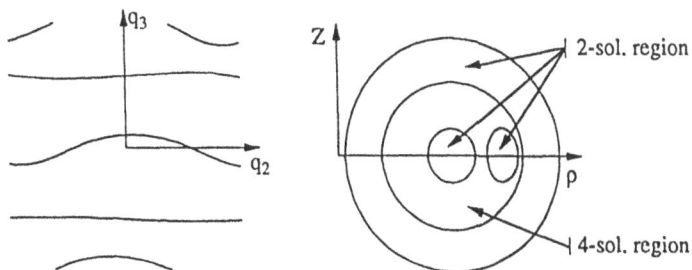

Figure 6.

4.4. TYPE-1 MANIPULATORS WITH TWO C-SHEETS

Under some geometric conditions, a general manipulator may have only two inverse kinematic solutions. Because of the sufficient condition stated in §3.1, such manipulators are type-1. Fig. 7 shows such a manipulator with $\alpha_1 = -70°, \alpha_2 = 60°, a_1 = 2, a_2 = 1, a_3 = 0.5, d_1 = 0, d_2 = 1, d_3 = 0.2$.

5. CONCLUSION

In this paper, we have derived a new necessary and sufficient condition for a 3-DOF serial manipulator to be type-2, i.e. nonsingular posture changing. This condition is based on the existence of cusps in the workspace. A type-2 manipulator can be easily recognized when tracing the boundary surfaces in a cross section of the workspace. Several illustrative examples have been provided. The main important notions in the kinematics of serial manipulators were recalled, stressing the difference between CS and

162

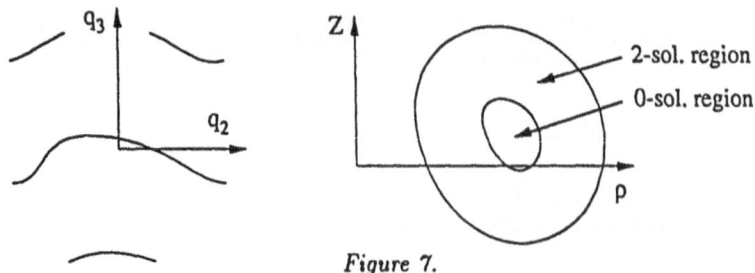

Figure 7.

pseudo-singular manifolds. Future work includes the search for general analytical conditions on the DH parameters for a manipulator to be type-2. For general 6-DOF serial manipulators, application of the previous condition is not straightforward since all joints must be considered. However, it seems that the existence of cusps should, again, play an important role.

References

1. Parenti C.V. and Innocenti C., (Yug., 1988) *"Position Analysis of Robot Manipulator: Regions and Subregion"* ARK'88, pp. 150–158, Ljubljana.
2. Burdick J. W., (1988) *"Kinematic Analysis and Design of Redundant Manipulators"* PhD Diss., Stanford.
3. Wenger P., (May 1992) *"A new General Formalism for the Kinematic Analysis of all Non-redundant Manipulators"* IEEE Int. Conf. on Rob. and Aut., Nice, France, pp. 442–447.
4. Paï D.K. and Leu M.C., (Oct. 1992) *"Genericity and Singularities of Robot Manipulators"* it Proc. IEEE Trans. on Rob. and Aut., Vol. 8, pp. 545-559.
5. Smith D.R., (1990) *"Design of Solvable 6R Manipulators"* PhD Thesis, Georgia Institute of Technology, Atlanta.
6. Burdick J. W.,(1991) *"A classification of 3R regional manipulator singularities and geometries"* IEEE Int. Conf. on Rob. and Aut., pp. 2670–2675, Cal.
7. Wenger P., (Nov. 1993) *"A Classification of Manipulators Geometries Based on Singularity Avoidance Ability"*, ICAR'93, pp. 649–654, Tokyo, Japan.
8. Wenger P., El Omri J., (July 1994) *"On the Kinematics of Singular and Nonsingular Posture changing Manipulators"* Proc. ARK'94, pp. 29–38, Ljubjana, Slovenia.
9. Wenger P., El Omri J., *"Postures and Singularities in Serial Manipulators"* proposed to the IEEE'95 Int. Conf on Rob. and Aut.
10. Borrel P. and Liegeois A., (1986) *"A Study of Manipulator Inverse Kinematic Solutions with Application to Trajectory Planning and Workspace Determination"*, IEEE Int. Conf. Rob. and Aut., pp. 1180–1185.
11. Tsai K. Y., Kholi D., Arnold J., (Dec. 93) *"Trajectory Planning in Joint Space for Mechanical Manipulators"* ASME J. of Mech. Design, Vol. 115, pp. 909–914.
12. Ranjbaran F., Angeles J., (June 94) *"On Positioning Singularities of 3-Revolute Robotic Manipulators"* Proc. 12th Symp. on Engineering Applications in Mechanics, pp. 273–282, Montreal, Canada.
13. Innocenti C., Parenti-Castelli V., *"Singularity-free Evolution from one Configuration to Another in Serial and Fully-parallel Manipulators"* ASME92 DE-Vol. 45, Rob., Spatial Mech., and Mechanical Systems.
14. Kholi D. and Hsu M.S., (1987) *"The jacobian analysis of workspaces of mechanical manipulators"* J. of Mech. and Mach. Theory, Vol. 22, No. 3, pp. 265–275.
15. Tsai K. Y., Kholi D., (Dec. 93) *"Trajectory Planning in Task Space for General Manipulators"* ASME J. of Mech. Design, Vol. 115, pp. 915–921.

IDENTIFICATION AND CLASSIFICATION OF THE SINGULAR CONFIGURATIONS OF MECHANISMS

D. ZLATANOV, R.G. FENTON AND B. BENHABIB
Computer Integrated Manufacturing Laboratory
Department of Mechanical Engineering
University of Toronto
Toronto, Ontario, Canada
beno@me.utoronto.ca

Abstract

This paper presents a novel method for finding and classifying all the singularities of an arbitrary non-redundant mechanism. The proposed technique is based on the velocity-equation formulation of kinematic singularity and the singularity classification first introduced in (Zlatanov et al., 1994–1,2). Criteria for singularity are derived and applied to formulate procedures for computing the singularity set and revealing its division into singularity classes. Further development of methods for automatic singularity analysis is discussed.

1. Introduction

The identification of singularities has been investigated extensively in the literature for open kinematic chains (Hunt 1986, Wang and Waldron 1987, and Burdick 1992). Classifications and conditions for singularity have also been developed for parallel manipulators (Agrawal 1990, Kumar 1992, Merlet 1989). Gosselin and Angeles (1990), however, were the first to address the singularity analysis of general closed-loop chains.

A common feature to most of these studies is the definition of singularity in terms of only the input and output velocities of the mechanism, (i.e., the passive-joint velocities are not considered). In (Zlatanov et al. 1994–1,2), it was shown that approaches based solely on input-output equations may fail to detect certain singularities in the general closed-loop case. In the first of these papers, singularity of non-redundant mechanisms was defined by means of a velocity equation including all the joint velocities of the mechanism. A comprehensive classification, based on six singularity types was introduced. This classification, reviewed below in Section 2, is valid for arbitrary kinematic chains.

In the present paper, the problem of singularity identification is addressed. The objective is to find all existing singularities for a given mechanism and determine their classification. Conditions for the occurrence of singularity are derived and used as tools in two methods for singularity identification and classification.

J.-P. Merlet and B. Ravani (eds.), Computational Kinematics, 163–172.
© 1995 *Kluwer Academic Publishers.*

2. Preliminaries

2.1. DEFINITION OF SINGULARITY VIA THE VELOCITY EQUATION

In this paper, the singular configurations of an *arbitrary kinematic chain* are studied. For simplicity, it is assumed that all the N kinematic pairs have 1 dof. The *full-cycle mobility* of the mechanism (Hunt 1978) is denoted by n. It is assumed that only n of the N joints (the input joints) are *active*, i.e., their joint parameters can be actively changed. The remaining $N-n$ joints are *passive*. We restrict our attention to *non-redundant* input-output devices, i.e., it is assumed that the number of input parameters, as well as the dimension of the output space are equal to the general mobility, n. By default, the n active joint velocities will be referred to as *input*, and the n differential output parameters (the output velocities), specifying the instantaneous motion of the output link, as *output*.

Let us denote by $m = \left[T^T, \Omega^{aT}, \Omega^{pT} \right]^T$ the vector of the velocity parameters of the mechanism. T, Ω^a and Ω^p are the arrays of the output, input and passive-joint velocities, respectively. For any configuration, q, there exist an $N \times (N+n)$ matrix, $L(q)$, such that m is a feasible instantaneous motion of the mechanism if and only if:

$$L(q)m = 0. \tag{1}$$

Herein, we assume that $L(q)$ is a known continuous function of q. To obtain $L(q)$, one would first write the linear equations of loop closure for the joint twists of a system of independent loops (Davies 1981). We shall assume that this process yields $(N-n)$ equations. These equations together with the n output velocity equations form the system (1), the matrix $L(q)$ being a known continuous function of q. If more than $(N-n)$ equations are obtained, as can be the case for over-constrained mechanisms, a matrix with a greater number of rows may be replaced in (1) and used instead of L in the methods which are described in the following sections.

A configuration, q, is defined as non-singular when Eq.(1) can be solved both in terms of Ω^a and T, for that q.

2.2. CLASSIFICATION OF SINGULARITIES OF A GENERAL MECHANISM

According to (Zlatanov et al., 1994–1,2) there exist six singularity types and each singular configuration belongs to at least two types. The six singularity types are: (i)–(ii) *redundant input/output* (RI/RO respectively), which occur when a non-zero input/output is possible with zero output/input; (iii)–(iv) *impossible input/output* (II/IO respectively), which occurs when a certain input/output is not feasible for any output/input; (v) *redundant passive motion* (RPM), which occurs when a non-zero instantaneous motion is possible with both the input and output being equal to zero, and (vi) *increased instantaneous mobility* (IIM), which occur when the transitory or instantaneous mobility is higher than the full-cycle mobility of the kinematic chain.

As shown in (Zlatanov et al. 1994–1,2), the loss of output dof (IO type) is not always accompanied by the acquisition of extra input dof (RI type), and, dually, the RO type (extra output dof) is not equivalent to the II type (loss of input dof). The interdependence of the singularity types is given by Table 1.

Each cell of the table denotes a combination of certain singularity types. Only the cells marked by "Y" correspond to possible combinations of singularity types. Thus, the singularity set of any mechanism can be divided into up to 21 *classes*. Two singularities belong to the same class when they belong to exactly the same singularity types.

TABLE 1. Possible combinations of singularity types for non-redundant mechanisms.

	IO	II	IIM	IO and II	IO and IIM	II and IIM	IO and II and IIM
RI	Y						
RO		Y					
RPM			Y	Y			Y
RI and RO				Y	Y	Y	Y
RI and RPM				Y	Y		Y
RO and RPM				Y		Y	Y
RI and RO and RPM			Y	Y	Y	Y	Y

The purpose of an exhaustive singularity analysis process is to obtain not only the singularity set as a whole, but also its partition into classes. Knowing the class of a singularity is of significant practical importance, since this determines how the instantaneous-kinematics properties of the mechanism degenerate at this singularity.

2.3. FEASIBLE CONFIGURATIONS

A mechanism configuration, q, is an N-tuple of values of all joint parameters. For closed-loop mechanisms not all such N-tuples correspond to feasible configurations. The configuration space is given by the solution set of a system of nonlinear equations, $l(q) = 0$, referred to as the loop equations of the kinematic chain. Only $(N–n)$ of the loop equations are independent, but the dimension of $l(q)$ may be greater for some mechanisms.

When attempting to find the singularities of a given mechanism it must be assured that the values obtained for q are compatible with the loop equations. If only parts of the configuration space need to be considered, additional inequality constraints on the joint parameters are imposed. The set determined by the joint constraints will be denoted by Q. Thus, the set of feasible configurations is $\{q \in Q \mid l(q) = 0\}$.

3. Conditions for Singularity

The singularity of a given configuration, q, can be determined by examining the matrix $L(q)$ of the velocity equation. Let L_I, L_O and L_p be the submatrices of L obtained by removing the columns corresponding to the input, output, and both the input *and* output, respectively. The following general singularity condition then holds:

Proposition. For any mechanism, a configuration, q, is non-singular if and only if both the matrices L_I and L_O are non-singular at q.

This statement is implied by the definition of singularity and the fact that the velocity equation (1) is a sufficient condition for the feasibility of an instantaneous motion, m.

The criteria for the separate singularity types are given by the following proposition:

Proposition.

(i)	$q \in \{RI\} \Leftrightarrow \operatorname{rank} L_O < \operatorname{rank} L_p + n$	
(ii)	$q \in \{RO\} \Leftrightarrow \operatorname{rank} L_I < \operatorname{rank} L_p + n,$	
(iii)	$q \in \{RPM\} \Leftrightarrow \operatorname{rank} L_p < N - n,$	
(iv)	$q \in \{II\} \Leftrightarrow \operatorname{rank} L_I < \operatorname{rank} L,$	
(v)	$q \in \{IO\} \Leftrightarrow \operatorname{rank} L_O < \operatorname{rank} L,$	
(vi)	$q \in \{IIM\} \Leftrightarrow \operatorname{rank} L < N,$	
(vii)	$q \in \{RO\}$ or $q \in \{RPM\} \Leftrightarrow q \in \{II\}$ or $q \in \{IIM\} \Leftrightarrow L_O$ is singular ,	
(viii)	$q \in \{RI\}$ or $q \in \{RPM\} \Leftrightarrow q \in \{IO\}$ or $q \in \{IIM\} \Leftrightarrow L_I$ is singular.	

The above eight conditions can be derived from the velocity-equation definitions of the singularity types given in (Zlatanov et al. 1994–1).

4. Singularity Identification and Classification

When a feasible configuration, q, is given, the rank of the matrices L_I, L_O, L_p and L can be computed to check for singularity, and the type of singularity be determined by simply reviewing the conditions (i) to (viii) listed in Section 3.

However, when q is unknown, to obtain the singularities of the mechanism, the conditions must be interpreted as systems of equations for q, and the singularity set and its subsets be obtained as solutions of these equations. This process is described below.

4.1. DETERMINATION OF THE SINGULARITY SET

For the goals of singularity identification of closed-loop mechanisms, the matrices L_I and L_O play a role analogous to the one of the Jacobian in the case of a serial chain. The singularities of a non-redundant mechanism with known kinematic chain, link parameters and joint constraints, can be determined by solving the following two systems of non-linear equations:

$$\det L_I(q) = 0,$$
$$l(q) = 0, \tag{2}$$

and

$$\det L_O(q) = 0,$$

$$l(q) = 0,$$

(3)

subject to the joint constraints Q.

Therefore, the problem of singularity identification can be solved by (a) deriving the loop equations, (b) writing the velocity equation, and then (c) solving (2) and (3) and taking the union of the two solutions.

For a non-redundant mechanism each of the two subsets of the singularity set, obtained in Systems (2) and (3), is the solution of a system of $(N - n + 1)$ equations. Therefore, the singularity set will be typically of dimension $(n - 1)$ or equivalently of co-dimension in the n-dimensional configuration space of the mechanism. Thus, mechanisms with mobility of 1 usually have a finite number of isolated singularities, while for higher values of n the singularity set will be continuous with ∞^{n-1} points.

Though the solution of Systems (2) and (3) identifies all the singularities of a mechanism, however, it does not classify them. In general, by using only the matrices L_I and L_O, it is not possible to classify all the singularities of a mechanism. This can be done, however, for some mechanisms, and for a large part of the singularities of other mechanisms. The statements (vii) and (viii) imply that if for a given configuration L_I is singular but L_O is non-singular, the configuration is a singularity of class (RI & IO). Conversely, when a configuration satisfies condition (viii) but not (vii) it must be of the (RO & II) class. It is only when both L_I and L_O are singular that conditions other than (vii) and (viii) need to be considered. Singularities that satisfy both (vii) and (viii) may have substantially different kinematic features, e.g., they may lead to either a loss or a gain in output/input dof. In fact, a configuration where both L_I and L_O are singular could belong or not belong to any of the six types.

4.2. DETERMINATION OF THE SINGULARITY TYPES

If it were known that there are no singularities of the IIM or RPM types, the identification and classification process could be completed by examining only conditions (vii) and (viii). The main strategy of the method described below is to first identify and classify the IIM and RPM singularities and then analyze the remaining configurations using the determinants of L_I and L_O.

As in Section 4.1, it is understood that the singularity equations are solved subject to the joint constraints and the loop equations. To simplify the presentation these operations are not explicitly included in the description of the algorithm. Below, $\{k\}$ stands for "all configurations obtained in Step k of the algorithm."

(1)   Find all feasible q satisfying condition (vi).

(2)   Find all feasible q satisfying condition (iii).

(3)   Classify $\{1\} \cup \{2\}$:

       (3.1)   For $\{1\}$, check (iv) and (v). Obtain 4 sets:
                IIM; IIM&II; IIM&IO; IIM&II&IO.

       (3.2)   For $\{2\}$, check (i) and (ii). Obtain 4 sets:
                RPM; RPM&RI; IIM&RO; RPM&RI&RO.

       (3.3)   Find all the intersections of each set in $\{3.1\}$ and each set in $\{3.2\}$. Obtain **10 classes**. (These are the 10 classes that belong to the

IIM *and* RPM types, see Table 1)

(3.4)   Subtract {2} from each set in {3.1}. Obtain **4 classes.**
(The 4 classes of IIM, but *not* RPM singularities, see Table 1).

(3.5)   Subtract {1} from each set in {3.2}. Obtain **4 classes.**
(The 4 classes of RPM, but *not* IIM singularities, see Table 1).

(4)   Find all q satisfying condition (vii). From these subtract $\{1\} \cup \{2\}$.
(5)   Find all q satisfying condition (viii). From these subtract $\{1\} \cup \{2\}$.
(6)   Intersect {4} and {5}. Obtain **3 classes.**
(Singularities that are neither IIM nor RPM).

Thus, the singularities that belong to each of the 21 classes in Table 1 are identified.

The operations in Steps (1) and (2) require the identification of the points x for which some rectangular matrix $M(x)$ is singular. This can be done by finding all x for which all sub-matrices of maximum dimension have zero determinants, i.e., by solving a system of simultaneous non-linear equations. In Steps (3.1) and (3.2) it is required to find sets of the type $R = \{x \mid \text{rank } A(x) < \text{rank } B(x)\}$. This can be done by presenting R as the union of the sets $R_i = \{x \mid \text{rank } A(x) < i \leq \text{rank } B(x)\}$. The sets R_i can be obtained by solving systems of equations.

It can be noted that since the condition for RPM (or IIM) singularity requires the rank-deficiency of a rectangular matrix, a larger number of equations must be satisfied and the dimension of the solution set will be typically lower than the dimension of the singularity set as a whole. In practice, IIM singularities occur only for mechanisms with specially proportioned link parameters. RPM singularities, when they exist, form sets of low dimensions. The algorithm is organized in such a way that the conditions for RI, RO, II and IO, which may involve the examination of multiple cases, are solved only together with the conditions for IIM (RPM), i.e., for a comparatively small subset of singularities.

4.3 EXAMPLE

Consider the mechanism shown in Figure 1 ($N = 8$, $n = 2$). The inputs are the joint velocities at A and E, the output is the motion of point G. The link dimensions are $AB = AD = BC = DE = 1$, $CD = FG = 2$, $CG = 1.5$, $EF = 3$. The 8×10 L matrix is:

$$L = \begin{bmatrix} I_2 & 0 & \mathbf{m}_{EG} & 0 & 0 & 0 & 0 & 0 & \mathbf{m}_{FG} \\ 0 & S_A & 0 & S_B & S_C & S_D & 0 & 0 & 0 \\ 0 & 0 & S_E & 0 & 0 & S_D & S_C & S_G & S_F \end{bmatrix} \tag{4}$$

where S_P, $P = A, B,..., G$, are 3-dimensional planar screws, $S_P = (1, y_P, -x_P)^T$, $\mathbf{m}_{PG} = (y_P - y_G, x_G - x_P)^T$ and I_2 is the 2×2 unit matrix. To find all singularities and establish their types, the procedure described in Subsection 4.2 is followed:

(1) Check for IIM singularities. For the given mechanism, it is established that condition (vi) has no solution compatible with the given link lengths.

(2) Check for RPM singularities. The condition (iii) is satisfied only when the determinants of both $[S_B S_C S_D]$ and $[S_C S_G S_D]$ vanish. This gives 8 distinct singular configurations (one of them is shown in Figure 2).

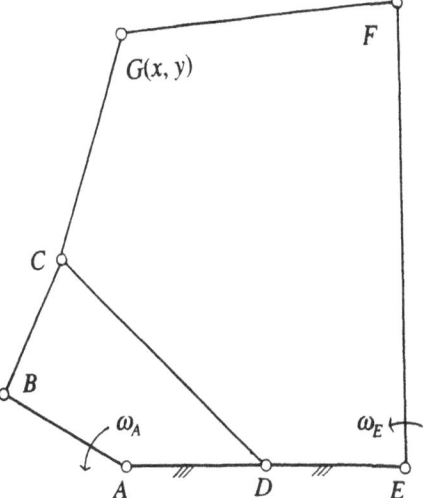

Figure 1. A 2-dof planar linkage.

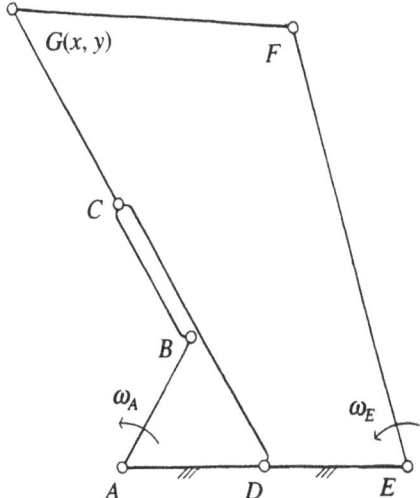

Figure 2. A singular configuration of class (RPM & IO *and* II).

(3) (3.2) For each of 8 the configurations in {2}, conditions (i) and (ii) are checked and it is found that neither is satisfied.

(3.5) It is concluded that the the (RPM & IO *and* II) class consists of the 8 elements of {2}.

(4) The condition (vii) is applied. (vii) is equivalent to the singularity of at least one of the matrices $[\mathbf{S}_B\mathbf{S}_C\mathbf{S}_D]$ or $[\mathbf{S}_C\mathbf{S}_G\mathbf{S}_F]$. The solution of each of these equations (combined with the loop equations) is a 1-dimensional submanifold of the 2-dimensional configuration space. The first manifold has 4 connected components, and the second one has 3 components. All elements of the union of these manifolds,

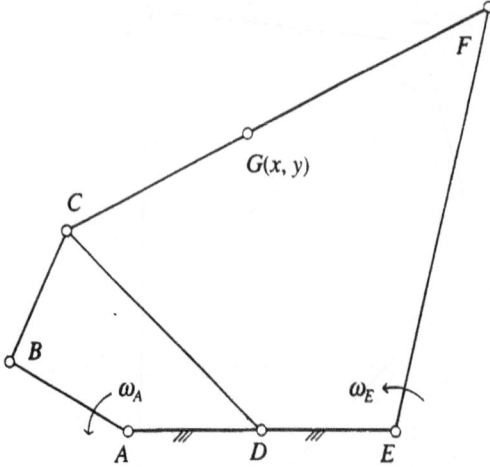

Figure 3. A singular configuration of class (RO & II).

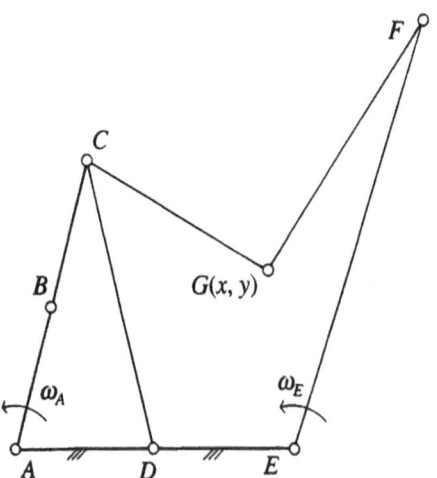

Figure 4. A singular configuration of class (RI & IO).

except the 8 elements of {2} found in Step 2, are of the (RO & II) class. One such singularity is shown in Figure 3. The corresponding connected component is obtained by moving the linkage while keeping the joint angle at G constant.

(5) The condition (viii) is applied. (viii) is equivalent to the singularity of at least one of the matrices $[S_A S_B S_C]$, $[S_G S_C S_D]$ or $[S_E S_G S_F]$. The solution for each of these equations (combined with the loop equations) is a 1-dimensional submanifold of the 2-dimensional configuration space. The first and third manifolds have each 2 connected components, while the second one has 4. All elements of the union of these manifolds, *except* the 8 elements of {2}, belong to the (RI & IO) class. Figure 4 provides an example. The connected component corresponding to the shown configuration is obtained by moving the linkage while keeping the points B and C fixed.

(6)   The intersection of the sets obtained in Steps 4 and 5 consists of 16 configurations Apart from the 8 configurations classified in Step 3.5 as (RPM & II *and* IO)-singularities, the others are (RI *and* RO & IO *and* II)-class singularities. The remaining configurations obtained in Step 4 /5 belong to the class (RO & II)/(RI & IO).

Thus, four different classes of singularities are obtained for the given mechanism: 8 (RPM & II *and* IO) singularities, Step (3.5); 8 (RI *and* RO & IO *and* II) singularities, Step (6); ∞^1 (RO & II) configurations, Steps (4) and (6); and, ∞^1 (RI & IO) configurations, Steps (5) and (6).

5.   Automatic   Singularity   Analysis

For all but the simplest mechanisms the singularity set contains infinitely many configurations and therefore to locate the singularities implies the task of obtaining a good description of a multi-dimensional subspace of the mechanism's configuration space. This could be done by either obtaining simplified symbolic equations for the singularity set or by providing an algorithm able to trace numerically and represent graphically the projections and cross-sections of this set.

The procedures in Section 4 describe an algorithm for the automatic identification of the singularity set, however significant kinematic and computational problems remain to be solved before a "black box" can emerge for singularity analysis. Some of these issues are briefly outlined below.

The first step in an algorithm for singularity analysis must be the automatic generation of the loop equations. It is desirable to make use of symbolic methods designed to take advantage of possible closed form solutions (Kecskeméthy 1993). On the other hand, since an algebraic solution cannot be guaranteed, a representation that is suitable for numerical iterative solution should be preferred. In particular, the position parameters should be chosen in such a way that the resulting equations are polynomial.

The next step is the (automatic) formulation of the singularity conditions. As it was shown in Sections 3 and 4 these conditions involve the rank-deficiency of some (polynomial) matrix function of q. According to Davenport et al. (1993) for such matrices (functions of multiple variables) the Cramer rule is a more efficient way for symbolic computation of the determinants than any process of Gaussian elimination (transforming the matrix into a triangular form). However, if the kinematic nature of $L(q)$ is taken into account, the matrix could be simplified and the computation of singularity conditions for the submatrices be made easier. The strategies for passive-joint screw elimination by reciprocal screws developed for parallel and hybrid chains could be helpful (Zanganeh and Angeles 1994, Zlatanov et al., 1994–3).

Finally, once the systems of algebraic equations have been generated, the goal would be to extract maximum information about their solution sets. These sets (algebraic varieties) are subsets of the singularity set. This investigation may involve symbolic simplification of the equations or their numerical solution. (On the other hand, some interesting properties of the solution set may be deduced without solving the equations by applying algebraic-geometry tools (Merlet, 1993)). Ideally, one would like to obtain a stratification of the singularity set, which would decompose the set into non-intersecting manifolds consisting of singularities of the same class.

172

6. Conclusions

This paper presents a solution technique to the singularity-identification problem. The singular configurations of an arbitrary non-redundant mechanism with a given geometry, all its can be determined and classified by solving the singularity criteria proposed in the paper. The conditions for singularity are derived using a velocity-equation formulation of singularity. Two methods of singularity analysis are presented. The first one reduces the identification of all singular configurations to the solution of two systems of nonlinear equations. The second method, a more complex procedure in nature, is proposed for simultaneous singularity identification and classification.

7. References

Agrawal, S.K (1990) Rate kinematics of in-parallel manipulator systems, *IEEE, Int. Conf. on Robotics and Automation,* 104–109.

Burdick, J.W. (1992) A recursive method for finding revolute-jointed manipulator singularities, *IEEE, Int. Conf. on Robotics and Automation,* 448–453.

Davies, T.H. (1981) Kirchhoff circulation law applied to multi-loop kinematic chains, *Mech. and Mach. Theory,* **16**, 171–183.

Davenport, J.H., Siret, Y. and Tournier, E., (1993) *Computer Algebra,* Academic Press, London.

Gosselin, C, and Angeles, J.(1990) Singularity analysis of closed-loop kinematic chains, *IEEE, Trans. on Robotics and Automation* **6**, 281–290.

Hunt, K.H. (1986) Special configurations of robot-arms via screw theory. Part 1, *Robotica* **4**, 171–179.

Hunt, K.H. (1978) *Kinematic geometry of mechanisms,* Clarendon press.

Kumar, V. (1992) Instantaneous kinematics of parallel-chain robotic mechanisms, *J. of Mechanical Design* **114**, 349–358.

Kecskeméthy (1993) On closed form solutions of multiple-loop mechanisms, in J. Angeles, G. Hommel and P Kovács (eds.), *Computational Kinematics,* Kluwer Academic Publishers, Dordrecht, pp.263–274

Merlet, J-P. (1989) Singular configurations of parallel manipulators and Grassman geometry, *Int. J. of Robotics Res.* **8**, 45-56.

Merlet, J-P. (1993) Algebraic geometry tools for the study of kinematics of parallel manipulators, in J. Angeles, G. Hommel and P Kovács (eds.), *Computational Kinematics,* Kluwer Academic Publishers, Dordrecht, pp.183–194

Wang, S.L. and Waldron, K.J. (1987) A study of the singular configurations of serial manipulators, *Trans. of the ASME, J. of Mech., Trans., and Aut. in Design,* **109**, 14–20.

Zanganeh, K.E. and Angeles, J., (1994) Instantaneous kinematics of modular parallel manipulators, *ASME, 23rd Mechanisms Conf.,* **DE–72**, 271–277.

Zlatanov, D., Fenton, R.G. and Benhabib, B. (1994–1) Singularity analysis of mechanisms and manipulators via a velocity-equation model of the instantaneous kinematics, *IEEE, Int. Conf. on Robotics and Automation,* **2**, 986–991.

Zlatanov, D., Fenton, R.G. and Benhabib, B. (1994–2), Singularity analysis of mechanisms and manipulators via a motion-space model of the instantaneous kinematics, *IEEE, Int. Conf. on Robotics and Automation,* **2**, 980–985.

Zlatanov, D., Fenton, R.G. and Benhabib, B. (1994–3), Analysis of the instantaneous kinematics and singular configurations of hybrid-chain manipulators, *ASME, 23rd Mechanisms Conf.,* **DE–72**, 467-476.

Acknowledgements

The financial support of the Natural Sciences and Engineering Research Council of Canada is gratefully acknowledged.

ANALYTICAL CONSTRAINTS FOR A WORKSPACE DESIGN
OF 2R MANIPULATORS

Marco CECCARELLI and Gianbattista SCARAMUZZA

Dipartimento di Ingegneria Industriale
Università di Cassino
Via Zamosch 43, 03043 CASSINO (FR) , Italy

Abstract - In this paper an algebraic formulation is used to deduce a closed-form design algorithm for 2R manipulators. The general expression of a torus has been inverted for synthesis purposes and the algebraic nature has been preserved to deduce both solving formulas and design constraints. These constraints have been used to draw a chart of feasible region for a prescribed workspace through some given points. Once structural parameters are found, dimensional sizes of the 2R chain have been solved by inverting the expressions of definition for the structural parameters. A third order algebraic equation has been obtained and the solutions discussed, so that it turns out that a torus can be generated by two or four different 2R manipulators only. Numerical examples illustrate a design procedure and some peculiar charts of feasible regions.

1. Introduction

Designing 2R manipulators for a prescribed workspace can be considered of great relevance since workspace characteristics are of primary importance for manipulating purposes, and 2R chains are widely used in industrial robots.

The geometry of a torus workspace of a 2R chain has been described and geometrically analyzed by Fichter E. F. and Hunt K. H., [1], with the aim of deducing theorems useful for both analysis and synthesis procedures. They have also demonstrated by means of geometric arguments the existence of four cognate 2R dyads which trace the same toroidal surface. The design problem has been discussed in detail by Roth B., [2], who developed a design algorithm based on the analytical expression of a torus written in terms of an arbitrary coordinate system so that the position and the orientation of the robot base have been included into the design parameters. Although the expression of a torus has been used, the design algorithm is numerical in nature. In this paper we propose an analytical design for 2R manipulators by preserving the algebraic nature of the torus expression. This allow us

J.-P. Merlet and B. Ravani (eds.), Computational Kinematics, 173–182.

to deduce analytically design constraints for workspace prescribed points. Moreover, the synthesis problem has been solved by an analytical inversion of the design equations, giving the possibility to have at once all the feasible solutions.

2. Formulation of a Design Problem

From the geometry of the 2R manipulator chain, Fig. 1, the parametric equations for the locus of a reference point H can be expressed with respect to a fixed reference frame XYZ as its Cartesian coordinates

$$x = a_2 \cos\theta_1 \cos\theta_2 - a_2 \cos\alpha_1 \sin\theta_1 \sin\theta_2 - d_2 \sin\alpha_1 \sin\theta_1 + a_1 \cos\theta_1$$

$$y = a_2 \sin\theta_1 \cos\theta_2 + a_2 \cos\alpha_1 \cos\theta_1 \sin\theta_2 + d_2 \sin\alpha_1 \cos\theta_1 + a_1 \sin\theta_1 \quad (1)$$

$$z = - a_2 \sin\alpha_1 \sin\theta_2 + d_2 \cos\alpha_1$$

where the kinematical parameters for a design procedure are:
a_1, the link length giving the common normal distance between joint axes Z_1 and Z_2;
a_2, the distance of a reference point H from Z_2;
α_1, the twist angle between Z_1 and Z_2, measured positive clockwise about X_1;
d_2, the link offset giving the distance between the joint centers along Z_2.
The angle θ_1 and θ_2 are the revolution angles of the joints about the joint axes measured between X axis and a_1 direction, a_1 and a_2 directions, respectively. $d_1 = 0$ is assumed.

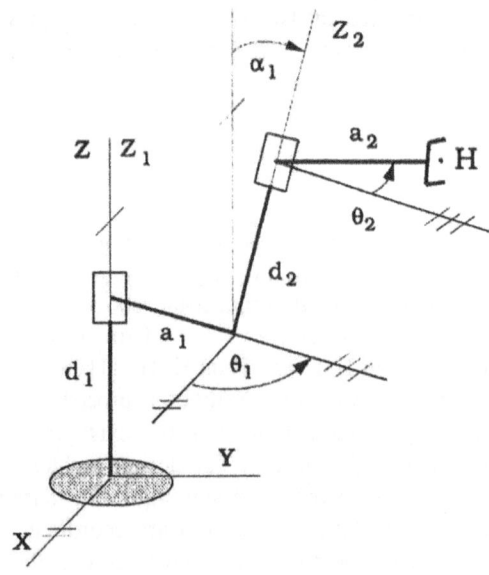

Fig. 1. A 2R manipulator and its kinematical parameters.

The angles θ_1 and θ_2 can be eliminated from Eqs. (1) by solving the last equation for $\sin\theta_2$ and substituting this into the sum of the squares of the first two equations. Then, a rationalization of the resulting equation with the hypotheses $a_1 \neq 0$, $a_2 \neq 0$, $\alpha_1 \neq 0$, gives the equation of a general torus in the form

$$\left(r^2 + z^2 - A\right)^2 + (Cz + D)^2 + B = 0 \tag{2}$$

when the structural parameters A, B, C, D are defined as

$$A = a_1^2 + a_2^2 + d_2^2$$

$$B = -4a_1^2 a_2^2 \tag{3}$$

$$C = \frac{2a_1}{\operatorname{sen}\alpha_1}$$

$$D = -2a_1 d_2 \frac{\cos\alpha_1}{\sin\alpha_1}$$

A design problem for a prescribed workspace can be formulated by properly inverting Eqs. (2) and (3).

A workspace can be given by prescribing a certain number of points through their Cartesian coordinates. For a toroidal workspace surface there can be at most four points to equal the number of the unknowns, [2], when the manipulator base is given with respect to a fixed frame.

Thus, we may write four equations of the form of Eq. (2), i.e.

$$\left(r_i^2 + z_i^2 - A\right)^2 + (Cz_i + D)^2 + B = 0 \qquad i = 1,...,4 \tag{4}$$

as a function of the unknown structural parameters A, B, C and D. Of course, the given four points must have different coordinates, that is

$$(r_i, z_i) \neq (r_j, z_j) \qquad\qquad i, j = 1,...,4, \; i \neq j \tag{5}$$

to write independent design equations (4). By inverting Eqs. (4) it is possible to obtain

$$B = -(\rho_1 - A)^2 - (Cz_1 + D)^2 \tag{6}$$

and

$$-2A(\rho_i - \rho_1) + C^2\left(z_i^2 - z_1^2\right) + 2CD(z_i - z_1) = -\left(\rho_i^2 - \rho_1^2\right) \qquad i = 2,...,4 \tag{7}$$

where $\rho_i = r_i^2 + z_i^2$ is a prescribed robot reach at point i.

The design problem requires the solution of the system which is composed by Eqs. (7) in the form

$$U x = w \tag{8}$$

in which the unknowns A, C^2, and CD can be expressed in a vector $x=(A, C^2, CD)^t$, (t is the transpose operator); U is the matrix of the coefficients and w is a vector of the known terms given as

$$U = \begin{bmatrix} -2(\rho_2 - \rho_1) & \left(z_2^2 - z_1^2\right) & 2(z_2 - z_1) \\ -2(\rho_3 - \rho_1) & \left(z_3^2 - z_1^2\right) & 2(z_3 - z_1) \\ -2(\rho_4 - \rho_1) & \left(z_4^2 - z_1^2\right) & 2(z_4 - z_1) \end{bmatrix} \tag{9}$$

$$w = \begin{bmatrix} -\left(\rho_2^2 - \rho_1^2\right) \\ -\left(\rho_3^2 - \rho_1^2\right) \\ -\left(\rho_4^2 - \rho_1^2\right) \end{bmatrix} \tag{10}$$

Provided that the condition

$$\det U \neq 0 \tag{11}$$

holds a solution of the equations system (8) can be obtained.

In particular, a design problem can be solved algebraically, and after some algebra the structural parameters can be expressed in the form

$$A = \frac{\rho_1^2 k_1 - z_1^2 k_2 + z_1 k_3 - k_4}{2K_U} \tag{12}$$

$$C^2 = \frac{\rho_1^2 k_5 - \rho_1 k_2 + z_1 k_6 - k_7}{K_U} \tag{13}$$

$$CD = -\frac{z_1^2 k_6 + \rho_1^2 k_8 - \rho_1 k_3 - k_9}{2K_U} \tag{14}$$

where

$$k_1 = z_3 z_4 \Delta z_{43} - z_2 z_4 \Delta z_{42} + z_2 z_3 \Delta z_{32} \quad, \quad k_2 = \rho_2^2 \Delta z_{43} - \rho_3^2 \Delta z_{42} + \rho_4^2 \Delta z_{32} \quad,$$

$$k_3 = \rho_2^2 \Delta z_{43}^2 - \rho_3^2 \Delta z_{42}^2 + \rho_4^2 \Delta z_{32}^2 \quad, \quad k_4 = \rho_2^2 z_3 z_4 \Delta z_{43} - \rho_3^2 z_2 z_4 \Delta z_{42} + \rho_4^2 z_2 z_3 \Delta z_{32}$$

$$k_5 = \rho_2 \Delta z_{43} - \rho_3 \Delta z_{42} + \rho_4 \Delta z_{32} \quad, \quad k_6 = \rho_3 \rho_4 \Delta \rho_{43} - \rho_2 \rho_4 \Delta \rho_{42} + \rho_2 \rho_3 \Delta \rho_{32}$$

$$k_7 = z_2 \rho_3 \rho_4 \Delta \rho_{43} - z_3 \rho_2 \rho_4 \Delta \rho_{42} + z_4 \rho_2 \rho_3 \Delta \rho_{32} \quad, \quad k_8 = \rho_2 \Delta z_{43}^2 - \rho_3 \Delta z_{42}^2 + \rho_4 \Delta z_{32}^2$$

$$k_9 = z_2^2 \rho_3 \rho_4 \Delta \rho_{43} - z_3^2 \rho_2 \rho_4 \Delta \rho_{42} + z_4^2 \rho_2 \rho_3 \Delta \rho_{32} \tag{15}$$

In addition, the analytical expression of Eq. (9) is useful to give the det U in an algebraic form as

$$K_U = r_1^2 c_1 + z_1^2 c_2 + z_1 c_3 + c_4 \tag{16}$$

where

$$c_1 = z_3 z_4 \Delta z_{43} + z_2 z_4 \Delta z_{24} + z_2 z_3 \Delta z_{32} \quad , \quad c_2 = -r_2^2 \Delta z_{43} - r_3^2 \Delta z_{24} - r_4^2 \Delta z_{32} \tag{17}$$

$$c_3 = r_2^2 \Delta z_{43}^2 + r_3^2 \Delta z_{24}^2 + r_4^2 \Delta z_{32}^2 \quad , \quad c_4 = -r_2^2 z_3 z_4 \Delta z_{43} - r_3^2 z_2 z_4 \Delta z_{24} - r_4^2 z_2 z_3 \Delta z_{32}$$

Both in Eqs. (15) and (17) the coefficients are expressed as a function of the given coordinates which have been grouped in the form

$$\Delta \rho_{ij} = \rho_i - \rho_j \quad , \quad \Delta z_{ij} = z_i - z_j \quad , \quad \Delta z_{ij}^2 = z_i^2 - z_j^2 \quad . \quad i, j = 2, \dots, 4 \tag{18}$$

Eqs. (6) and (12) to (18) give the algebraic solution for a design problem for 2R manipulators with prescribed workspace through four given points.

3. Workspace Design Constraints

A fundamental question for a successful design of 2R manipulator with prescribed workspace can be considered: may the given points be prescribed arbitrarily?
As we already point out, the conditions (5) need to be met to give independent and solvable design equations (4), and a given workspace can be described at the most through four points.
A further condition is expressed through Eq. (11) to check if the given points can be prescribed arbitrarily. In fact since the determinant of U can be expressed in the form of K_U in Eq. (16) it may vanish if and only if the c_i i=1,...,4 coefficients vanish simultaneously. This may occur when $z_2 = z_3 = z_4$ so that it is not possible to prescribe three or more points on a horizontal straight line. This constraint condition could be obtained also from a geometrical analysis of the general torus shape, as well as from the fourth order torus equation. However, because of the "banana shape" general form of a torus cross-section, three or four points on a vertical straight line are permitted.
Moreover, when three points P_2, P_3, P_4 are properly assigned, further workspace design analytical constraints for the fourth given point P_1 can be obtained by using the design formulas (12), (13), and (16). In fact for given P_2, P_3, P_4 the feasible design region may be found for P_1 by using the following considerations:

- the structural parameter A must be positive because of its definition in Eqs. (3) so that a feasible design region R_A can be characterized as

$$R_A(P_1) = (r_1, z_1) : A > 0 \tag{19}$$

The boundary is determined from the condition $A = 0$ by using Eqs. (12) and (15);

- the structural parameter C must be real, so that a feasible design region R_C can be characterized as

$$R_C(P_1) = (r_1, z_1) : C^2 > 0 \tag{20}$$

178

The boundary is determined from the condition $C = 0$ by using Eqs. (13) and (15);
- the determinant K_U cannot be zero so that a feasible design region R_K can be characterized as

$$R_K(P_1) = (r_1, z_1) : K_U \neq 0 \qquad (21)$$

The boundary is determined from $K_U = 0$ by using Eqs. (16) and (17).
Summarizing, once the three points P_2, P_3, P_4 are given, not on an horizontal straight-line, a fourth prescribed point P_1 must be selected inside a feasible region $R_W(P_1)$ simultaneously satisfying the above mentioned constraints. The feasible region can be expressed as an intersection between these in the form

$$R_W(P_1) = R_A \cap R_C \cap R_K \qquad (22)$$

Fig. 2 illustrates an example of a general topology of the feasible region $R_W(P_1)$ given by Eq. (21) for a workspace design of 2R manipulators with assigned base location, when the distances are expressed in unit length and the angles in degrees.

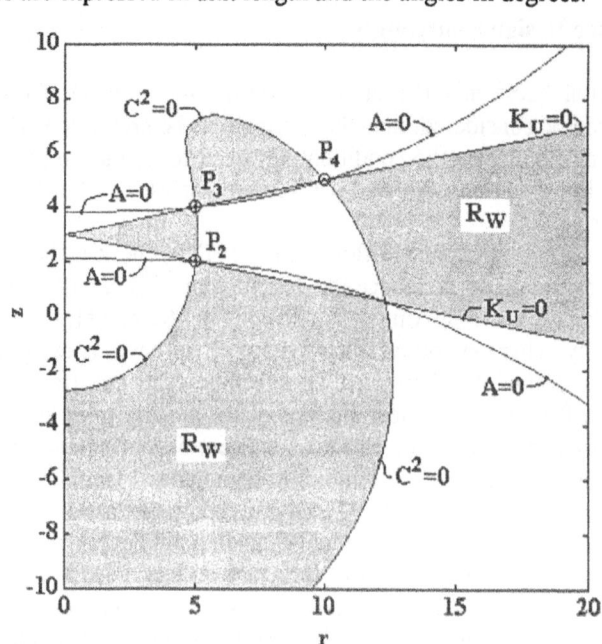

Fig. 2 Characteristic topology of feasible design region R_W for P_1.
This example is for $P_2=(5, 2)$, $P_3=(5, 4)$, and $P_4=(10, 5)$.

4. A Design Algorithm and Cognate Manipulators

A design algorithm for 2R manipulators with prescribed workspace through four suitable points can be developed in a closed algebraic form by using Eqs. (6), (12), (13), and (14) to find the structural parameters A, B, C, and D and successively by properly inverting the expressions of definition in Eqs. (3) to give the dimensional sizes of the chain a_1, a_2, d_2, α_1.

In particular, a set of meaningful values for the structural parameterscan be obtained only for a positive value of C^2. This means that only one torus may be traced crossing over the four given points P_1, P_2, P_3, P_4. Now the problem is to compute cognate manipulators having this torus as workspace. By inverting the definition Eqs. (3) of the structural parameters, the chain sizes can be given as

$$a_2^2 = - \frac{B}{4a_1^2}$$

$$d_2^2 = - \frac{D^2}{4a_1^2 - C^2}$$

$$\alpha = \text{atan2}\left(\frac{2a_1}{C} , - \frac{D}{d_2 C} \right)$$

(23)

when the a_1 length has been previously found and atan2 is the atan operator for oriented angles. By using the expressions (23) in the A formula of Eqs. (3) we obtain the following design equation

$$\left(a_1^2\right)^3 - \frac{4A + C^2}{4}\left(a_1^2\right)^2 + \frac{AC^2 - D^2 - B}{4}\left(a_1^2\right) + \frac{BC^2}{16} = 0 \qquad (24)$$

where the only unknown is a_1^2, since A, B, C, D are given from a previous design step by Eqs. (6), (12), (13), (14). This cubic equation can be reduced to standard form by using the substitution $a_1^2 = \xi + (4 A + C^2)/12$ to give

$$\xi^3 + p\xi + q = 0 \qquad (25)$$

where

$$p = - 3\left(\frac{4A + C^2}{12} \right)^2 + \frac{AC^2 - D^2 - B}{4}$$

(26)

$$q = - 2\left(\frac{4A + C^2}{12} \right)^3 + \frac{AC^2 - D^2 - B}{4} \cdot \frac{4A + C^2}{12} + \frac{BC^2}{16}$$

The solutions can be expressed as

$$\xi_1 = u_0 + v_0 , \quad \xi_2 = u_0 e^{i\frac{2\pi}{3}} + v_0 e^{-i\frac{2\pi}{3}} , \quad \xi_3 = u_0 e^{-i\frac{2\pi}{3}} + v_0 e^{i\frac{2\pi}{3}} \qquad (27)$$

where

$$u_0 = \sqrt[3]{- \frac{q}{2} + \sqrt{\Delta}} \qquad \qquad v_0 = - \frac{p}{3u_0} \qquad (28)$$

with the discriminant Δ

$$\Delta = \frac{p^3}{27} + \frac{q^2}{4} \qquad (29)$$

and i is the imaginary unit. Depending on the value of Δ, we may have one, two or three solutions for ξ. Nevertheless, since the coefficient $- B\ C^2 /16$ is positive and a positive value for a_1^2 is required because of its geometrical definition, only two real and distinct solutions at the most can be found. Consequently, because of the design Eqs. (23) one or two 2R chains can be obtained and two or four cognate manipulators may be synthesized, when the joint revolution direction is considered meaningful.

5. Numerical Examples

Some numerical examples are reported to illustrate the characteristic geometry of feasible region R_W, and its usefulness for manipulator design with prescribed workspace. The coordinates and the manipulator lengths are expressed in unit length and the angle in degrees.

Fig. 2 shows the general topology of the feasible region R_W, which is composed of four different regions, each of one is joined to the neighbour one at one point. Three of these points are always defined by the prescribed three points P_2, P_3, P_4 and the fourth cannot a prescribed one, although all designed tori will cross over it. This fact is illustrated in Figs 3 and 4. Each region has a typical shape although its size may vary considerably depending on the relative position of the given points P_2, P_3, P_4, as a comparison of Figs. 2, 3 and 4 shows. Particularly, the regions are strongly affected by the curves $C^2=0$ and $K_U=0$, which may or may not determine nozzle regions far away from the z axis, giving the possibility of a hole. Indeed, it has been experienced that, although it is impossible to have a feasible region R_W disconnected from the z axis because of the $C^2 = 0$ curve, nozzle shape regions may give hole existence since a feasible torus must be adapted to the geometry imposed by the permitted tangent direction at the abovementioned four critical points. Thus the toroidal surface is forced to have some tangent direction within a limited range indicated by R_W at the point where it crosses one of the four critical points, and therefore it may occur that although P_1 can be in R_W a torus cannot be designed: this occurs for P_1 positions very far from the other given P_2, P_3, P_4, as happens if in Fig. 2 P_1 has, for example, coordinates r=1, z=-10. Fig. 3 together with Table 1 stress the fact that for a given workspace through four points there may be only one torus, but several manipulators may trace it. Particularly, Table 1 lists the four different cognate manipulators for the designed workspace shown in Fig. 3.

We note once more that although there are differences in sign and twist angles, these cognate manipulators have only two 2R chains at the most, as it was pointed out in [1], but from a practical viewpoint there are four different manipulators because of the orientation of Z_2 axis, i.e. the revolute joints may rotate in either direction. Specifically, solutions n.1 and 2 of Table 1 are one 2R chain and n.3 and 4 the latter. Finally, Fig.4 shows how different solutions for the P_1 location may give different tori both in size and shape depending on the region of selection. Infact depending on whether P_1 is placed in the region $K_U>0$ or $C^2>0$ the torus may have respectively an egg-shaped or a banana shaped cross-section. Greatest size variation can be obtained when the P_1 position varies in the $K_U > 0$ region.

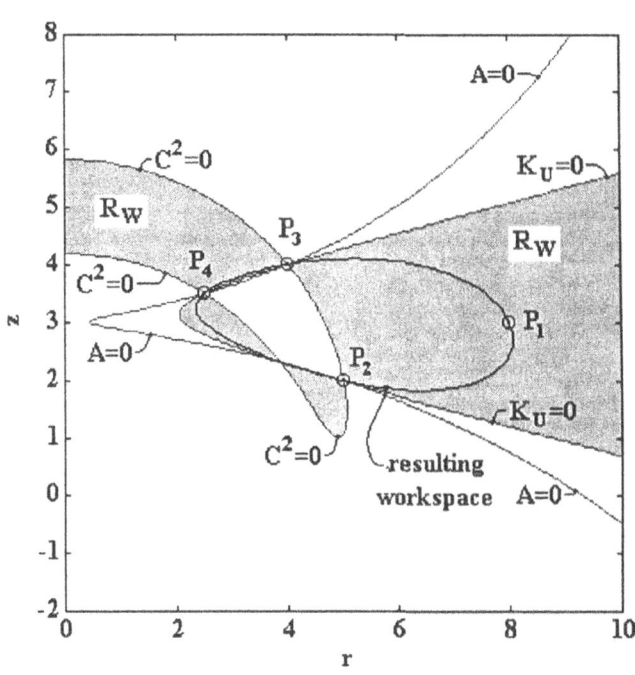

Fig. 3 A numerical example of feasible region R_W for P_1=(8, 3), P_2=(5, 2), P_3=(4, 4) and P_4= (2.5, 3.5). The reported toroidal workspace is given by the designed manipulators of Table 1.

Table 1. Design results of cognate manipulators for A= 44.0714, B= -837.4994, C^2= 650.4285, CD=-75.7123 with the workspace shown in Fig.3

Solution n.	a_1	a_2	d_2	α_1
1	5.04	2.87	3.23	23.27
2	5.04	2.87	-3.23	156.73
3	2.78	5.20	3.04	12 59
4	2.78	5.20	-3.04	167.41

6. Conclusion

A design problem for 2R manipulators with prescribed workspace has been formulated by means of an algebraic approach for a toroidal workspace. Thus a design algorithm has been proposed to find the structural parameters and the chain sizes of 2R manipulators sequentially. Analytical expressions have been deduced and they have

182

been used to formulate interesting workspace design constraints. In fact it has been found that, for the case of given robot base location, a workspace cannot be prescribed with four points arbitrarily located. Feasible design regions have been investigated and some numerical examples have been useded to draw and remark on these design constraints. Moreover, the algebraic approach has been also useful for discussing the multiplicity of solutions and to give an analytical justification of cognate 2R manipulators, previously determined through a descriptive geometric reasoning, [1], or numerically, [2].

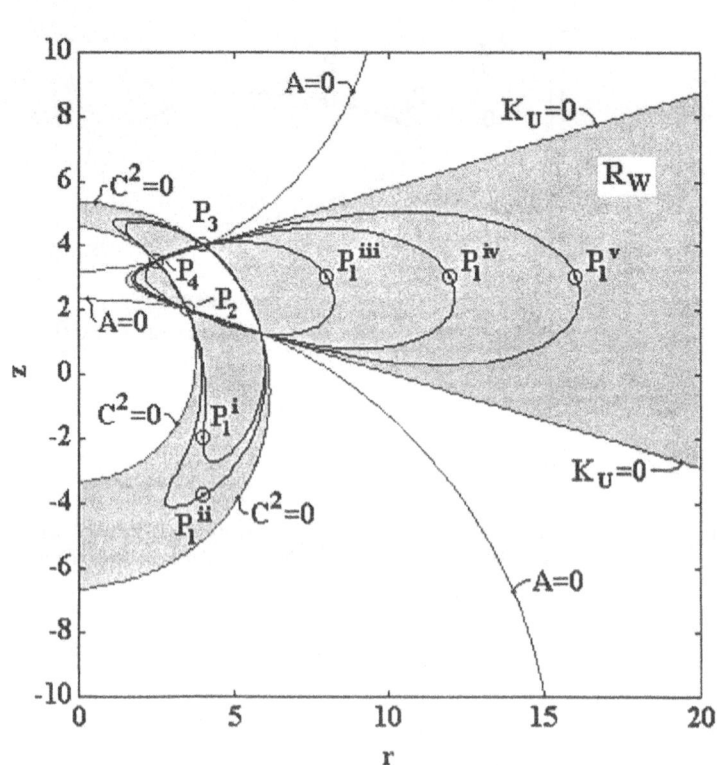

Fig. 4 Workspaces as a function of P_1 position with respect to a feasible design region for prescribed P_2, P_3 and P_4.

7. References

1. Fichter, E.F. and Hunt, K.H.: The Fecund Torus, its Bitangent-Circles and Derived Linkages, *Mechanism and Machine Theory*, 10 (1975), 167-176.
2. Roth B.: Analytical Design of Two-Revolute Open Chains, *Preprints of the Sixth CISM-IFToMM Symposium on Theory and Practice of Robots and Manipulators*, Cracow, 1986, 180-187.

Acknowledgements

A financial support of the Italian Ministry for the University and the Scientific and Technological Research through funds MURST ex 40% and 60% is gratefully acknowledged.

A HIERARCHICAL REPRESENTATION OF THE SPACE OCCUPANCY OF A ROBOT MECHANISM

R. FEATHERSTONE
Dept. of Engineering Science
Oxford University
Parks Road, Oxford, OX1 3PJ, England

Abstract.
   Given a kinematic model of a mechanism and geometric models of its individual parts, it is possible to define a space-occupancy function that calculates the set of points occupied by the mechanism as a function of its configuration. This paper describes a method of representing the graph of this function with a hierarchy of hyper-cylinders. The hierarchy is generated automatically from data structures called swept bubbles, and is suitable for use in collision detection, path planning and related problems.

1. Introduction

A mechanism consists of a collection of rigid bodies, called links, which are connected together by joints. If we characterize each link with a geometric model describing its shape and a kinematic model giving its location as a function of the joint variables, then it is possible to define a space-occupancy function for the mechanism: a function that calculates the set of points occupied by the mechanism at any given configuration. This function, or its graph (the space-occupancy graph), is useful in path planning, collision detection, and related problems.

This paper presents a method of approximating the space-occupancy graph (SOG) with a hierarchy of hyper-cylinders (high-dimensional versions of ordinary cylinders), such that each hyper-cylinder is a node in the hierarchy, each terminal node encloses a piece of the SOG, and each non-terminal node encloses the union of the pieces enclosed by its children. The idea is analogous to the use of sphere hierarchies to represent 3-D shape

183

J.-P. Merlet and B. Ravani (eds.), Computational Kinematics, 183–192.
© *1995 Kluwer Academic Publishers.*

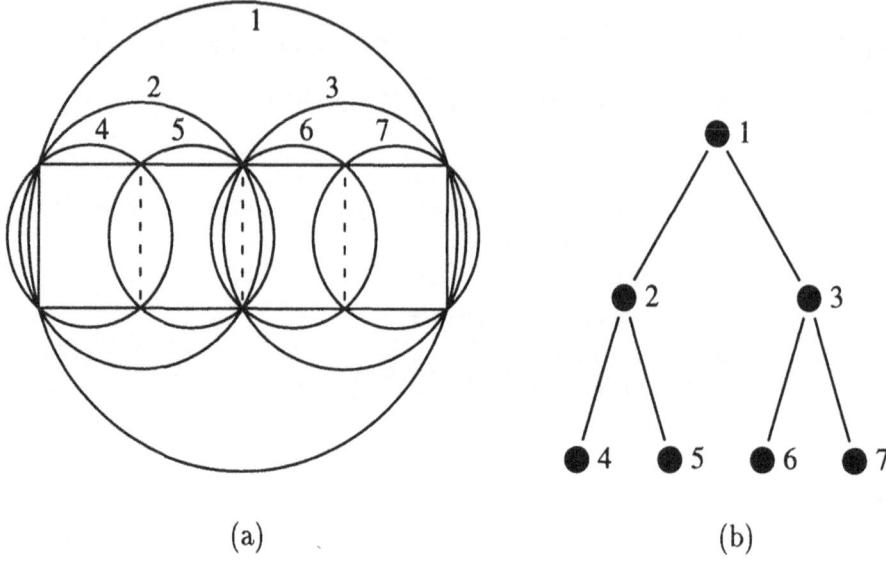

Figure 1. A sphere hierarchy constructed by recursive binary subdivision of a rectangle. Both the geometrical arrangement (a) and the hierarchical arrangement (b) are shown.

[2, 3, 4]; but there are two key differences, both of which are motivated by considering that the SOG is a complicated, high-dimensional object:

1. the hierarchy is generated in the absence of an explicit description of the SOG; and
2. any part of the hierarchy can be generated on demand (so it is not necessary to precompute the whole hierarchy).

The hierarchy is generated by an efficient algorithm based on the use of data structures called swept bubbles. This algorithm works intrinsically with hyper-cylinders, which is why they are used in preference to other possible shapes. Swept bubbles were originally described in [1, 5].

The remaining sections describe sphere hierarchies, the SOG, how to generate a hyper-cylinder hierarchy from swept bubbles, and finally some applications and experimental results.

2. Sphere Hierarchies

It is well known that the shape of a rigid body can be approximated to arbitrary accuracy as the union of a collection of spheres, and that a hierarchical data structure can be built around these spheres, producing a representation that is useful for performing efficient intersection tests and related operations [2, 3, 4].

There are many different ways to construct a sphere hierarchy. Figure 1 illustrates one of the simplest: the method of binary subdivision of the object. In this method, the hierarchy takes the form of a binary tree with a sphere stored at each node. Each sphere encloses part of the object; each sphere at a non-terminal node encloses the union of the parts enclosed by its children; and the sphere at the root node encloses the whole object.

A point is considered to be inside a sphere hierarchy if it is inside one of the leaf-node spheres, as determined by the following recursive algorithm.

function point_in_sh(point, sh)
begin
   if not point_in_sphere(point, sh.sphere) **then**
      return *false*;
   else if leaf(sh) **then**
      return *true*;
   else
      for each child **of** sh **do**
         if point_in_sh(point, child) **then**
            return *true*;
      return *false*;
end

A similar algorithm can be devised for testing the intersection of two sphere hierarchies.

Any point outside the sphere hierarchy is also outside the shape it represents; but the points inside the hierarchy aren't necessarily inside the shape. A sphere hierarchy is said to have an accuracy of d if every point inside the hierarchy is either inside the shape it represents or the minimum distance between the point and the shape is less than or equal to d. This definition also applies to sub-hierarchies and individual spheres.

3. The Space-Occupancy Graph

The space-occupancy function of a mechanism is defined by $\sigma : Q \mapsto P(E)$, where Q is the mechanism's configuration space, E is Euclidean space (2-D or 3-D), $P(E)$ is the power set of E (the set of all subsets of E), and $\sigma(q)$ is the set of points occupied by the mechanism in configuration q. The SOG is the graph of this function, defined by:

$$\Gamma(\sigma) = \{\langle q, p \rangle \mid q \in Q, p \in \sigma(q)\}, \quad \Gamma(\sigma) \subseteq Q \times E.$$

Figure 2 shows the SOG of a rectangle in $E(2)$ that is free to rotate about the origin by an angle $q \in [0, \frac{\pi}{2}]$, which is the configuration variable. The SOG is therefore an object in the 3-D space $[0, \frac{\pi}{2}] \times E(2)$. Figure 2 also shows the enclosing hyper-cylinder, which looks like an ordinary cylinder in this example, and it illustrates the two main uses of a SOG (which can be implemented by set intersection and projection):

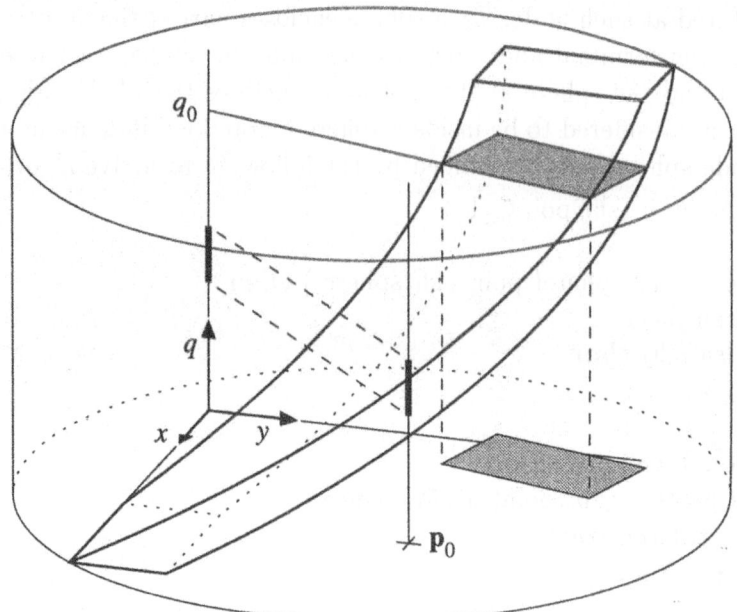

Figure 2. The SOG of a rectangle free to rotate about the origin in the x–y plane. Also shown are the intersections of the SOG with the line $\mathbf{p} = \mathbf{p}_0$ and the plane $q = q_0$, and the hyper-cylinder enclosing the SOG.

1. Given a configuration $\mathbf{q}_0$, or a set of configurations, find the set of points that are occupied by the mechanism in at least one of the given configurations.
2. Given a point $\mathbf{p}_0 \in \mathsf{E}$, or a set of such points, find the set of configurations satisfying $\mathbf{p}_0 \cap \sigma(\mathbf{q}) \neq \emptyset$.

The hyper-cylinders used in this paper are generated by the swept-bubble algorithm described below, and are defined as the product of a box in Q and a sphere in E (a box being the product of a set of closed intervals, one for each dimension of Q). Such a shape is invariant with respect to rotations in E, and it decouples the dimensions of Q (allowing each dimension to be treated separately).

The SOG shown in Figure 2 has only three dimensions. In practice, a typical SOG will have between four and ten. This high dimensionality presents certain difficulties:

1. The number of hyper-cylinders required to achieve a given accuracy grows exponentially with the dimension of the SOG, so it may not always be practical to precompute the whole of the hierarchy.
2. It may not be feasible to construct an explicit geometric model of the SOG as a first step to constructing the hierarchy.

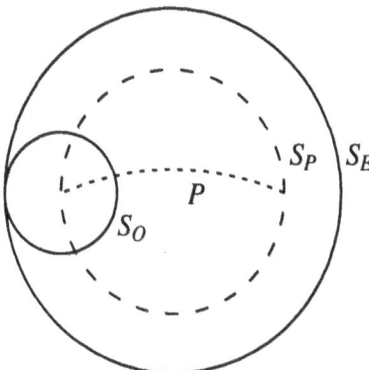

Figure 3. A swept bubble, showing the object sphere (S_O), the path followed by S_O (P), the sphere enclosing the path (S_P) and the sphere enclosing S_O at every point along P (S_E).

The swept-bubble algorithm overcomes these difficulties by allowing any part of the hierarchy to be constructed on demand, and by constructing it directly from the kinematic model and link shape descriptions (sphere hierarchies) rather than from an explicit model of the SOG.

4. Swept Bubbles

A swept bubble is a data structure that forms a node in a self-generating hierarchy (a bubble hierarchy) that describes a hyper-cylinder represent-ation of a SOG. This section describes the swept bubble itself and the method of generating a bubble hierarchy.

Figure 3 shows the basic idea. Suppose we have an object with one degree of motion freedom, so that its configuration space is one-dimensional and consists solely of the motion variable q; and suppose also that its shape is described by a sphere hierarchy, and that the sphere S_O is a member of that hierarchy. If we allow q to vary over an interval $Q = [q_0, q_1]$ then the centre of S_O will follow a path segment P. We can construct a sphere S_P that encloses the path, and hence an enclosing sphere, S_E, that encloses S_O at every point along P. Together, the motion interval Q and the sphere S_E define a hyper-cylinder in $Q \times E$ that encloses the SOG of the piece of object represented by S_O in the region of configuration space satisfying $q \in Q$.

A swept bubble contains all the necessary data to perform this calcula-tion: a description of the object's motion in a form suitable for calculating S_P, explicit values for Q, S_P and S_E, and a pointer to a separate data structure describing S_O. It also contains pointers to its children, for use in constructing the hierarchy.

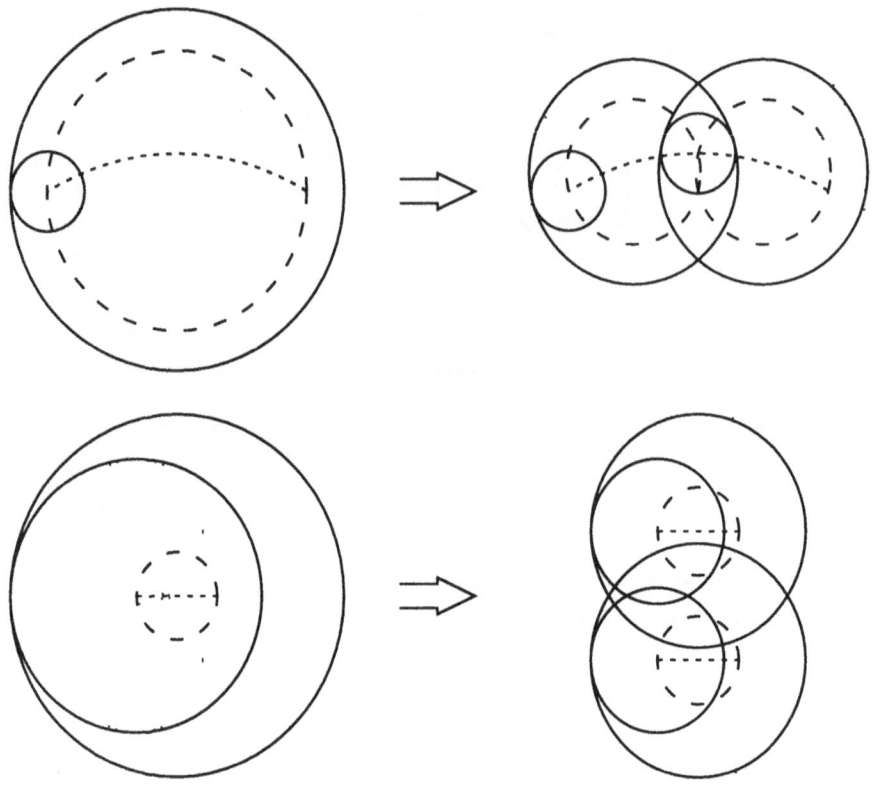

Figure 4. Generating the children of a swept bubble by cutting the motion interval (top), or by cutting the object (bottom).

A bubble hierarchy is generated using an operation called 'expansion', which creates the children of a swept bubble. (See Figure 4.) There are two ways to perform an expansion:

1. Cut the motion interval into two pieces. Each child inherits one piece, defining a shorter path, and new values of S_P and S_E are calculated accordingly.
2. Cut the object; *i.e.*, replace S_O with its children, which represent smaller pieces of the object. Each child inherits one of the children of S_O, and new values of S_P and S_E are calculated accordingly.

A swept bubble is expanded by one or other of these two operations, never both. The choice is made by considering which operation will produce the bigger improvement in accuracy. Any swept bubble can be expanded, unless its motion interval has reached a predetermined minimum size and its object sphere is a leaf node, in which case the swept bubble is a leaf node in the bubble hierarchy.

In terms of the hyper-cylinder hierarchy, the effect of expansion is to generate the hierarchy by binary subdivision of the SOG. The two different

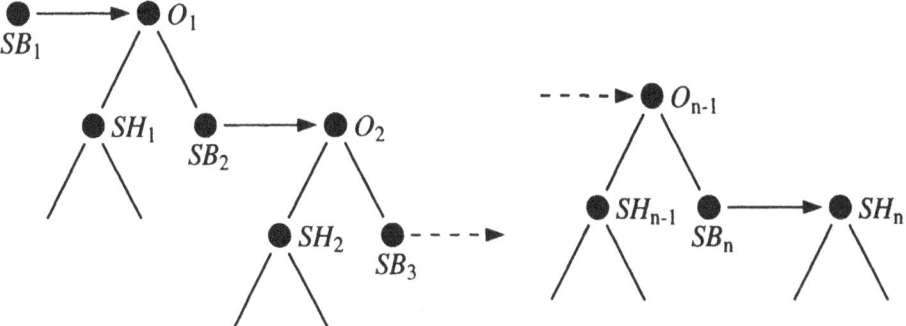

Figure 5. A swept-bubble hierarchy for a serial n-link mechanism.

expansion operations correspond to cutting the SOG along dimensions in Q or in E. Expansion can be used to pre-compute the hierarchy, or to generate any part of it on demand.

Now consider an object with n degrees of motion freedom, such that its location is the product of n 1-DoF displacements. We can construct a SOG for this object using a chain of n swept bubbles, one for each displacement, which are labelled $SB_1 \ldots SB_n$ and connected together so that SB_{i+1} is the object structure of SB_i. The root node of the sphere hierarchy is the object structure of SB_n. If we associate motion variable q_i with SB_i, and define Q_i as the $(n+1-i)$-dimensional configuration space constructed from variables $q_i \ldots q_n$, then SB_i defines a hyper-cylinder in $Q_i \times E$ that is the product of the enclosing sphere of SB_i with the box defined by the motion intervals of $SB_i \ldots SB_n$.

It follows that only SB_1 and its descendents define a hyper-cylinder representation of the SOG of the original system. The other swept bubbles and their descendents define lower-dimensional hyper-cylinders that are used as intermediate results.

The expansion procedure is applied to SB_1, and proceeds in exactly the same manner as for a 1-DoF system, with one exception: if a swept bubble is to be expanded by cutting its object, and its object is a swept bubble with no children, then the expansion procedure must first apply itself (recursively) to expanding the object bubble before continuing with the original expansion. This causes $SB_2 \ldots SB_n$ to be expanded as required to support the expansion of SB_1.

The SOG of a multibody system can be calculated from a data structure like that shown in Figure 5. $SH_1 \ldots SH_n$ are sphere hierarchies describing the shapes of the links; $SB_1 \ldots SB_n$ are swept bubbles describing the motions of the joints; and $O_1 \ldots O_{n-1}$ are link nodes that connect the various pieces together into a single data structure. The sphere at node O_i encloses the spheres at nodes SH_i and SB_{i+1}. The expansion procedure is applied to SB_1, and swept-bubble chains will emerge as the link nodes are cut.

5. Applications

Most swept-bubble applications involve performing an intersection test. The basic algorithm for doing this is:

```
function bh_sh_intersect( bh, sh )
begin
    if not sphere_intersect( bh.encl_sph, sh.sphere ) then
        return false;
    else if leaf(bh) and leaf(sh) then
        return true;
    else if cut_bh_first(bh,sh) then
        expand(bh);
        for each child of bh do
            if bh_sh_intersect( child, sh ) then return true;
    else
        for each child of sh do
            if bh_sh_intersect( bh, child ) then return true;
    return false;
end
```

This function intersects a bubble hierarchy with a sphere hierarchy, and returns a truth value indicating whether or not an intersection occurs. Except for the statement 'expand(bh);' this function is almost identical to the corresponding function for intersecting two sphere hierarchies.

With appropriate modifications, this function can be made to report where in configuration space the intersection occurs, and to continue searching until it has found every intersection. This modified algorithm can be used to build a map of obstacle regions in configuration space [1], or to find solution regions for inverse kinematics problems (in which case 'sh' represents a target location rather than an obstacle). In all cases, the algorithm respects joint motion limits.

Another useful modification is to let the joint variables be functions of other variables, and to calculate their motion ranges accordingly. This allows collision checking along a given path to be performed. Figure 6 illustrates this application: a planar 3R robot moves along a straight-line path in joint space from $(\pi/2, 0, -\pi/2)$ to $(\pi/6, -\pi/3, -\pi/3)$, and the problem is to determine whether or not the robot hits an obstacle along the way. In this example, each link is a 1×0.2 rectangle and the obstacle is a 0.5×0.5 square. These shapes are modelled by sphere hierarchies that are accurate to better than 0.01 units: 39 spheres are required for each link, and 31 for the obstacle. The independent variable varies uniformly from 0 to 1 along

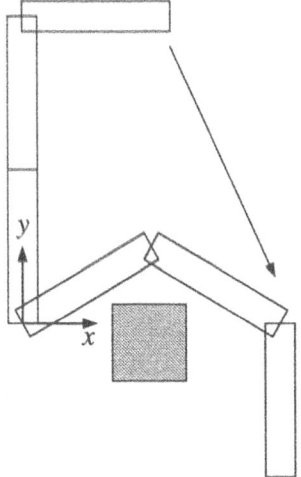

y	coll	sph	exp
−1.0	no	10	7
−0.5	no	25	25
−0.2	no	36	34
0.0	no	45	41
0.1	no	60	52
0.2	no	91	73
0.22	no	118	85
0.23	yes	38	15
0.3	yes	55	19
0.5	yes	49	21

Figure 6. Collision checking along a given trajectory.

the path, and the expansion algorithm is allowed to divide it into intervals no smaller than 1/128.

The top left corner of the obstacle is placed at coordinates $(0.6, y)$ for various values of y shown in the table. For each y value, the table shows the outcome of the collision test, the number of sphere intersections performed, and the number of swept-bubble expansions performed. (If the bubble hierarchy is precomputed then no expansions are required.) The correct answer to this problem is that a collision will occur for $y \geq 0.231$; so the results in the table are accurate to better than 0.01 units. Notice that the computational effort is greatest in the case of a near miss ($y = 0.22$), and that it declines rapidly as the obstacle is moved away from the path of the robot. This is a general property of the sphere and hyper-cylinder hierarchies. In the cases where a collision is detected, the computational effort varies erratically and depends on the search strategy used.

My current implementation of this algorithm performs 3-D calculations throughout; so a sphere-intersection test takes 11 arithmetic operations, and an expansion takes 80 plus 2 square roots and 4 sin/cos calculations. (There is scope for optimization here.) According to these figures, a 1-Megaflop computer could solve the worst-case example in Figure 6 in about 10 milliseconds, or 1.3 milliseconds with a precomputed bubble hierarchy.

In general, the computational effort required by this and other swept-bubble algorithms rises exponentially with the dimension of the search space, and, for a given dimension, polynomially with desired accuracy. Swept bubbles are best suited to problems where low accuracy is acceptable, such as gross motions that do not pass close to obstacles. I have performed experiments with 4-DoF robots, and found that the number of sphere tests rises above 10,000 in a cluttered environment. Verwer has implemented a path planner for a 5-DoF robot using swept bubbles, and reports execution times of 20 seconds (on a SUN-4/280) to find an initial path [5].

6. Conclusion

This paper has presented a technique for representing the space-occupancy graph (SOG) of a mechanism using data structures called swept bubbles, and algorithms that build a hierarchical hyper-cylinder representation of the SOG. The swept-bubble technique is useful in applications where it is necessary to search both Euclidean space and configuration space looking for intersections between solids. Such applications include path planning, collision checking and inverse kinematics.

The main advantages of swept bubbles are that they work with general geometry and kinematics; they search Euclidean and configuration spaces simultaneously, homing in quickly on regions of interest; they respect joint motion limits; they are easy to implement; and they use sphere-intersection tests, which are very cheap.

The main disadvantages are that they suffer from the curse of dimensionality; they require shapes to be represented by sphere hierarchies; and they deliver only approximate answers.

Acknowledgements

This work was supported by SERC Advanced Research Fellowship number B92/AF/1466.

References

1. Featherstone, R., Swept Bubbles: A Method of Representing Swept Volume and Space Occupancy, Document no. MS-90-069, Philips Labs, Briarcliff Manor, NY, 1990.
2. Liu, Y., Noborio, H. and Arimoto, S., Hierarchical Sphere Model (HSM) and its Application for Checking an Interference between Moving Robots, *Proc. IEEE Int. Workshop on Intelligent Robots and Systems (IROS '88)*, pp. 801–806, Tokyo, 1988.
3. O'Rourke, J. and Badler, N., Decomposition of Three-Dimensional Objects into Spheres, *IEEE Trans. Pattern Analysis and Machine Intelligence*, vol. PAMI-1, no. 3, pp. 295–305, 1979.
4. Sawatzky, G. and El-Zorkany, H., Using an Efficient Collision Detector in the Solution of the Find-Path Problem of Industrial Robots, *Proc. SPIE, vol. 579, Intelligent Robots and Computer Vision*, pp. 131–141, 1985.
5. Verwer, B. J. H., A Multiresolution Work Space, Multiresolution Configuration Space Approach to Solve the Path Planning Problem, *Proc. IEEE Int. Conf. Robotics and Automation*, pp. 2107–2112, Cincinnati, 1990.

DOMAINS OF OPERATION AND INTERFERENCE FOR BODIES IN MECHANISMS AND MANIPULATORS

E.J. HAUG, F.A. ADKINS, AND C.M. LUH
Center for Computer-Aided Design
and
Department of Mechanical Engineering
The University of Iowa
Iowa City, Iowa 52242

Abstract. Domains associated with one or more working bodies of a mechanism or manipulator are defined that characterize the range of operation of the bodies and interference that may occur between the equations defining domains are derived in terms of equations defining the kinematics of the mechanism and the geometry of the working bodies. Analytical criteria defining the boundaries of such domains are derived and numerical methods for computing families of generators on boundaries are outlined. The criteria and numerical methods presented are applied to planar mechanism and manipulator examples.

1. Introduction

Criteria for boundaries of workspaces of mechanisms and manipulators, using conditions associated with singularity of constraint Jacobian or velocity transformation matrices, have been developed by a number of authors in the recent past. Litvin [6] used the implicit function theorem to define singular configurations of mechanisms as criteria for boundaries of workspaces. Analytical conditions associated with special features of the geometry of specific manipulators have been used by a number of authors to obtain explicit criteria for boundaries of workspaces (see [2], [7], [8], and [10]). Singularity of the velocity transformation between input and output coordinates has likewise been used to characterize singular surfaces of manipulators (see [5] and [9]). Numerical methods for mapping boundaries of workspaces of

193

J.-P. Merlet and B. Ravani (eds.), Computational Kinematics, 193–202.
© *1995 Kluwer Academic Publishers.*

mechanisms and manipulators have recently been presented (see [3], [4], and [11]). These methods are summarized here, with accompanying numerical methods.

2. Analytical Conditions For Working Domains And Boundaries

In order to define working domains in bodies that move with an underlying mechanism, whose configuration is defined by a generalized coordinate $\mathbf{q}$, the shape of the domain of the working bodies is parameterized by a vector μ. Defining the nz-vector $\mathbf{z} = \left[\mathbf{q}^T, \mu^T\right]^T$, the nu-vector $\mathbf{u}$ to a working point on a working body is given analytically in the form $\mathbf{u} = \mathbf{g}(\mathbf{z})$, where the vector function $\mathbf{g}(\mathbf{z})$ is twice continuously differentiable. The kinematic constraint equations for the mechanism and equations involving parameter-ization of the shapes of working bodies, which may account for inequalities with slack variables [4] to define domains, are written in the form $\boldsymbol{\Phi}(\mathbf{z}) = \mathbf{0}$, where the m-vector of functions $\boldsymbol{\Phi}(\mathbf{z})$ is twice continuously differentiable. The system of $nu + m$ equations that define points in a domain specified by the associated criteria is

$$\boldsymbol{\Psi}(\mathbf{u},\mathbf{z}) \equiv \begin{bmatrix} \mathbf{u} - \mathbf{g}(\mathbf{z}) \\ \boldsymbol{\Phi}(\mathbf{z}) \end{bmatrix} = \mathbf{0} \tag{1}$$

Thus, the domain of interest is the set

$$D = \{\mathbf{u} \in \Re^{nu} : \boldsymbol{\Psi}(\mathbf{u},\mathbf{z}) = \mathbf{0}, \; for \; some \; \mathbf{z}\} \tag{2}$$

To characterize this domain, criteria for points on its boundary are needed. Let $\mathbf{u}^* \in D$. Then, there is a $\mathbf{z}^*$ such that $\boldsymbol{\Psi}(\mathbf{u}^*,\mathbf{z}^*) = \mathbf{0}$ and $\mathbf{u}^* = \mathbf{g}(\mathbf{z}^*)$. If $\boldsymbol{\Psi}_{\mathbf{z}}(\mathbf{u}^*,\mathbf{z}^*)$ is full row rank then, by the implicit function theorem, there exists a twice-continuously differentiable function $\mathbf{h}(\mathbf{u})$ and neighborhoods $\mathbf{U}$ and $\mathbf{Z}$ of $\mathbf{u}^*$ and $\mathbf{z}^*$, respectively, and $\mathbf{z} = \mathbf{h}(\mathbf{u})$ such that $\boldsymbol{\Psi}(\mathbf{u},\mathbf{h}(\mathbf{u})) = \mathbf{0}$, for all $\mathbf{u} \in \mathbf{U}$ and $\mathbf{z} \in \mathbf{Z}$. This implies that $\mathbf{u}^*$ is in the interior of D. A necessary condition for $\mathbf{u}^*$ to be on the boundary of D, denoted ∂D, is thus that $\mathbf{u}^* \in D$ and $\boldsymbol{\Psi}_{\mathbf{z}}(\mathbf{u}^*,\mathbf{z}^*)$ is row-rank deficient. A necessary and sufficient condition for row-rank deficiency of $\boldsymbol{\Psi}_{\mathbf{z}}(\mathbf{u}^*,\mathbf{z}^*)$ is that there exists a vector $\boldsymbol{\xi}$ such that $\boldsymbol{\xi}^T\boldsymbol{\xi} = 1$ and

$$\boldsymbol{\Psi}_{\mathbf{z}}^T(\mathbf{u}^*,\mathbf{z}^*)\boldsymbol{\xi} = \mathbf{0} \tag{3}$$

A necessary condition for points on the boundary of the domain D is, therefore,

$$\partial D \subset \left\{\mathbf{u} \in D : \boldsymbol{\Psi}_{\mathbf{z}}^T(\mathbf{u},\mathbf{z})\boldsymbol{\xi} = \mathbf{0}, \boldsymbol{\xi}^T\boldsymbol{\xi} = 1, \boldsymbol{\Psi}(\mathbf{u},\mathbf{z}) = \mathbf{0}, \; for \; some \; \mathbf{z} \; and \; \boldsymbol{\xi}\right\} \tag{4}$$

3. Numerical Methods For Mapping Boundaries

A brief summary of calculations involved in mapping one-dimensional generators for ∂D [3] is given here.

Finding an Interior Point of D: Simulations can often be carried out to find a configuration of the mechanism and points on working bodies satisfying criteria of Eq. 1, yielding a point $\mathbf{u}^0$ interior to D. Alternatively, an iterative method can be used to find such a point, as follows. Let $\mathbf{s} = \left[\mathbf{u}^T, \mathbf{z}^T\right]^T$ and $\mathbf{s}^{(0)}$ be an estimate of a point in D. The generalized Newton method [3] defined by the recursive relation

$$\mathbf{\Psi_s}\left(\mathbf{s}^{(j)}\right)\Delta\mathbf{s}^{(j)} = -\mathbf{\Psi}\left(\mathbf{s}^{(j)}\right), \;\; \mathbf{s}^{(j+1)} = \mathbf{s}^{(j)} + \Delta\mathbf{s}^{(j)} \tag{5}$$

where $j = 1, 2, \ldots$ is the iteration counter and the first of Eqs. 5 is solved using the Moore-Penrose generalized inverse [1]. If the method converges, the resulting $\mathbf{u}^{(j+1)}$ is in D and is denoted $\mathbf{u}^0$.

Finding an Initial Point on ∂D: Finding an initial point on ∂D, as the basis for initiating continuation computation to locate all candidate points on ∂D is a non-trivial task. A unit vector $\mathbf{c}$ is selected to define a ray emanating from the point $\mathbf{u}^0$ interior to D that is to be traversed using a small step-size h, until a point outside D is encountered. From a candidate point $\mathbf{u}^{i-1}$, a new point is defined by $\mathbf{u}^i = \mathbf{u}^{i-1} + h\mathbf{c}$, $i = 1, 2, \ldots$. Defining $\mathbf{a} = \left[\mathbf{q}^T, \boldsymbol{\mu}^T\right]^T$, at a predicted point $\mathbf{u}^i$, the recursive relation of Eq. 5, with $\mathbf{s}$ replaced by $\mathbf{a}$, is solved using the Moore-Penrose generalized inverse. If the iterative process converges to a solution of Eq. 1, then a point interior to D has been determined and another step is taken along ray $\mathbf{c}$. This process is continued until a point is reached at which the iterative process fails to converge, signaling that the point is outside D. An interval halving technique is employed to subdivide the interval containing the boundary point, until a boundary point is found.

Mapping One-Dimensional Solution Curves on ∂D: Defining the complete set of variables involved in the formulation as $\mathbf{x} = \left[\mathbf{u}^T, \mathbf{z}^T, \boldsymbol{\xi}^T\right]^T$, the conditions of Eq. 4 for points on ∂D are

$$\mathbf{G}^*\left(\mathbf{x}\right) = \begin{bmatrix} \mathbf{\Psi}\left(\mathbf{u}, \mathbf{z}\right) \\ \mathbf{\Psi}_\mathbf{z}^T\left(\mathbf{u}, \mathbf{z}\right)\boldsymbol{\xi} \\ \boldsymbol{\xi}^T\boldsymbol{\xi} - 1 \end{bmatrix} = \mathbf{0} \tag{6}$$

If **u** is in the plane, then Eq. 6 defines a one-dimensional curve that is a candidate for a segment of ∂D. If **u** is in three-dimensional space, then a cutting plane in the output set is defined as $\mathbf{L}^T \mathbf{u} = b$, where the vector **L** serves as the normal to the plane intersecting the solution set of Eq. 1 and the parameter b defines the location of the plane along the vector **L**. Selecting a grid of values of b yields a family of one-dimensional curves on ∂D that are generators of the surface. Numerical methods developed for mapping one-dimensional boundary segments of accessible output sets [3] are directly applicable for numerous working domain applications.

4. Accessible Output Sets

The simplest domain of operation associated with a working body is the accessible output set, denoted A. It consists of the set of all points in space that can be reached by a working point P that is fixed in the working body, for some value of the mechanism general coordinate vector **q**. Working point P is fixed in a working body with an x'-y'-z' body-fixed reference frame, located by the body-fixed vector $\mathbf{s}'^P$. It is transformed to the global x-y-z frame by the orthogonal orientation transformation matrix $\mathbf{A}(\mathbf{q})$ that is a function of the generalized coordinates; i.e., $\mathbf{s}^P = \mathbf{A}(\mathbf{q})\mathbf{s}'^P$. The vector **r** that locates the origin of the body-fixed reference frame is a function of the vector **q** of generalized coordinates of the mechanism that controls motion of the working body. The output vector **u** in this case is $\mathbf{u} = \mathbf{r}^P = \mathbf{r}(\mathbf{q}) + \mathbf{A}(\mathbf{q})\mathbf{s}'^P \equiv \mathbf{g}(\mathbf{q})$. The system of equations of Eq. 1 for the accessible output set is thus

$$\mathbf{\Psi}(\mathbf{u},\ \mathbf{q}) \equiv \begin{bmatrix} \mathbf{u} - \mathbf{r}(\mathbf{q}) + \mathbf{A}(\mathbf{q})\mathbf{s}'^P \\ \mathbf{\Phi}(\mathbf{q}) \end{bmatrix} = \mathbf{0} \tag{7}$$

The accessible output set is $A = \{\mathbf{u} : \mathbf{\Psi}(\mathbf{u},\ \mathbf{q}) = \mathbf{0},\ for\ some\ \mathbf{q}\}$.

The planar Stewart platform shown in Fig. 1 consists of a moving upper platform and three actuators that control the three degrees of freedom of the platform. The working point is chosen at the center of the top platform, so $\mathbf{s}'^P = \mathbf{0}$. The actuator lengths ℓ_1, ℓ_2, and ℓ_3 are input coordinates that control the position $\mathbf{r} = [x,\ y]^T$ of point P and the orientation ϕ of the top platform. Actuator length unilateral constraints are $0 < \ell_i^{\min} \le \ell_i \le \ell_i^{\max}$, $i = 1,\ 2,\ 3$, where $\ell_1^{\min} = \ell_2^{\min} = \sqrt{2}$, $\ell_3^{\min} = 1$, $\ell_1^{\max} = \ell_2^{\max} = 2$, and $\ell_3^{\max} = \sqrt{3}$. Introducing new input coordinates v_1, v_2, and v_3 so that $\ell_i = a_i + b_i \sin v_i$, $i = 1,\ 2,\ 3$, where $a_1 = a_2 = \left(\sqrt{2} + 2\right)/2$, $a_3 = \left(1 + \sqrt{3}\right)/2$, $b_1 = b_2 = \left(2 - \sqrt{2}\right)/2$, and $b_3 = \left(\sqrt{3} - 1\right)/2$, the actuator unilateral

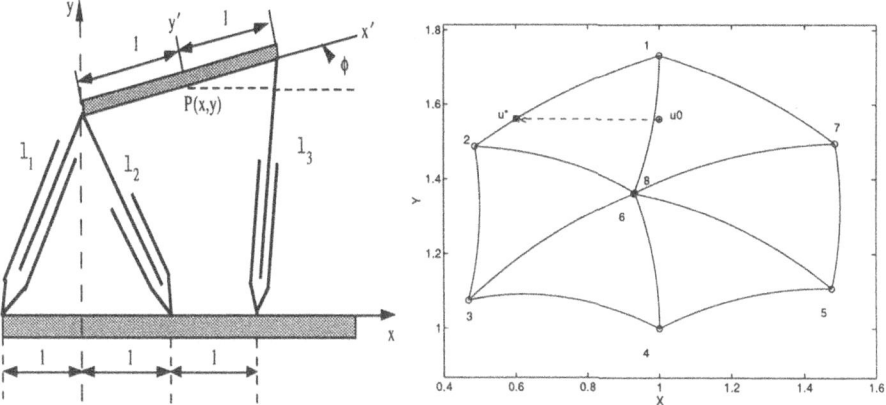

Figure 1. Planar Stewart Platform   Figure 2. Accessible Output Set

constraints are automatically satisfied. The generalized coordinate vector is thus $\mathbf{q} = [x,\ y,\ v_1,\ v_2,\ v_3,\ \phi]^T$ and the constraint equations for this manipulator are

$$\boldsymbol{\Phi}(\mathbf{q}) \equiv \begin{bmatrix} \ell_1^2 - (x - \cos\phi + 1)^2 - (y - \sin\phi)^2 \\ \ell_2^2 - (1 - x + \cos\phi)^2 - (y - \sin\phi)^2 \\ \ell_3^2 - (x + \cos\phi - 2)^2 - (y + \sin\phi)^2 \end{bmatrix} = \mathbf{0} \qquad (8)$$

where ℓ_i is a function of v_i. Since $\mathbf{s}'^P = \mathbf{0}$, $\mathbf{u} = [x,\ y]^T \equiv \mathbf{g}(\mathbf{q})$.

To obtain an initial point to begin numerical calculations, the platform is raised to a mid-point in which the upper platform is horizontal; i.e., $\phi = 0$. The initial values for the generalized coordinates for this point interior to A are thus $\mathbf{q}^0 = [1,\ 1.5607,\ 0.5236,\ 0.5236,\ 0.5236,\ 0.0]^T$. Selecting a search direction parallel to the x-axis; i.e., $\mathbf{c} = [-1,\ 0]^T$, the search process of Section 3 for an initial point on ∂A is carried out, yielding the point $\mathbf{u}^*$ on ∂A shown in Fig. 2. The final configuration on ∂A obtained in this process is $\mathbf{q}^* = [0.6018,\ 1.5607,\ -0.5209,\ 1.5694,\ 1.5658,\ 0.1235]^T$. With this approximate point on ∂A, a least squares approximate solution of Eq. 3, with $\mathbf{e}_j^T \boldsymbol{\xi}^j = 1$, $\mathbf{e}_j = [\delta_{ij}]$, and j set such that $\boldsymbol{\Phi}_\mathbf{z} \boldsymbol{\Phi}_\mathbf{z}^T + \mathbf{e}_j \mathbf{e}_j^T$ has smallest condition number [3], is obtained as $\boldsymbol{\xi}^2 = [0.0,\ 1.0,\ 0.9284]^T$. With this starting point, candidate curves on ∂A are mapped using the continuation method outlined in Section 3. The curves obtained are shown in Fig. 2, where numbered points are bifurcation points. Points 6 and 8, while appearing to be close together, are actually the projections onto the $\mathbf{u}$-space of two distinctly different bifurcation points. The generalized coordinate vector for point 6 is $\mathbf{q} = [x,\ y,\ \phi]^T = [0.928571,\ 1.36090,\ -0.38075]^T$ and that for point 8 is $[0.931663,\ 1.366332,\ 0.37183]^T$.

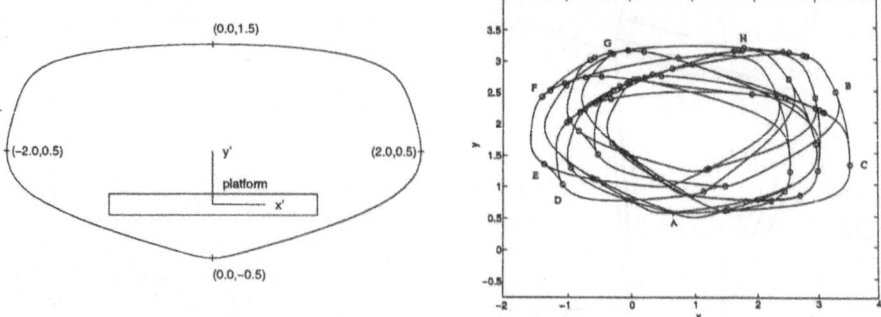

Figure 3. Projection Dome   **Figure 4.** Singular Boundary Curves

5. Operational Envelopes

A point P in the domain of the working body is located in space by the vector $\mathbf{u} = [x,\ y,\ z]^T = \mathbf{r}(\mathbf{q}) + \mathbf{A}(\mathbf{q})\,\mathbf{s}'^P$. In the present case, however, it is intended that point P is free to vary over the domain of the working body. A typical point P in the domain of the working body can be defined, relative to the body-fixed x'-y'-z' reference frame, as a function of a parameter vector $\boldsymbol{\mu}$, $\mathbf{s}'^P = \mathbf{f}(\boldsymbol{\mu})$. The vector $\boldsymbol{\mu}$ may be comprised of independent parameters, or it may be required to satisfy an equation of the form $\mathbf{B}(\boldsymbol{\mu}) = 0$. This equation is combined with purely kinematic constraint equations in $\mathbf{q}$, to form the composite set of constraint equations $\boldsymbol{\Phi}(\mathbf{q},\ \boldsymbol{\mu}) \equiv \boldsymbol{\Phi}(\mathbf{z}) = 0$. The operational envelope, denoted OE, of the working body is defined as the set of all points in space that can be occupied by some point P on the working body, for some admissible value of the generalized coordinate vector $\mathbf{q}$. Conditions for points in space that are occupied by points in the domain of the working body, as it moves under the influence of kinematic constraints defining the underlying mechanism, are

$$\boldsymbol{\Psi}(\mathbf{u},\ \mathbf{z}) \equiv \begin{bmatrix} \mathbf{u} - \mathbf{r}(\mathbf{q}) - \mathbf{A}(\mathbf{q})\,\mathbf{f}(\boldsymbol{\mu}) \\ \boldsymbol{\Phi}(\mathbf{q},\ \boldsymbol{\mu}) \end{bmatrix} = 0 \tag{9}$$

where the convention that $\mathbf{z} = \begin{bmatrix} \mathbf{q}^T,\ \boldsymbol{\mu}^T \end{bmatrix}^T$ has been used. Thus, the operational envelope of the working body is $OE = \{\mathbf{u} : \boldsymbol{\Psi}(\mathbf{u},\ \mathbf{z}) = 0\ for\ some\ \mathbf{z}\}$.

Consider the operational envelope of the projection dome of Fig. 3 that is attached to the planar Stewart platform of Fig. 1 to form a driving simulator. The generalized coordinate vector is $\mathbf{q} = [x,\ y,\ v_1,\ v_2,\ v_3,\ \phi]^T$ and the constraint equations for the manipulator are given in Eq. 8. The equation for the boundary of the dome is $\mu_1^2 + (2 + 1.3(\mu_2 - 0.5)^3)^2((\mu_2 - 0.5)^2 -$

1) $= 0$. The domain of the dome is thus characterized by the inequality $\mu_1^2 + (2 + 1.3(\mu_2 - 0.5)^3)^2((\mu_2 - 0.5)^2 - 1) \leq 0$. To represent the domain of the dome in equation form, a slack variable μ_3 is introduced and the equation for the domain is

$$B(\boldsymbol{\mu}) = \mu_1^2 + (2 + 1.3(\mu_2 - 0.5)^3)^2((\mu_2 - 0.5)^2 - 1) + \mu_3^2 = 0 \qquad (10)$$

A typical point in the domain of the dome is thus given, in the platform body-fixed x'-y' frame, as $\mathbf{s}'^P = [\mu_1, \mu_2]^T \equiv \mathbf{f}(\boldsymbol{\mu})$. The equation defining the operational envelope of the domain of the dome of the planar Stewart platform is

$$\Psi(\mathbf{u}, \mathbf{q}, \boldsymbol{\mu}) \equiv \begin{bmatrix} u_1 - x - \mu_1 \cos\phi + \mu_2 \sin\phi \\ u_2 - y - \mu_1 \sin\phi - \mu_2 \cos\phi \\ \mu_1^2 + (2 + 1.3(\mu_2 - 0.5)^3)^2((\mu_2 - 0.5)^2 - 1) + \mu_3^2 \\ \ell_1^2 - (x - \cos\phi + 1)^2 - (y - \sin\phi)^2 \\ \ell_2^2 - (1 - x + \cos\phi)^2 - (y - \sin\phi)^2 \\ \ell_3^2 - (x + \cos\phi - 2)^2 - (y + \sin\phi)^2 \end{bmatrix} = 0$$

$$(11)$$

To obtain a point interior to OE, the mechanism legs were set to their maximum length, which determines the values of x, y and ϕ. The values $\mu_1 = 0.0$ and $\mu_2 = 0.5$, and $\mu_3 = 2.0$ were chosen, and the search direction was selected as $\mathbf{c} = [0, 1]^T$. Searching in this direction, from the configuration $\left[\mathbf{u}^{0T}, \mathbf{q}^{0T}, \boldsymbol{\mu}^{0T}\right]^T = \left[1.0, \sqrt{3}, 1.0, \sqrt{3}, \frac{\pi}{2}, \frac{\pi}{2}, \frac{\pi}{2}, 0, 0, 0.5, 2\right]^T$, using the method of Section 3, yielded the following point on OE: $\mathbf{x}^* = \left[\mathbf{u}^{*T}, \mathbf{q}^{*T}, \boldsymbol{\mu}^{*T}\right]^T = [1.0, 3.232, 1.0, 1.732, 1.571, 1.571, 1.571, 0., 0., 1.5, 0.]^T$. Starting from $\mathbf{x}^*$, the boundary is traced using the method outlined in Section 3. The resulting set of boundary segments is shown in Fig. 4. The outer curves define the boundary of the operational envelope, OE. In Fig. 4, several points of interest are marked. There are points at which the projections of boundary curves cross, but the curves computed do not intersect in the higher dimensional space. An example of such points is point A. Points such as these pose a problem when attempting to calculate the boundary of the operational envelope without also calculating the interior singular curves. Many segments on the boundary of the operational envelope arise from tracing the shape of the dome with the platform in a fixed configuration. Examples of this behavior are segments EF and GH. There are, however boundary segments that result from moving the platform and the dome, sweeping out a curve from one configuration of the platform to another. Examples of such segments are BC and DE. On segment BC the mechanism is moving by changing the third leg from its maximum to its minimum length, hence v_3 changes from $\pi/2$ to $-\pi/2$. On segment DE the mechanism is moving by changing the second leg from its minimum to its maximum length, hence v_2 changes from $-\pi/2$ to $\pi/2$.

6. Domains Of Interference

Consider an underlying mechanism that consists of a number of kinematically constrained moving bodies, with a vector of generalized coordinates that are subjected to kinematic constraint equations. Interference consideration is focused on a pair of working bodies of the underlying mechanism, denoted bodies 1 and 2. The origins of body-fixed x'-y'-z' reference frames that are used to characterize the position of each of the bodies are defined by vectors $\mathbf{r}_1(\mathbf{q})$ and $\mathbf{r}_2(\mathbf{q})$, which are functions of the generalized coordinates of the underlying mechanism. Orthogonal orientation transformation matrices $\mathbf{A}_1(\mathbf{q})$ and $\mathbf{A}_2(\mathbf{q})$, which are functions of the mechanism generalized coordinates, define the orientation of the body-fixed reference frames on bodies 1 and 2, respectively, relative to the global x-y-z frame. Points P in the domains of bodies 1 and 2 are characterized in terms of parameter vectors $\boldsymbol{\mu}_1$ and $\boldsymbol{\mu}_2$, in their respective body-fixed reference frames, as $\mathbf{s'}_1^P = \mathbf{f}_1(\boldsymbol{\mu}_1)$ and $\mathbf{s'}_2^P = \mathbf{f}_2(\boldsymbol{\mu}_2)$. Any equations involving $\boldsymbol{\mu}_1$ and $\boldsymbol{\mu}_2$ are combined with the kinematic constraint equations. Denoting the combined set of geometry parameterization vectors as $\boldsymbol{\mu} = \left[\boldsymbol{\mu}_1^T, \boldsymbol{\mu}_2^T\right]^T$, kinematic and geometric constraint equations are of the form $\boldsymbol{\Phi}(\mathbf{q}, \boldsymbol{\mu}) = \mathbf{0}$. The condition that some point P in space is occupied by a material point in each of bodies 1 and 2 is formulated as $\mathbf{r}_2(\mathbf{q}) + \mathbf{A}_2(\mathbf{q})\mathbf{f}_2(\boldsymbol{\mu}_2) - \mathbf{r}_1(\mathbf{q}) - \mathbf{A}_1(\mathbf{q})\mathbf{f}_1(\boldsymbol{\mu}_1) = \mathbf{0}$. The domain of points in body 1 that interfere with body 2 is denoted $I_1'D_2$. Let $\mathbf{u}_1'$ be a point on body 1, characterized in its body-fixed reference frame; i.e., $\mathbf{u}_1' = \mathbf{f}_1(\boldsymbol{\mu}_1)$. If $\mathbf{u}_1'$ is in $I_1'D_2$, then there exist $\mathbf{q}$ and $\boldsymbol{\mu}$ that satisfy

$$\boldsymbol{\Psi}(\mathbf{u}_1', \mathbf{z}) \equiv \begin{bmatrix} \mathbf{u}_1' - \mathbf{f}_1(\boldsymbol{\mu}_1) \\ \mathbf{r}_2(\mathbf{q}) + \mathbf{A}_2(\mathbf{q})\mathbf{f}_2(\boldsymbol{\mu}_2) - \mathbf{r}_1(\mathbf{q}) - \mathbf{A}_1(\mathbf{q})\mathbf{f}_1(\boldsymbol{\mu}_1) \\ \boldsymbol{\Phi}(\mathbf{q}, \boldsymbol{\mu}) \end{bmatrix} = \mathbf{0} \quad (12)$$

Thus the domain of interference is $I_1'D_2 = \{\mathbf{u}_1' : \boldsymbol{\Psi}(\mathbf{u}_1', \mathbf{z}) = \mathbf{0}, \; for \; some \; \mathbf{z}\}$.

Consider the planar Stewart platform and dome of Section 5, operating in a room that is 3.0 units high and 5.4 units wide. The global reference frame is used as the local frame for the room and the center of the room is at $(0.7, 1.5)$. The room is selected so that the dome will interfere with the ceiling and the right wall, but not the floor and the left wall. The coordinates of points outside of the room are $\mathbf{s'}_2^P = [\mu_{21}, \quad \mu_{22}]^T$. Equations are thus needed to represent the right wall and the ceiling. Points to the right of the right wall can be given as $\mu_{21} = 3.4 + \mu_{23}^2$, and points above the ceiling can be given as $\mu_{22} = 3.0 + \mu_{24}^2$, where μ_{23} and μ_{24} are slack variables. The interference equations are $\mu_{21} - x - \mu_{11}\cos(\phi) + \mu_{12}\sin(\phi) = 0$ and

$\mu_{22} - y - \mu_{11}\sin(\phi) - \mu_{12}\cos(\phi) = 0$. The governing equations for $I_1' D_2$ relative to the right of the right wall are

$$\Psi(\mathbf{u},\mathbf{q},\boldsymbol{\mu}) \equiv \begin{bmatrix} u_1 - \mu_{11} \\ u_2 - \mu_{12} \\ \mu_{21} - x - \mu_{11}\cos\phi + \mu_{12}\sin\phi \\ \mu_{22} - y - \mu_{11}\sin\phi - \mu_{12}\cos\phi \\ \ell_1^2 - (x - \cos\phi + 1)^2 - (y - \sin\phi)^2 \\ \ell_2^2 - (1 - x + \cos\phi)^2 - (y - \sin\phi)^2 \\ \ell_3^2 - (x + \cos\phi - 2)^2 - (y + \sin\phi)^2 \\ \mu_{11}^2 + (2 + 1.3(\mu_{12} - 0.5)^3)^2((\mu_{12} - 0.5)^2 - 1) + \mu_{13}^2 \\ \mu_{21} - 3.4 - \mu_{23}^2 \end{bmatrix} = \mathbf{0}$$

(13)

The governing equations for $I_1' D_2$ above the ceiling are obtained by replacing the last of Eqs. 13 with $\mu_{22} - 3.0 - \mu_{24}^2 = 0$. Initial points on the boundaries of these sets were determined by geometry, and the method of Section 2 was used to map boundary segments. Results for $I_1' D_2$ are presented in Fig. 5.

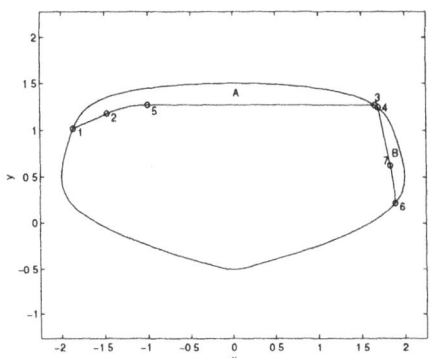

Figure 5. $I_1' D_2$ for Stewart Platform with Dome and Room

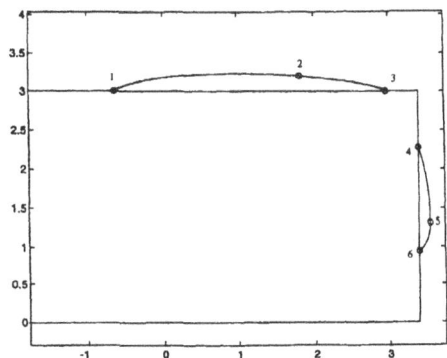

Figure 6. $I_2' D_1$ for Stewart Platform with Dome and Room

Points in region A of Fig. 5 result from the intersection of the dome with the ceiling and points in region B result from intersection with the right wall. The boundary segment from point 3 to point 5 in Fig. 5, is due to the edge of the ceiling penetrating the dome, with all legs of the mechanism at full extension. The boundary segment from point 2 to point 5 is the result of the mechanism moving by reducing the length of the third actuator, with the other two legs held at their maximum lengths. The boundary segment from point 6 to point 7 is a result of varying ℓ_3 to its minimum length, with ℓ_1 at maximum and ℓ_2 at minimum length. The segment from point 7 to point 4 is the result of the edge of the ceiling being traced with ℓ_1 at maximum, and both ℓ_2 and ℓ_3 at minimum length. Similarly, results for $I_2'D_1$ are presented in Fig. 6. The boundary segment from point 1 to point 2 in Fig. 6 is due to the edge of the dome penetrating into the ceiling, with all legs of the mechanism at full extension. The boundary segment from point 2 to point 3 is the result of the mechanism moving by reducing the length of the second actuator, with the other two legs held at their maximum lengths. The boundary segment from point 4 to point 5 is a result of varying ℓ_3 to its minimum length, with ℓ_1 at maximum and ℓ_2 at minimum length. The segment from point 5 to point 6 is the result of the edge of the dome being traced with ℓ_1 at maximum and both ℓ_2 and ℓ_3 at minimum length.

Research supported by US Army Automotive Research Grant No. DAAE07–94–C–R094 and ARPA Augmentation Awards for Science and Engineering Research Training Grant No. EEC 92-22784.

References

1. Allgower, E.L., and K. Georg. 1990. *Numerical Continuation Methods*. Berlin: Springer-Verlag.
2. Freudenstein, F. and E.J. Primrose. 1984. On the Analysis and Synthesis of the Workspace of a Three-Link, Turning-Pair Connected Robot Arm. *Journal of Mechanisms, Transmissions, and Automation in Design* 106:365-370.
3. Haug, E.J., C.M. Luh, F.A. Adkins, and J.Y Wang. 1994. Numerical Algorithms for Mapping Boundaries of Manipulator Workspaces. *Advances in Design Automation* ASME DE-69:2:447-459.
4. Jo, D.Y., and E.J. Haug. 1989. Workspace Analysis of Multibody Mechanical Systems Using Continuation Methods. *Journal of Mechanisms, Transmissions, and Automation in Design* 3(4):581-589.
5. Kumar, A., and K.J. Waldron. 1981. The Workspaces of a Mechanical Manipulator. *Journal of Mechanical Design* 103:665-672.
6. Litvin, F.L. 1980. Application of Theorem of Implicit Function System Existence for Analysis and Synthesis of Linkages. *Mechanism and Machine Theory* 15:115-125.
7. Spanos, J., and D. Kohli. 1985. Workspace Analysis of Regional Structures of Manipulators. *Journal of Mechanisms, Transmissions, and Automation in Design* 107:216-222.
8. Tsai, Y.C., and A.H. Soni. 1981. Accessible Region and Synthesis of Robot Arms. *Journal of Mechanical Design* 103:803-811.
9. Waldron, K.J., S. Wang, and S.J. Bolin. 1985. A Study of the Jacobian Matrix of Serial Manipulators. *Journal of Mechanisms, Transmissions, and Automation in Design* 107:230-238.
10. Yang, D.C.H., and Lee, T.W. 1983. On the Workspace of Mechanical Manipulators. *Journal of Mechanisms, Transmissions, and Automation in Design* 105:62-69.
11. Yang, F.C., and E.J. Haug. 1994. Numerical Analysis of the Kinematic Working Capability of Mechanisms. To appear in *Journal of Mechanical Design*.

DESIGNING A PARALLEL MANIPULATOR FOR A SPECIFIC WORKSPACE

J-P MERLET
INRIA
BP 93, 06902 Sophia-Antipolis, France

Abstract. We present an algorithm to determine all the possible locations of the attachment points of a Gough-type parallel manipulator which has to reach a desired workspace. This workspace is described by a set of segments which defines the location of a specific point of the moving platform, the platform orientation being kept constant. This algorithm takes into account the leg length limits, the mechanical limits on the passive joints and interference between links.

1. Introduction

In this paper we consider a 6 d.o.f. Gough-type parallel manipulator constituted by a fixed base plate and a mobile plate connected by 6 extensible links. For a parallel manipulator, workspace limits are due to the bounded range of linear actuators, mechanical limits on passive joints and interference between links. One important step in the design of a parallel manipulator is to define its geometry according to the desired workspace. Various geometrical algorithms for computing the workspace boundary when the platform's orientation is kept constant, either in 2D or 3D, have been described by Gosselin [3],[5] and Merlet [7]. The problem which will be addressed in this paper is to find all the possible locations of the passive joints such that the robot workspace includes the desired workspace. The topic of this paper has been addressed by very few authors. Claudinon [1] uses a numerical method to find the optimal values of the design parameters which optimize some kinematic and dynamic features of a robot. Stoughton [9] uses a numerical procedure for optimizing the workspace of a specific parallel manipulator whose length limits are known. Liu [6] characterizes some

J.-P. Merlet and B. Ravani (eds.), Computational Kinematics, 203–212.
© 1995 *Kluwer Academic Publishers.*

extremal positions as a function of the geometry of the robot and of the extremal link lengths. Gosselin [4] establishes a design rule for getting a spherical 3 d.o.f parallel manipulator with full rotation and has studied the optimization of the workspace of planar three d.o.f parallel manipulators [2].

Let A_i, B_i denote the attachment points of the link on the base and on the platform. For a set of A_i, B_i we attach a reference frame $O(x, y, z)$ to the base such that the z coordinate of A_i is equal to 0. In the same manner we attach to the platform a mobile frame $C(x_r, y_r, z_r)$ such that the z_r coordinate of B_i is equal to 0. A subscript r will denote a point or a vector whose coordinates are written in the mobile frame. Let α_i be the angle between the Ox axis and OA_i and let β_i be the angle between the Cx_r axis and CB_{i_r} (figure 1).

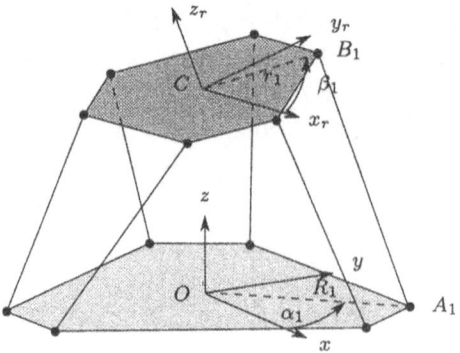

Figure 1. The design parameters for a set of points A_i, Bi are the distances R_1, r_1 between the points and the origins O, C of their frames.

Under these assumptions we have:

$$\mathbf{A_i O} = \begin{pmatrix} -R_1 \cos \alpha_i \\ -R_1 \sin \alpha_i \\ 0 \end{pmatrix} = R_1 \mathbf{u_i} \quad \mathbf{CB_{ir}} = \begin{pmatrix} r_1 \cos \beta_i \\ r_1 \sin \beta_i \\ 0 \end{pmatrix} = r_1 \mathbf{v_i} \quad (1)$$

where $\mathbf{u_i}, \mathbf{v_i}$ are constant unit vectors. The purpose of this paper is to determine the possible values of R_1, r_1 such that the workspace of the corresponding robot contains a given workspace. The following assumptions are made:

- the specific workspace is defined for a constant orientation of the platform.
- the specific workspace is defined by a set of segments.
- the minimum and maximum values of the leg lengths are known.
- the angles α_i, β_i are known.

The purpose of the next sections is to determine the boundary of the region in the R_1, r_1 plane (denoted $\mathcal{P}_{R_1 r_1}$) which defines the allowable values for R_1, r_1. Such regions will therefore be called the *allowable regions*.

2. Allowable region for the length constraints

2.1. ALLOWABLE REGION FOR ONE LEG

In this section we will assume that the constraints limiting the workspace are only the leg lengths. The minimum and maximum values of these lengths will be denoted $\rho_{min_i}, \rho_{max_i}$. The position of the robot is defined by the coordinates of C in the fixed frame and the rotation matrix R between the fixed and mobile frame. A trajectory is defined by two points $M_1(x_1, y_1, z_1)$, $M2(x_2, y_2, z_2)$, and for any C belonging to the trajectory we may write:

$$\mathbf{OC} = \mathbf{OM_1} + \lambda \mathbf{M_1 M_2} \tag{2}$$

where λ is a parameter in the range [0,1]. Let us calculate the leg length for a point C on the segment $M_1 M_2$. The leg length ρ is the norm of the vector $\mathbf{AB}$. We have:

$$\rho^2 = \|\mathbf{AO} + \mathbf{OC} + \mathbf{CB}\|^2 \tag{3}$$

Using equations (2,1) we can rewrite this equation as:

$$\begin{aligned}
\rho^2 =\ & R_1^2 + r_1^2 + R_1 r_1\ R\mathbf{v}.\mathbf{u} + R_1 \mathbf{u}.(\mathbf{OM1} + \lambda \mathbf{M_1 M_2})^T + \\
& r_1 R\mathbf{v}.(\mathbf{OM_1} + \lambda \mathbf{M_1 M_2})^T + \lambda^2 \mathbf{M_1 M_2}.\mathbf{M_1 M_2}^T + \\
& \mathbf{OM_1}.\mathbf{OM_1}^T + 2\lambda \mathbf{OM_1}.\mathbf{M_1 M_2}^T
\end{aligned}$$

This equation can also be written as:

$$\mathcal{E}(R_1, r_1, \lambda, \rho) = 0 \tag{4}$$

The structure of equation (4) leads to the following theorems:

Theorem 1 *For given values of ρ, λ the equation defines an ellipse in the $\mathcal{P}_{R_1 r_1}$ plane or has no solution in R_1, r_1.*

Theorem 2 *Let M be a point in the $\mathcal{P}_{R_1 r_1}$ plane. If we have $\mathcal{E} \leq 0$ (i.e. the point M is inside the ellipse) for a given λ (i.e. for a fixed position of the platform) then the corresponding length of the leg is less than or equal to ρ.*

2.1.1. *Computing the allowable region for the maximum length constraint* Consider now the functions $\mathcal{E}_{max}(\lambda) = \mathcal{E}(R_1, r_1, \lambda, \rho_{max})$ obtained for all the λ in [0,1]. These functions define a set of ellipses in the $\mathcal{P}_{R_1 r_1}$ plane,

206

each of which is called a *maximal ellipse*. If for any λ in the range $[0,1]$ we have $\mathcal{E}(R_1, r_1, \lambda, \rho_{max}) \leq 0$ then for any position of the platform on the trajectory, the leg length is less than or equal to ρ_{max}. Consequently the set of points $M(R_1, r_1)$ such that $\mathcal{E}_{max}(\lambda) \leq 0$ for any λ in $[0,1]$ defines the allowable region for the maximum length constraint. This means that any such point M must be inside all the ellipses in the set; therefore the allowable region with respect to the constraint $\rho \leq \rho_{max}$ is the intersection $\mathcal{I}$ of all the ellipses of the set. We denote by $\mathcal{E}_{max}(0)$ and $\mathcal{E}_{max}(1)$ the two ellipses obtained for $\lambda = 0$ and $\lambda = 1$. We have proved the following theorems:

Theorem 3 *As λ varies, the center of the corresponding ellipse lies on a segment which in some cases may be reduced to a point. The angle between the main axis of the ellipse and the R_1 axis is $\pi/4$.*

Theorem 4 *If the ellipses $\mathcal{E}_{max}(0)$, $\mathcal{E}_{max}(1)$ exist then either an ellipse exists for each value of λ in the range $[0,1]$ or the intersection of all the ellipses in the set is empty.*

Theorem 5 *The intersection $\mathcal{I}$ of all the ellipses in the set is equal to:*

$$\mathcal{E}_{max}(0) \cap \mathcal{E}_{max}(1)$$

Therefore the allowable region is simply the intersection of the ellipses computed for the extreme points of the trajectory.

2.1.2. *Computing the allowable region for the minimum length constraint*
Consider now the functions $\mathcal{E}_{min}(\lambda) = \mathcal{E}(R_1, r_1, \lambda, \rho_{min})$ in the $\mathcal{P}_{R_1 r_1}$ plane. These functions define a set of ellipses, each of which is called a *minimal ellipse*. If for a given point M and a given λ we have $\mathcal{E}_{min}(\lambda) > 0$, then the corresponding leg length is greater than ρ_{min}. Therefore this relation has to hold for any point belonging to the allowable region and for all λ in $[0,1]$. Consequently any point in the allowable region must lie outside the region $\mathcal{U}$ defined by $\mathcal{E}_{min}(\lambda) = 0$ i.e. the union of the minimal ellipses (a simple method to compute this union is described in [8]).

2.1.3. *Computing the allowable region for all the leg length constraints*
The computation of the allowable region for the leg length constraints is done using the following algorithm:

1. compute the two maximal ellipses for the extreme points of the trajectory
2. compute the intersection $\mathcal{I}$ of these two ellipses. If there is none then there is no allowable region.
3. compute the union $\mathcal{U}$ of the minimal ellipses.
4. subtract $\mathcal{U}$ from $\mathcal{I}$ to obtain the allowable region

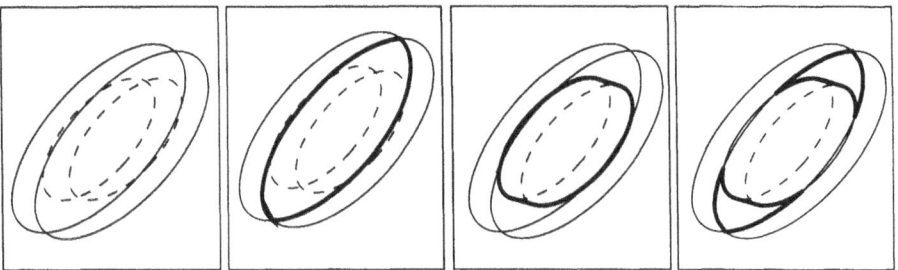

Figure 2. The computation of the allowable region. On the left side are drawn the maximal ellipse for the extreme point of the trajectory (in thin line) and the minimal ellipse for the same points (in dashed line). The intersection of the maximal ellipse is shown on the second drawing. The union of the forbidden ellipse is shown on the third drawing. The final allowable region is shown in thick line on the last drawing.

Figure 2 shows the result of this algorithm.

3. Allowable region for the mechanical limits on the joints

3.1. A MODEL FOR THE MECHANICAL LIMITS

We have described in [7] a method for modeling the mechanical limits on the passive joints. The mechanical limits of a joint can be described by a pyramid whose apex is the joint center and whose planar faces are such that if the joint constraints are satisfied, then the link will be inside the interior of the pyramid. For the joints attached to the base, the center of this pyramid is located at point A (Figure 3).

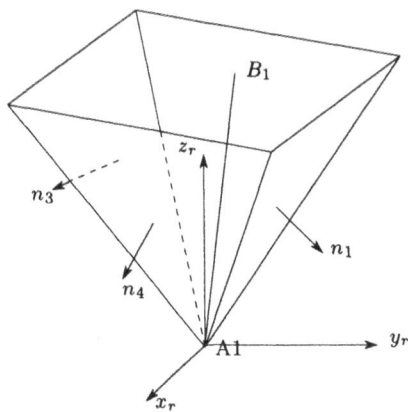

Figure 3. An example of constraint modeling for a passive joint located at A_1. If the mechanical limits of the joints are satisfied then link A_1B_1 is inside the volume bounded by the pyramid.

For the constraints on the passive joints attached to the platform a similar model may be used.

3.2. ALLOWABLE REGION FOR ONE LEG

Let $\mathbf{n_j^i}, j \in [1, k]$ denote the normals to the faces of the pyramid for the joint i. For a given position of the platform the leg $A_i B_i$ will lie inside the pyramid (which means that the position of the leg respects the mechanical limits of the joints) if $\mathbf{A_i B_j} . \mathbf{n_j^i}^T \leq 0 \quad \forall j \in [1, k]$. We consider a specific leg and a specific face of a pyramid. Using equation (2), we can rewrite the previous inequality as:

$$R_1 \mathbf{u} . \mathbf{n}^T + r_1 R \mathbf{v} . \mathbf{n}^T + \lambda \mathbf{M_1 M_2} . \mathbf{n}^T + \mathbf{OM_1} . \mathbf{n}^T \leq 0 \qquad (5)$$

Let us denote by $\mathcal{L}(R_1, r_1, \lambda)$ the left side of this inequality. The equation $\mathcal{L}(R_1, r_1, \lambda) = 0$ defines a pencil of lines in the $\mathcal{P}_{R_1 r_1}$ plane. All these lines have a constant slope. This pencil of lines defines a region in the plane whose boundary is constituted by the lines $\mathcal{L}(R_1, r_1, 0) = L_0$, $\mathcal{L}(R_1, r_1, 1) = L_1$. One of the lines L_0, L_1 separates the plane into two half-planes such that on one side of the line, each point $M(R_1, r_1)$ satisfies $\mathcal{L}(R_1, r_1, \lambda) \leq 0$ for each λ in [0,1], and on the other side, $\mathcal{L}(R_1, r_1, \lambda) > 0$ at least for one value of λ in [0,1]. Therefore this line defines a half-plane which is the allowable region for the joint and for this face of the pyramid.

This process is repeated for each face of the pyramid, leading to a set of half-planes. The intersection of these half-planes will be a closed region which defines the allowable region determined by the mechanical limits on the joint. An example of this computation is shown in figure 4.

As a specific example suppose that all the joints lie on two horizontal circles and we want to determine the possible radii of these two circles. The process is to compute the sets of allowed half-planes for all the joints and all the trajectories and compute their intersection.

4. Forbidden region determined by links interference

4.1. PRINCIPLE OF THE COMPUTATION OF THE ALLOWABLE REGION

In this section we will assume that the links have no thickness. We want to determine the zone in the $\mathcal{P}_{R_1 r_1}$ plane for which there is no interference between any pair of links, i.e. the zone for which the intersection point M between the lines i, j, if any, does not belong either to $A_i B_i$ or to $A_j B_j$. We will consider the case where $i = 1$ and $j = 2$ without loss of generality. If the two lines intersect then:

$$\mathbf{A_1 A_2} . (\mathbf{A_1 B_1} \times \mathbf{A_2 B_2})^T = 0 \qquad (6)$$

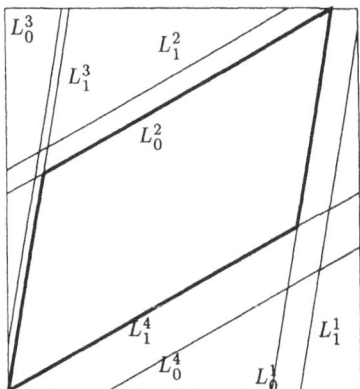

Figure 4. The mechanical limits of this particular joint are described by a four-faced pyramid. We have computed the separating half-plane for each of the faces by computing the L_0, L_1. The intersection of these half-planes defines a closed region (in thick line) which is the allowable region for this joint.

which can be written as:

$$- R_1 r_1 (b_1 \lambda + b_2 R_1 + b_3 r_1 + b_4) = 0 \tag{7}$$

where the b_i's are constants given by the geometry and the trajectory of the platform. Various cases can now be considered:

1. $b_1 = b_2 = b_3 = b_4 = 0$: the lines intersect no matter what the dimension of the robot is for any point on the trajectory.
2. $b_1 = 0$: the lines will intersect if the R_1, r_1 are on a line in the $\mathcal{P}_{R_1 r_1}$ plane
3. in the general case the lines will intersect if the R_1, r_1 are on a pencil of lines in the $\mathcal{P}_{R_1 r_1}$ plane

Each of these cases defines a region $\mathcal{R}$ in the $\mathcal{P}_{R_1 r_1}$ plane for which there is intersection of the two lines. In the first case this region is the full plane, in the second case the region is the line $b_2 R_1 + b_3 r_1 + b_4 = 0$ and in the last case the region is the zone bounded by the lines $b_1 + b_2 R_1 + b_3 r_1 + b_4 = 0$ and $b_2 R_1 + b_3 r_1 + b_4 = 0$, which are the extremal lines of the pencil. Consequently computing this region is easy. But we are interested only in the sub-region where the links intersect. To compute this sub-region we project the points A_1, A_2, B_1, B_2 onto the plane $O(x, y)$ of the reference frame and denote the projected points by a superscript ^{P}. If the segments $A_1 B_1, A_2 B_2$ intersect then their projection will also intersect. The intersection point will belong to the link in three cases. The first one occurs when:

$$(\mathbf{A_1 B_1}^P \times \mathbf{B_1 B_2}^P)_z > 0 \qquad (\mathbf{B_2 A_2}^P \times \mathbf{B_2 B_1}^P)_z < 0 \tag{8}$$

$$(\mathbf{A_1 A_2}^P \times \mathbf{A_1 B_1}^P)_z < 0 \qquad (\mathbf{A_2 A_1}^P \times \mathbf{A_2 B_2}^P)_z > 0 \tag{9}$$

where the subscript $_z$ denotes the z component of the vector. The second case occurs when:

$$(\mathbf{A_1B_1}^P \times \mathbf{B_1B_2}^P)_z < 0 \qquad (\mathbf{B_2A_2}^P \times \mathbf{B_2B_1}^P)_z > 0 \qquad (10)$$
$$(\mathbf{A_1A_2}^P \times \mathbf{A_1B_1}^P)_z > 0 \qquad (\mathbf{A_2A_1}^P \times \mathbf{A_2B_2}^P)_z < 0 \qquad (11)$$

The last case occurs when $A_1^P, A_2^P, B_1^P, B_2^P$ are collinear (in which case the previous inequalities become equalities). This may happen at one point on the trajectory or all along the trajectory, in which case the points A_1, A_2, B_1, B_2 lie in the same horizontal plane.

The quantities which appear on the left side of the inequalities can be expressed as functions of λ, R_1, r_1. They all have the same generic form:

$$R_1(e_1\lambda + e_2R_1 + e_3r_1 + e_4) \quad \text{or} \quad r_1(e_1\lambda + e_2R_1 + e_3r_1 + e_4)$$

which defines a pencil of lines. Note that the inequalities may define unbounded regions. As the other constraints on the workspace lead to bounded region we will consider only a limited portion of the $\mathcal{P}_{R_1r_1}$ plane, for example a square whose sides are equal to the maximum dimension of the rectangle which encloses all the maximal ellipses. After the computation of $\mathcal{R}$ we consider each set of inequalities. We then divide the square into four regions Q_i defined by

$$R_1 \geq 0 \quad r_1 \geq 0 \qquad R_1 \geq 0 \quad r_1 \leq 0 \qquad R_1 \leq 0 \quad r_1 \geq 0 \qquad R_1 \leq 0 \quad r_1 \leq 0$$

In each region Q_i the sign of the inequalities is now fully defined by four inequalities:

$$e_1^i\lambda + e_2^iR_1 + e_3^ir_1 + e_4^i \leq 0$$

By dealing with these four inequalities we are then able to determine the region of the square for which interference of links will occur. By repeating this process for each pairs of links and then taking the union of the results, we get the region of the $\mathcal{P}_{R_1r_1}$ plane for which at least one pair of links will interfere. Figure 5 shows an example of such a computation.

5. Verifying all the constraints

If we want to determine the geometries of the robots whose attachment points lie on two circles and which satisfy all the constraints, we compute the allowable regions defined for the leg length constraints and then compute the intersection of the result with the allowable region defined for the mechanical limits on the joints. Then we compute the forbidden region arising from the link interferences constraint and subtract it from the previous region. The result defines the location of all the possible attachment points.

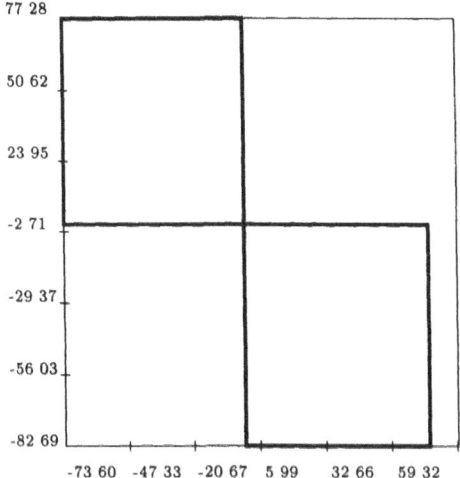

Figure 5. The region in the $\mathcal{P}_{R_1 r_1}$ plane for which link 0 will interfere with some other link.

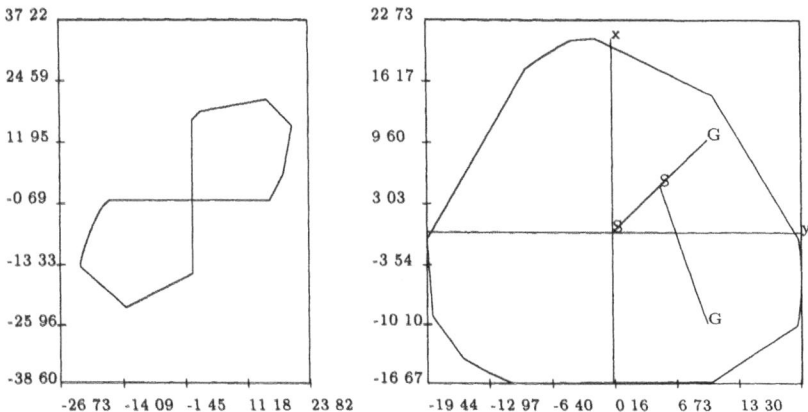

Figure 6. An allowable zone for the whole set of constraints. A robot has been defined by taking the values of R_1, r_1 inside the allowable region. The figure presents the workspace boundary and the trajectories for this particular robot.

Figure 6 shows an example of such a zone. This figure shows that the two trajectories lie inside the workspace of a robot whose parameters lie inside the allowable region.

6. Conclusion

The algorithm presented in this paper enables one to compute all the possible locations of the attachment points for any Gough-type parallel robot which can reach a specified workspace. It may be noted that although the algorithm has been described for a specific type of parallel robot, similar results hold for other mechanical architectures. After having determined the robots which can reach the desired workspace, we can use some other criterion to determine an "optimal" robot using a numerical algorithm with a search domain which is now considerably restricted. Possible criterion might be, for example, to minimize the maximum of the articular forces when the robot moves a given load in the specified workspace, or to minimize the maximum of the positioning errors for the platform for a given error of the sensors measuring the leg length. One objective of our future research will be to extend the description of the desired workspace to more complex geometrical objects like curves, polygons or polyhedra.

References

1. Claudinon B. and Lievre J. Test facility for rendez-vous and docking. In *36th Congress of the IAF*, pages 1–6, Stockholm, October 7-12, 1985.
2. Gosselin C. *Kinematic analysis optimization and programming of parallel robotic manipulators.* Ph.D. Thesis, McGill University, Montréal, June 15, 1988.
3. Gosselin C. Determination of the workspace of 6-dof parallel manipulators. *J. of Mechanical Design*, 112(3):331–336, September, 1990.
4. Gosselin C. and Angeles J. The optimum kinematic design of a planar three-degree-of-freedom parallel manipulator. *J. of Mechanisms, Transmissions and Automation in Design*, 110(1):35–41, March, 1988.
5. Gosselin C., Lavoie E., and Toutant P. An efficient algorithm for the graphical representation of the three-dimensional workspace of parallel manipulators. In *22nd Biennial Mechanisms Conf.*, pages 323–328, Scotsdale, September 13-16, 1992.
6. Liu K., Fitzgerald M.K., and Lewis F. Kinematic analysis of a Stewart platform manipulator. *IEEE Trans. on Industrial Electronics*, 40(2):282–293, April, 1993.
7. Merlet J-P. Détermination de l'espace de travail d'un robot parallèle pour une orientation constante. *Mechanism and Machine Theory*, 29(8):1099–1113, November, 1994.
8. Merlet J-P. Designing a parallel robot for a specific workspace. Research Report 2527, INRIA, April, 1995.
9. Stoughton R. and Kokkinis T. Some properties of a new kinematic structure for robot manipulators. In *ASME Design Automation Conf.*, pages 73–79, Boston, June 28, 1987.

ON THE ISOTROPIC DESIGN OF GENERAL SIX-DEGREE-OF-FREEDOM PARALLEL MANIPULATORS

K.E. ZANGANEH AND J. ANGELES
McGill Centre for Intelligent Machines &
Department of Mechanical Engineering, McGill University
817 Sherbrooke Street West, Montreal, Canada H3A 2K6
etemadi@cim.mcgill.ca, angeles@cim.mcgill.ca

Abstract.

Dexterity is considered as an important factor in the design of all robotic manipulators, whether serial or parallel. If the focus is on kinematics, rather than dynamics, one can quantify this factor using the condition number of the Jacobian matrix. In this paper, we study the optimum kinematic design of general six-degree-of-freedom (6-DOF) parallel manipulators, through the minimization of the condition number of their Jacobian matrices. Moreover, an isotropic architecture is shown to exist and an example of the isotropic design is included for illustration.

1. Introduction

Many factors are involved in the design of a robotic manipulator, among which dexterity is considered of the utmost importance. From a physical point of view, different concepts for manipulator dexterity have been proposed in the literature. For example, dexterity can be considered as a measure of the kinematic extent over which a manipulator can reach all orientations (Gupta and Roth, 1982; Vijaykumar et al., 1985), or as a measure of the goodness of grasping (Kobayashi, 1985). In terms of manipulator dynamics, dexterity can be a specification of the dynamic response (Yoshikawa, 1985a), or the ability of a manipulator to move and apply forces in arbitrary directions as easily as possible (Park and Brockett, 1994). On the other hand, these physical concepts can be classified as either *global* or *local*. In general, the former measures the overall performance of a ma-

213

J.-P. Merlet and B. Ravani (eds.), Computational Kinematics, 213–220.

nipulator in an averaged sense (Yoshikawa, 1985a; Gosselin and Angeles, 1991; Park and Brockett, 1994), the latter provides a measure in a particular configuration, which makes it applicable in redundancy resolution schemes, in the determination of the optimal workpiece location, and in the design of manipulators for accurate motions in a specific region of the workspace (Paul and Stevenson, 1983; Yoshikawa, 1985b; Klein and Blaho, 1987; Angeles et al., 1992).

If the focus is on the kinematics, rather than dynamics, one can quantify dexterity by defining a mathematical measure of it in terms of the Jacobian matrices that map the Cartesian velocities into the joint rates and vice versa. Moreover, the most commonly used measures of local dexterity are the *determinant*, the *condition number* and the *minimum singular value* of the Jacobian matrix (Klein and Blaho, 1987). Among these measures, the condition number appears as the most suitable index from an accuracy point of view, serial manipulators whose Jacobian matrix is capable of attaining a condition number of unity, being called *isotropic* (Salisbury and Craig, 1982). It should be noted that the actual value of the determinant cannot be used for quantitative accuracy estimates (Golub and Van Loan, 1990), as illustrated by the example below: Let us consider two $n \times n$ matrices B and D in the forms

$$B = \begin{bmatrix} 1 & -1 & \cdots & -1 \\ 0 & 1 & \cdots & -1 \\ \vdots & \vdots & \ddots & \vdots \\ 0 & 0 & \cdots & 1 \end{bmatrix} ; \qquad D = \mathrm{diag}(\, 10^{-1}, \ldots, 10^{-1}\,)$$

One can readily verify that D is isotropic while its determinant can approach zero as n increases. This is in contrast to B, whose determinant is unity, while its condition number increases with n according to $n2^{n-1}$.

The concept of kinematic isotropy has been used as a criterion in the design of planar and spherical parallel manipulators (Gosselin and Angeles, 1988; Gosselin and Lavoie, 1993; Mohammadi-Daniali and Zsombor-Murray, 1994). It has been pointed out that the kinematic analysis of parallel manipulators leads to two Jacobian matrices (Gosselin and Angeles, 1990). Based on the role that these matrices play in the kinetostatic transformations between joint and Cartesian variables, they are commonly referred to as the *forward* or *direct* and the *inverse* Jacobians. Hence, the optimization process, in general, involves two condition numbers that should be minimized simultaneously.

In this paper, we study the kinematic design of a general 6-DOF parallel manipulator, often referred to as the generalized Stewart platform (Innocenti and Parenti-Castelli, 1993), that can attain an isotropic configuration. Because of the number of design variables involved, this is done through the numerical minimization of the condition number. Moreover, an example of the isotropic design is included for illustration.

2. A General 6-DOF Parallel Manipulator

Figure 1 depicts a general 6-DOF parallel manipulator that comprises six legs, each having a prismatic actuator and connecting a moving platform (MP) to a base platform (BP). The legs at the ends are joined to the BP and MP by two sets of universal and spherical joints, respectively, the centers of these joints being the non-coplanar attachment points of the legs to the platforms. The attachment points of the ith leg on the BP and MP are denoted by $\{R_i\}_1^6$ and $\{P_i\}_1^6$, respectively. Moreover, two coordinate frames $\mathcal{B}$ and $\mathcal{M}$ are fixed to the BP and MP at points O and P.

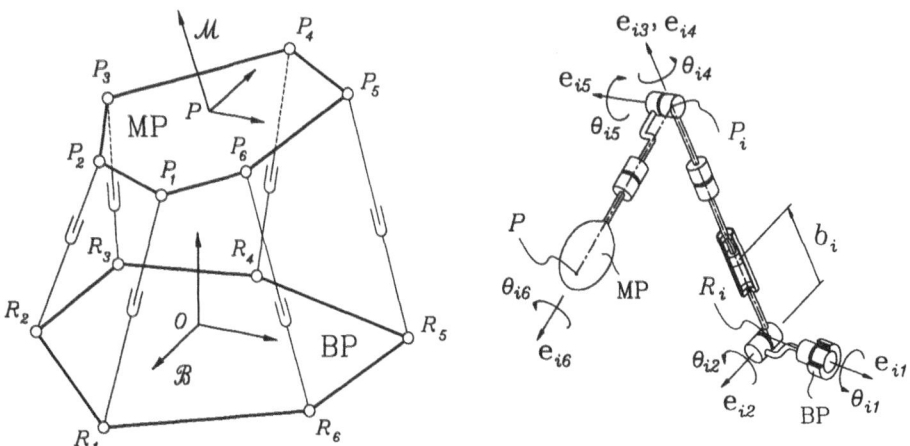

Figure 1. A general parallel manipulator   *Figure 2.* A detailed drawing of a leg

2.1. VELOCITY ANALYSIS

Figure 2 depicts a detailed drawing of the ith leg. In this figure, the jth joint variable of the ith leg is represented by θ_{ij}, while e_{ij} denotes the unit vector parallel to the axis of the jth joint of the ith leg. Note that both i and j vary from 1 to 6. To perform the velocity analysis of the manipulator, we follow the procedure introduced in (Zanganeh and Angeles, 1994). Thus, we consider the ith leg as a serial subchain and let $\hat{\pi}_i$ and π_i represent the vectors directed from point P to R_i and P_i, respectively. We can then write, for each leg,

$$t_P = J_i \dot{\theta}_i ; \qquad i = 1, \ldots, 6 \qquad (1)$$

where, $\dot{\boldsymbol{\theta}}_i \equiv [\dot{\theta}_{i1}, \ldots, \dot{\theta}_{i6}]^T$, $\boldsymbol{t}_P$ is the 6-dimensional twist vector of point P of the MP and

$$J_i \equiv \begin{bmatrix} e_{i1} & e_{i2} & 0 & e_{i4} & e_{i5} & e_{i6} \\ \hat{m}_{i1} & \hat{m}_{i2} & e_{i3} & m_{i4} & m_{i5} & m_{i6} \end{bmatrix}$$

$$\hat{m}_{ij} \equiv \hat{\pi}_i \times e_{ij}; \qquad m_{ik} \equiv \pi_i \times e_{ik}$$

In the above definitions, $\mathbf{0}$ denotes the 3-dimensional zero vector. Moreover, we note that the jth column of J_i is, in fact, the *Plücker* array of the screw axis of the jth joint in ray-coordinate form. Since the axes of the five unactuated joints in the ith leg intersect the line that passes through points R_i and P_i, for $i = 1, \ldots, 6$, those axes form a *linear complex*. Hence, the reciprocal screw to the screw of the unactuated joint of the ith leg is readily obtained in axis-coordinate form as

$$s_i = \begin{bmatrix} m_{i3} \\ e_{i3} \end{bmatrix}; \qquad i = 1, \ldots, 6 \qquad (2)$$

Upon multiplying both sides of eq.(1) by s_i^T, all unactuated-joint rates are eliminated, and we obtain

$$s_i^T t_P = \dot{b}_i; \qquad i = 1, \ldots, 6 \qquad (3)$$

where $\dot{b}_i$ is the actuated joint rate. The six equations within eq.(3), when combined into a single equation, take on the form

$$F t_P = \dot{b} \qquad (4)$$

In the above equation, $F \equiv [s_1, \ldots, s_6]^T$, $\dot{b} \equiv [\dot{b}_1, \ldots, \dot{b}_6]^T$ and the inverse Jacobian is simply the 6×6 identity matrix.

2.2. ISOTROPY CONDITIONS

In order to find an isotropic configuration of the manipulator, two matrices F_p and F_o are defined as

$$F_p \equiv \begin{bmatrix} m_1^T \\ \vdots \\ m_6^T \end{bmatrix} = \begin{bmatrix} (\pi_1 \times e_1)^T \\ \vdots \\ (\pi_6 \times e_6)^T \end{bmatrix}; \qquad F_o \equiv \begin{bmatrix} e_1^T \\ \vdots \\ e_6^T \end{bmatrix} \qquad (5)$$

where, in the above definitions, we let $m_i \equiv m_{i3}$ and $e_i \equiv e_{i3}$, for $i = 1, \ldots, 6$, to simplify the notation. Now, the manipulator can attain an isotropic configuration if the condition below holds for matrix F, namely,

$$F^T F \equiv \begin{bmatrix} \frac{1}{L_t^2} F_p^T F_p & \frac{1}{L_t} F_p^T F_o \\ \frac{1}{L_t} F_o^T F_p & F_o^T F_o \end{bmatrix} = \begin{bmatrix} \alpha^2 \mathbf{1} & \mathbf{O} \\ \mathbf{O} & \alpha^2 \mathbf{1} \end{bmatrix} \qquad (6)$$

where $\mathbf{1}$ and $\mathbf{O}$ are the 3×3 identity and zero matrices, respectively. Moreover, L_t is a scale factor that is introduced to dimensionally homogenize the vector space of the MP twists, and α is a real-valued constant. Thus, in terms of submatrices $\mathbf{F}_p$ and $\mathbf{F}_o$, the above condition can be expressed in the forms

$$\frac{1}{L_t^2} \mathbf{F}_p^T \mathbf{F}_p \equiv \frac{1}{L_t^2} \sum_{i=1}^{6} m_i m_i^T = \frac{1}{L_t^2} \sum_{i=1}^{6} (\pi_i \times e_i)(\pi_i \times e_i)^T = \alpha^2 \mathbf{1} \quad (7a)$$

$$\mathbf{F}_o^T \mathbf{F}_o \equiv \sum_{i=1}^{6} e_i e_i^T = \alpha^2 \mathbf{1} \quad (7b)$$

$$\frac{1}{L_t} \mathbf{F}_o^T \mathbf{F}_p \equiv \frac{1}{L_t} \sum_{i=1}^{6} e_i m_i^T = \frac{1}{L_t} \sum_{i=1}^{6} e_i (\pi_i \times e_i)^T = \mathbf{O} \quad (7c)$$

Moreover, by taking the trace of both sides of eqs.(7a & b), the relations below are readily derived:

$$\alpha^2 = 2 \, ; \qquad L_t^2 \equiv \frac{1}{6} \sum_{i=1}^{6} \|m_i\|^2 = \frac{1}{6} \sum_{i=1}^{6} \|\pi_i \times e_i\|^2 \quad (8)$$

where $\| \cdot \|$ represents the Euclidean norm of $(\cdot)$. On the other hand, the constraint on the magnitude of vector e_i is expressed in the form

$$\|e_i\| = 1 \, ; \qquad i = 1, \ldots, 6 \quad (9)$$

Equations (7a–c), (8) and (9) involve 25 constraint equations and 37 design variables, namely, L_t and the components of $\{e_i\}_1^6$ and $\{\pi_i\}_1^6$. Thus, in the absence of any other constraint, a twelve-parameter family of isotropic configurations can be exploited. Moreover, if we express the components of vector e_i in spherical coordinates, namely,

$$e_i \equiv [\, e_{ix}, \, e_{iy}, \, e_{iz} \,]^T = [\, \sin\phi_i \cos\theta_i, \, \sin\phi_i \sin\theta_i, \, \cos\phi_i \,]^T \, ; \, i = 1, \ldots, 6 \quad (10)$$

then, eq.(9) is identically verified and the number of design variables reduces to 31, while the number of constraints reduces to 19, thereby ending up with twelve free parameters. However, due to the large number of variables involved, their optimum values have to be determined numerically. This is done by formulating the problem as a constrained nonlinear optimization scheme, using conditions (7a–c).

3. Numerical Example

To start the numerical routine, a set of initial values for the design variables is needed. We obtained this set by considering the line-symmetric loop Bricard mechanism (Eddie Baker, 1980) as a platform-type parallel manipulator. The parameters of the manipulator thus obtained are then used to define the set of initial guesses, namely,

$$\pi_1 = [0, -.866, .320]^T; \quad \pi_2 = [-.750, -.433, 0]^T; \quad \pi_3 = [-.750, .433, .320]^T$$
$$\pi_4 = [0, .866, 0]^T; \quad \pi_5 = [.750, .433, .320]^T; \quad \pi_6 = [.750, -.433, 0]^T$$
$$e_1 = [0, -.707, -.707]^T; \quad e_2 = [-.612, -.354, .707]^T; \quad e_3 = [-.612, .354, -.707]^T$$
$$e_4 = [0, .707, .707]^T; \quad e_5 = [.612, .354, -.707]^T; \quad e_6 = [.612, -.354, .707]^T$$

Using the above set, we obtained an optimum set of design variables, corresponding to $L_t = 1$, namely,

$$\pi_1 = [.0122, -1.7494, .7547]^T; \quad \pi_2 = [-1.0271, -.5849, .0446]^T$$
$$\pi_3 = [-1.6859, 1.4652, 1.0133]^T; \quad \pi_4 = [-.0008, 1.2185, -.0119]^T$$
$$\pi_5 = [1.1436, .8126, .5827]^T; \quad \pi_6 = [1.1579, -.6472, .001]^T$$
$$e_1 = [-.4826, -.8609, .1611]^T; \quad e_2 = [-.3677, -.3690, .8536]^T$$
$$e_3 = [-.4818, .8608, .1638]^T; \quad e_4 = [-.0008, .5795, .8150]^T$$
$$e_5 = [1, 0, 0]^T; \quad e_6 = [.6322, -.2143, .7446]^T$$

Now, to define the geometry of the isotropic manipulator, we follow the procedure described below: First, let r_i and σ_i, for $i = 1, \ldots, 6$, represent the position vectors of vertices R_i and P_i in frames $\mathcal{B}$ and $\mathcal{M}$, respectively. Thus, we can readily derive the relations

$$\sigma_i = Q^T \pi_i; \qquad\qquad i = 1, \ldots, 6 \qquad (11a)$$
$$r_i = p + \pi_i - l_i e_i; \qquad\qquad i = 1, \ldots, 6 \qquad (11b)$$

where Q is the rotation matrix representing the orientation of frame $\mathcal{M}$ to frame $\mathcal{B}$, while p is the position vector of P in frame $\mathcal{B}$ and l_i represents the length of the ith leg. We now define the home configuration (HC) of the manipulator by assigning specific values to Q and p, namely,

$$Q = 1; \qquad p = [0, 0, p_z]^T$$

Then, from eqs.(11a-b), $\sigma_i = \pi_i$, for $i = 1, \ldots, 6$, and

$$r_1 = [.4948, -.8886, 1.5936]^T; \quad r_2 = [-.6593, -.2159, .1910]^T$$
$$r_3 = [-1.2040, .6044, 1.8494]^T; \quad r_4 = [0, .6391, .1731]^T$$
$$r_5 = [.1436, .8131, 1.5822]^T; \quad r_6 = [.5256, -.4329, .2565]^T$$

where, to obtain the above set, we let $\{l_i\}_1^6 = 1$ and $p_z = 1$. Figure 3 depicts the geometry of the synthesized isotropic manipulator. It should be noted that, to achieve a more practical architecture, one has to introduce extra constraints. This may, however, take the design away from isotropy

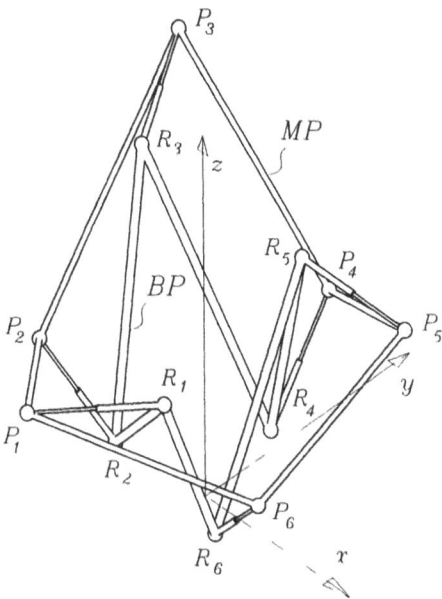

Figure 3. The isotropic parallel manipulator

(Pitten and Podhorodeski, 1993). Hence, depending on each application, the designer must specify a set of design requirements in order to attain a suitable architecture. A detailed discussion about the effects of introducing extra constraints on the optimum kinematic design of this type of manipulators can be found in (Zanganeh and Angeles, 1995).

4. Conclusions

In this paper, we introduced a local optimization scheme for the isotropic design of general 6-DOF parallel manipulators. This was done by imposing isotropy conditions on the velocity Jacobian of the manipulator. Moreover, using the theory of screws, we introduced a velocity analysis that readily led to the derivation of the foregoing Jacobian matrix. The numerical results showed that an isotropic design of the manipulator is possible. However, to achieve a practical design, one needs to introduce extra design constraints based on other requirements that arise from a specific application of the manipulator.

Acknowledgements

The research work reported here was made possible under NSERC (Natural Sciences and Engineering Research Council of Canada) Grant OGPIN 013.

220

References

Angeles, J., Ranjbaran, F., and Patel, R.V. (1992) On the Design of the Kinematic Structure of Seven-Axes Redundant Manipulators for Maximum Conditioning, *Proc. IEEE Int. Conf. on Robotics and Automation*, Nice, **Vol. 1**, pp. 494–499.

Eddie Baker, J. (1980) An Analysis of the Bricard Linkages, *Mech. Mach. Theory*, **Vol. 15**, pp. 267–286.

Golub, G.H., and Van Loan, C.F. (1990) *Matrix Computations.* The Johns Hopkins Univ. Press, Baltimore.

Gosselin, C., and Angeles, J. (1988) The Optimum Design of a Planar Three-Degree-of-Freedom Parallel Manipulator, *ASME J. Mech., Trans., Auto. in Design*, **Vol. 110**, pp. 35–41.

Gosselin, C., and Angeles, J. (1990) Singularity Analysis of Closed-Loop Kinematic Chains, *IEEE J. Robotics Automat.*, **Vol. 6 no. 3**, pp. 281–290.

Gosselin, C., and Angeles, J. (1991) A Global Performance Index for the Kinematic Optimization of Robotic Manipulators, *ASME J. Mech. Design*, **Vol. 113**, pp. 220–226.

Gosselin, C., and Lavoie, E. (1993) On the Kinematic Design of Spherical Three-Degree-of-Freedom Parallel Manipulators, *the Int. J. Robotics Res.*, **Vol. 12 no. 4**, pp. 394–402.

Gupta, K.C., and Roth, B. (1982) Design Considerations for Manipulator Workspace, *ASME J. Mech. Design*, **Vol. 104 no. 4**, pp. 704–712.

Innocenti, C., and Parenti-Castelli, V. (1993) Forward Kinematics of the General 6-6 Fully Parallel Mechanism: An Exhaustive Numerical Approach Via a Mono-Dimensional-Search Algorithm, *ASME J. Mech. Design*, **Vol. 115**, pp. 932–937.

Klein, C., and Blaho, B. (1987) Dexterity Measures for the Design and Control of Kinematically Redundant Manipulators, *the Int. J. Robotics Res.*, **Vol. 6 no. 2**, pp. 72–83.

Kobayashi, H. (1985) Control and Geometrical Considerations for an Articulated Robot Hand, *the Int. J. Robotics Res.*, **Vol. 4 no. 1**, pp. 3–12.

Mohammadi Daniali, H.R., and Zsombor-Murray, P.J. (1994) The Design of Isotropic Planar Parallel Manipulators, *Proc. 1st World Automation Congress*, Maui, **Vol. 2**, pp. 273–280.

Park, F.C., and Brockett, R.W. (1994) Kinematic Dexterity of Robotic Mechanisms, *the Int. J. Robotics Res.*, **Vol. 13 no. 1**, pp. 1–15.

Paul, R.P., and Stevenson, C.N. (1983) Kinematics of Robot Wrists, *the Int. J. Robotics Res.*, **Vol. 2 no. 1**, pp. 31–38.

Pitten, K.H., and Podhorodeski, R.P. (1993) A Family of Stewart Platforms with Optimal Dexterity, *J. Robotic Sys.*, **Vol. 10 no. 4**, pp. 463–479.

Salisbury, J.K., and Craig, J.J. (1982) Articulated Hands: Force Control and Kinematic Issues, *the Int. J. Robotics Res.*, **Vol. 1 no. 1**, pp. 4–12.

Vijaykumar, R., Tsai, M.J., and Waldron, K.J. (1986) Geometric Optimization of Manipulator Structures for Working Volume and Dexterity, *the Int. J. Robotics Res.*, **Vol. 5 no. 2**, pp. 99–111.

Yoshikawa, T. (1985a) Dynamic Manipulability of Robot Manipulators, *J. Robotic Sys.*, **Vol. 2 no. 1**, pp. 113–124.

Yoshikawa, T. (1985b) Manipulability of Robotic Mechanisms, *the Int. J. Robotics Res.*, **Vol. 4 no. 2**, pp. 3–9.

Zanganeh, K.E., and Angeles, J. (1994) Instantaneous Kinematics of Modular Parallel Manipulators, *Proc. ASME 23rd Mechanisms Conference*, Minneapolis, **DE-Vol. 72**, pp. 271–277.

Zanganeh, K. E., and Angeles, J. (1995) Kinematic Isotropy and the Optimum Design of Parallel Manipulators, *Centre for Intelligent Machines and Department of Mechanical Engineering Technical Report*, McGill University, Montreal.

STEWART PLATFORM BASED 6-AXIS FORCE AND TORQUE TRANSDUCERS

S.E. FENYI

Forschungszentrum Karlsruhe
Postfach 36 40
D – 76021 Karlsruhe

1. Introduction

Most 6-axis force and torque transducers (FTT) have a common feature: they are statically indeterminate. This fact induces some highly undesirable effects on the performance: the coupling of forces and moments, and the axial and lateral sensitivity differences. These drawbacks seriously impair the performance of the force control loop. A possible remedy is to choose determinate structures with the definitive payoff of better controllable performance indices. There is not much freedom to select such a design. It would have to be like a Stewart platform (SP). The only problem is the realizability. The central problem for the mechanical SP design is the choice of the bearings of the supporting structure. Due to these bearings we have only linear stress in the supporting limbs. Simple ball and socket joints are inapplicable. The backlash and stick slip caused by dry friction lead to intolerable nonlinearities as found in Gaillet and Reboulet [3]. They equipped the SP with a delicate bearingless isostatic supporting structure. This design is hard to miniaturize. Our proposal has SP geometry and is equipped with elastic joints, see Fig. 1. This monolithic design provides of two necks on the supporting Stewart limbs. The compliant behavior of the necks provides the partial torsional (rotary) and bending (cardanic) compensation of the limbs. The result of this compensation should be an approximate linear stress in the supporting framework. Our aim was to develop a linear elastic model with concentrated parameters so that we could examine the influence of the elastic joints. To cope with this task we have to solve a statically (or kinematically) indeterminate problem. The solution delivers the computed

221

J.-P. Merlet and B. Ravani (eds.), Computational Kinematics, 221–230.
© *1995 Kluwer Academic Publishers.*

stiffness matrix, and thus can be compared element by element with that of a SP based FTT with ideal frictionless joints.

Further we are making some other proposals for FTTs with multiple SP configuration, see [2]. These transducers are equipped with frictionless joints as is usual for the classic SP configurations regardless of their costs, see [11]. The supposed ideal joints make it possible to compute the stiffness matrix of the highly redundant structure, but it is also possible to use elastic joints which make the device technically more feasible. Now we summarize our proposals:

a) a statically determinate compliant version of FTT based on SP, with wrist and pedestal applications,

b) a statically indeterminate stiff version of FTT based on SP with elastic joints, with applications as in a).

c) a statically indeterminate compliant version of FTT based on SP in multiple configurations, with pedestal applications only.

For the proposal a) we apply geometrically nonlinear analysis. This enables us to check the limits of the linear analysis. We extensively applied results from line geometry and screw calculus.

2. Equilibrium of Forces

2.1 We need a concise vocabulary of line geometry. Lines are regarded as linear projective subvarieties of dimension 1 of the projective space $\mathbf{P}^3$. Lines can be represented by the following homogeneous Plcker vectors in $\mathbf{R}^6$

$$(\mathcal{L}) := (e_x, e_y, e_z; f_x, f_y, f_z)^T \qquad (1)$$

In this definition (e) is the unit vector of $\mathcal{L}$ and (f) its moment in a reference system of the Euclidean space $\mathbf{E}^3$ such that $(e) \perp (f)$. The lines of $\mathbf{P}^3$ correspond via def. (1) to points of $\mathbf{P}^5$. Because $(e) \perp (f)$, all these points belong to the Plcker quadric

$$\ll (e), (f) \gg = e_x f_x + e_y f_y + e_z f_z = 0 \qquad (2)$$

The elliptic polar of $(\mathcal{L})$ is the following row vector:

$$(\mathcal{L})^\tau := (f_x, f_y, f_z; e_x, e_y, e_z) \qquad (3)$$

The mutual invariant of two skew lines $\mathcal{L}_i$, $\mathcal{L}_j$ is the following inner product

$$\Omega_{ij} := (\mathcal{L})_i^\tau (\mathcal{L})_j = d_{ij} \sin \phi_{ij} \qquad (4)$$

with the shortest distance d_{ij} and the enclosed angle ϕ_{ij} between the oriented lines $(\mathcal{L})_i$, $(\mathcal{L})_j$. It is clear that $\Omega_{ij} = \Omega_{ji}$ and, formally, $\Omega_{ii} = \Omega_{jj} = 0$. The mutual invariants are reference free as (4) shows.

2.2 The parametric representation of line systems and the force equilibrium. Let us take $\nu = 3, 4, 5$ projectively independent lines in general positions and consider the following linear combinations:

$$(\mathcal{L}) = \sum_{\imath=1}^{\nu} \lambda_\imath (\mathcal{L})_\imath, \qquad (\nu = 3, 4, 5) \tag{5}$$

(5a) $\nu = 3$ lines define a regulus,
(5b) $\nu = 4$ lines define a linear congruence,
(5c) $\nu = 5$ lines define a linear complex.
The coefficients $\lambda_\imath$ are constrained: They have to satisfy the quadratic form

$$\sum_{\imath=1}^{\nu} \sum_{\jmath=1}^{\nu} \Omega_{\imath\jmath} \lambda_\imath \lambda_\jmath = 0 . \tag{6}$$

If (6) holds we say that $(\mathcal{L})$ through (5) depends linear-projectively on $\nu = 3, 4, 5$ independent lines. In the sequence (5) the lowest admissible value is $\nu = 3$, for the following reason: no other line can be linearly dependent on two projectively independent generally disposed (skew) lines, i.e. for any nonnull $\lambda_1, \lambda_2 : (\mathcal{L}) \neq \lambda_1(\mathcal{L})_1 + \lambda_2(\mathcal{L})_2$. This follows simply from the constraint (6).

THEOREM (2.2a) In $\mathbf{P}^3$ there are six projectively independent lines. For proof see [5, p. 325].

THEOREM (2.2b) (without proof): rank$[(\mathcal{L})_1, \ldots, (\mathcal{L})_6] = 6 \Leftrightarrow \mathcal{L}_1, \ldots, \mathcal{L}_6$ are projectively independent. We assemble the 6×6 matrix on the l.h.s. of this equivalence relation from the Plcker vectors (1).

THEOREM (2.2c) (without proof): every line of $\mathbf{P}^3$ is linearly dependent on 6 projectively independent lines

$$(\mathcal{L}) = \sum_{\imath=1}^{6} \lambda_\imath (\mathcal{L})_\imath. \tag{7}$$

The coefficients $\lambda_\imath$ $(1 \leq \imath \leq 6)$ in (7) can be regarded as *module* coordinates of a given L. They satisfy the projective quadratic (6) for $\nu = 6$ because the are constrained similary to (6). The decomposition (7) will play a major role in the partial calibration procedure of FTTs. Now we can briefly mention the challenging quest for full rank line systems. With (5a-c) we can give the following answer:

Six lines have full rank iff
(7a) not all six lines are members of the same regulus,
(7a) not all six lines are members of the same linear congruence,
(7a) not all six lines are members of the same linear complex.

These facts form the algebraic basis of the rank classes of line systems. We can therefore establish line systems with ranks 3, 4 and 5. To do this we can make the parametric representations (5) homogeneous by the following *involution* (Sylvester):

$$\sum_{i=1}^{\mu} R_i(\mathcal{L})_i = (0), \quad (\mu = 4, 5, 6). \tag{8}$$

If we interpret the coefficients R_i as the magnitudes of forces acting on the oriented lines $(\mathcal{L})_i$, we get the following classes of the *involutory* force equilibrium, see [16, p. 165]:

(8a) $\mu = 4$ forces are in equilibrium: all lines of action lie on the same hyperboloid, see [13, Vol. 1 p. 157] for details,

(8b) $\mu = 5$ forces are in equilibrium: all lines of action are members of the same linear congruence,

(8c) $\mu = 6$ forces are in equilibrium: all lines of action are members of the same linear complex.

In each case we can take the very last force as the equilibrant of the preceding ones, assuming that these preceding ones are projectively independent, corresponding to the conditions (5a-c); see [13, Vol. 1 p. 158].

Remark: the parametric representation of line systems (5) is equivalent with the representation by $6 - \nu$ linear homogeneous equations. In particular, if $\nu = 5$, a linear complex can be represented by a single equation and therefore the sequence (5a-c) can be replaced from the bottom up, by intersections of linear complexes.

To solve the homogeneous eq. (8) let us multiply them with the elliptic polars, see the definition (3)

$$\sum_{i=1}^{\mu} R_i(\mathcal{L})_j^T(\mathcal{L})_i = 0, \quad (1 \leq j \leq \mu) \quad \text{for} \quad \mu = 4, 5, 6 . \tag{9}$$

We get the following matrix form:

$$[\Omega]_\mu (R)_\mu = (0) \quad \text{for} \quad \mu = 4, 5, 6 \tag{10}$$

with the Gramian $[\Omega]_\mu := (\Omega_{ij})_{1 \leq i,j \leq \mu}$ and force magnitude vector $(R)_\mu = (R_i)_{1 \leq i \leq \mu} = (R_1, \ldots, R_\mu)^T$.

If the rank of the Gramian on the r.h.s. of (10) is $\mu - 1$, we will get one nontrivial solution for the vector of the force magnitudes $(R)_\mu$. This condition was fulfilled by the force systems (8a-c), i.e. they have exactly one nontrivial solution vector. It can be further shown that the constraint (6) for the inhomogeneous representation, and the condition for the nontrivial solution of the homogeneous eq. (10) are equivalent.

The solution of (10) is

$$(R_1, R_2, \ldots, R_\mu) = \rho(\sqrt{D_{11}}, \sqrt{D_{22}}, \ldots, \sqrt{D_{\mu\mu}}) \quad \forall \mu \qquad (11)$$

with the leading minors of the Gramian $D_{11}, \ldots, D_{\mu\mu}$ in (10) and a proportionality factor ρ. It is a remarkable fact that the solution (11) does not depend on the reference system like the equations (8). This invariance property corresponds intuitively to the fact that the force equilibrium cannot depend on the choice of the reference system.

2.3 Unconstrained line sums. If we drop the constraint (6) we introduce the *free* or *heteraptic* sum of lines

$$(\mathcal{D}) := \sum_{i=1}^{\nu} R_i(\mathcal{L})_i, \quad \nu \geq 2 \, . \qquad (12)$$

This definition is an admissible representation of a wrench

$$(\mathcal{D}) = (F_x, F_y, F_z; M_x, M_y, M_z)^T \qquad (13)$$

with its force and moment parts $(\mathcal{F}) := (F_x, F_y, F_z)^T$, $(\mathcal{M}) := (M_x, M_y, M_z)^T$. The definition (12) is sometimes called the reduction of force system acting on a rigid body. The action of ν forces is equivalent to the action of a wrench $(\mathcal{D})$ according to (12). The system of ν forces may comprise pure couples too. These represent the *Poinsot Pairs*.

The wrench $(\mathcal{D})$ has two invariants:

$$F_x^2 + F_y^2 + F_z^2 \, ; \quad F_x M_x + F_y M_y + F_z M_z \, , \qquad (14)$$

from which we can derive some others: the amplitude or intensity, and the pitch; see [5, p. 48]. The second part of (14) is the Plcker quadratic which is clearly nonnull because of the dropped constraint (6). The invariants (14) give us a clear hint that in $\mathbf{R}^6$ the isometric transformations preserve length and inner product ($\mathbf{R}^6$ is the associated vector space: $\mathbf{R}^6 \triangle \mathbf{P}^5$). To examine the invertibility of (12) let us treat it like (9):

$$\sum_{i=1}^{\nu} R_i(\mathcal{L})_j^\tau(\mathcal{L})_i = (\mathcal{L})_j^\tau(\mathcal{D}) \, , \quad (1 \leq j \leq \nu) \quad \text{for} \quad \nu \geq 2 \, . \qquad (15)$$

The entities on the r.h.s. are the line moments $M_j^l := (\mathcal{L})_j^\tau(\mathcal{D}) \, , \, (1 \leq j \leq \nu)$, originally defined by Chasles, see [12, p. 61].

We get the following systems of inhomogeneous linear equations:

$$[\Omega]_\nu(R)_\nu = (M^l)_\nu \quad \text{for} \quad \nu \geq 2 \qquad (16)$$

with the Gramian $[\Omega]_\nu := (\Omega_{ij})_{1 \le i, j \le \nu}$ and corresponding vectors $(R)_\nu := (R_i)_{1 \le i \le \nu}$ and $(M^l)_\nu := (M_i^l)_{1 \le i \le \nu}$.

The rank of the line system $[(\mathcal{L})_1, \ldots, (\mathcal{L})_\nu]$ now depends on the rank of the Gramian on the l.h.s. of (16) and therefore on its solvability. In the range $2 \le \nu \le 6$ we can make $\text{rank}[(\mathcal{L})_1, \ldots, (\mathcal{L})_\nu] = \nu$, therefore we can solve (16) for a given wrench. This solution is the decomposition of a given wrench in the line system $\mathcal{L}_1, \ldots, \mathcal{L}_\nu$ which we use now in a sense different from that in the preceeding section. The subcase $\nu = 2$ has paramount importance:

$$\begin{pmatrix} 0 & \Omega_{12} \\ \Omega_{21} & 0 \end{pmatrix} \begin{pmatrix} R_1 \\ R_2 \end{pmatrix} = \begin{pmatrix} M_1^l \\ M_2^l \end{pmatrix}. \tag{17}$$

This is the decomposition of a given wrench $(\mathcal{D})$ into two pure forces. This problem is immediately solvable; it was originally done by Chasles in other ways. The two skew lines $\mathcal{L}_1$ and $\mathcal{L}_2$ are the *reciprocal polars* of the linear complex defined by the wrench $(\mathcal{D})$ to be decomposed. For further details of Chasles' theorem see [12, p. 58]. The German term for this decomposition is *Kraftkreuz*; see for example [15]. The subcase $\nu = 6$ is central for FTTs because the rank of a line system is maximally 6. Let us consider eq. (12) for $\nu = 6$:

$$[(\mathcal{L})_1, (\mathcal{L})_2, (\mathcal{L})_3, (\mathcal{L})_4, (\mathcal{L})_5, (\mathcal{L})_6](R) = (\mathcal{D}) \tag{18}$$

with the vector $(R) := (R_1, R_2, R_3, R_4, R_5, R_6)^T$. The dual form of (18) is:

$$(\mathcal{D})^\tau [(\mathcal{L})_1, (\mathcal{L})_2, (\mathcal{L})_3, (\mathcal{L})_4, (\mathcal{L})_5, (\mathcal{L})_6] = (M_1^l, M_2^l, M_3^l, M_4^l, M_5^l, M_6^l) \tag{19}$$

with the elliptic polar $(\mathcal{D})^\tau := (M_x, M_y, M_z; F_x, F_y, F_z)$.

By assumption, rank $[(\mathcal{L})_1, (\mathcal{L})_2, (\mathcal{L})_3, (\mathcal{L})_4, (\mathcal{L})_5, (\mathcal{L})_6] = 6$, so that (18) leads to

THEOREM: (2.3a) (without proof): Every wrench can be decomposed into 6 pure forces.

This theorem is the theoretical fundament of the statically determinated FTTs. The formula (19) with the same rank condition leads to

THEOREM: (2.3b) (without proof): Every wrench is determined by 6 line moments.

Remark: there is a basic difference between a) $\nu = 2, 3, 4, 5$ and b) $\nu = 6$. The decomposition of a given wrench $(\mathcal{D})$ in subcase a) is possible with the Gramian (16) only. The subcase b) can be done directly without symmetrizing the eq. (12) as is shown by (18).

2.4 Compound equilibrium of forces. We will now give our geometric notions more physical content. Imagine a rigid body supported by κ limbs introduced by [5, p. 334]. These limbs cannot transmit couples from the foundation, but only exert a reactive force. We will call such limbs, Stewart limbs of fixed length (SLFL).

We consider ν forces acting on the supported rigid body. These forces induce κ reaction forces in the axes of the SLFLs. We can formally express the equilibrium conditions for the supported rigid body by

$$\sum_{i=1}^{\nu} R_i^a(\mathcal{L})_i = \sum_{j=1}^{\kappa} R_j^r(\mathcal{L})_j , \quad \nu \geq 1,\ \kappa \geq 1 . \tag{20}$$

The acting forces sum up to an acting wrench. We also consider a general acting wrench, i.e. $\nu \geq 2$, see Chasles' theorem (17).

Now (20) and (12) give

$$(\mathcal{D})^a = \sum_{j=1}^{\kappa} R_j^r(\mathcal{L})_j , \quad \kappa \geq 1 . \tag{21}$$

We suppose that in the range $1 \leq \kappa \leq 6$ rank $[(\mathcal{L})_1, \ldots, (\mathcal{L})_\kappa] = \kappa$.

THEOREM (2.4a): The equilibrium (21) is

(2.4.(i)) $\kappa < 6$ statically unstable,

(2.4.(ii)) $\kappa = 6$ statically determinate,

(2.4.(iii)) $\kappa > 6$ statically indeterminate.

We omit the proof. The literature on structural machanics shows some analogous theorems, see [19] for example. There is a kinematic counterpart of the above static theorem:

THEOREM (2.4b): With the same assumption as in theorem (2.4a) the supporting system is

(2.4.(i)') $\kappa < 6$ kinematically movable,

(2.4.(ii)') $\kappa = 6$ kinematically determinate,

(2.4.(iii)') $\kappa > 6$ kinematically indeterminate.

The subcase (2.4.(i)') can be proved as follows: It can be shown that a rigid body supported by κ SLFLs in the range $1 \leq \kappa \leq 6$ (and rank $[(\mathcal{L})_1, \ldots, (\mathcal{L})_\kappa] = \kappa$) has $6 - \kappa$ degrees of freedom (DOF); see [13, vol. 1 p. 50].

If $\kappa = 5$ the supported rigid body is allowed to move with 1 DOF. This constrained movement of SP with 5 SLFLs (and one active limb) is well known; see [9]. If $\kappa = 1$ the supported body is allowed to move with 5 DOF. It is well known that one SLFL is kinematicallly equivalent to one contacting point between 2 convex bodies. Such a constraint gives the contacting body exactly 5 DOF. The kinematic theorem (2.4b) fully explains why the subcase $\kappa = 2$ does not contradict Chasles' theorem (17): it cannot be expected that the center lines of the two SLFLs $\mathcal{L}_1$ and $\mathcal{L}_2$ fit the claim they should be reciprocal polars of the linear complex defined by the action wrench $(\mathcal{D})^a$. The only wrench that can be equilibrated is that for which the mentioned geometrical constraint holds. The supporting system for $\kappa = 2$ is therefore statically unstable as theorem (2.4a) states.

For $\kappa = 6$ we have obviously:

COROLLARY (2.4c): If a supporting system is statically determinate it is kinematically determinate too.

This corollary is the theoretical basis for the statics and kinematics of 6-axis FTTs.

Let us introduce the following abbreviated formulas for the statics of the FTT. With the definitions

$$[P] := [(\mathcal{L})_1, (\mathcal{L})_2, (\mathcal{L})_3, (\mathcal{L})_4, (\mathcal{L})_5, (\mathcal{L})_6] \quad \text{and}$$
$$(R^\ell) := (R_1^\ell, R_2^\ell, R_3^\ell, R_4^\ell, R_5^\ell, R_6^\ell)^T$$

(18) becomes concisely

$$[P](R^\ell) = (\mathcal{D}) , \tag{22}$$

this is the measurement equation of the FTT with rank $[P] = 6$ and (R^ℓ) is the vector of the supporting reaction forces ($l \simeq$ limb) which will be sensed. We must not forget the physical implications of the eq. (22), namely the force equilibrium of the mobile plate regarded as a rigid body. The action or load wrench $(\mathcal{D})$ will be equilibriated by precisely 6 reaction forces (R^ℓ). Because of the statical determinacy of this configuration the load determines the induced reaction forces in the supporting structure purely geometrically, without any elasticity theory. This fact is well known in classical statics, see the supports at [15, p. 36].

With the definitions preceding (22), eq. (7) becomes

$$[P](R^\ell) = (\mathcal{L}) \tag{23}$$

which is the partial calibration equation of the FTT. The calibrating load is now a pure unit force along $\mathcal{L}$.

Remark: eq. (22) admits the calibration with pure unit Poinsot couples. The reference invariance of eq. (23) was proved in the full report on this research: [2].

3. Conclusion and Outlook

In section 2 we collected a number of theorems widely scattered through classical mechanics and statics, which provide the theoretical basis of FTTs, regardless of their individual design details.

There is no place here to discuss our statically indeterminate proposals b) and c). We can find these details, especially the nonlinear analysis, in [2]. This report contains some details on the statics of degenerate or singular configurations of a redundant or nonredundant SLFL assembly, regardless of their use as FTT or parallel robot. The proposal in Fig. 1 was built as a laboratory probe about 30 mm in diameter for calibration purposes. The

9 mm version was also built, with very demanding methods of micromachining. The calibration of the 30 mm probe was not as successful as hoped. Especially the moment sensitivity was not satisfactory and a high scattering (about 20 %) in the sensitivity of the strain gages was also observed, which is basically not unusual.

We intend to build new probes with improved sensitivity.

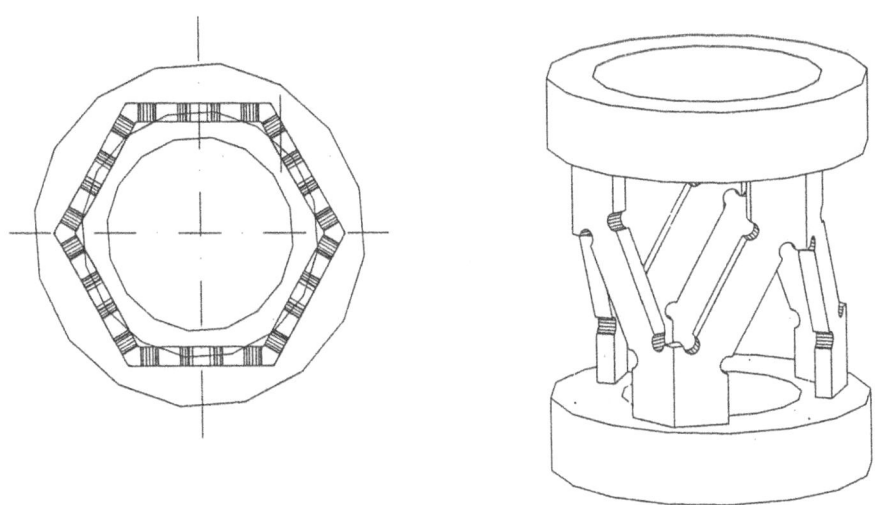

Fig. 1: Monolithic miniaturizable 6 – axis FTT design for microrobotics with elastic joints according to proposal b). The actual diameter of this device is about 9 mm. This design has an apparent similarity with the proposal discussed in [14] but the flexibility effect of the necks has quite another purpose. The necks actually compensate for the bending and torsion of the limbs.

References

1. Doyle, J.F. (1991) Static and Dynamic Analysis of Structures. Kluwer Acad. Publ., Dordrecht
2. Fenyi, S.E., Neisius, B. (1995) Miniaturizable and nonminiaturizable 6 – axis force and torque transducers. HIT Internal Report, Forschungszentrum Karlsruhe
3. Gaillet, A., Reboulet, C. (1983) An isostatic six component force and torque sensor. 13th International Symposium on Industrial Robots, Proceedings, Chicago, Illinois pp. 18 – 102, 18 – 111

4. Ghali, A., Neville, A.M. (1978) Structural Analysis. Chapman Hall, London
5. Hunt, K.H. (1978) Kinematic Geometry of Mechanisms. Clarendon Press, Oxford
6. Jessop, C.M. (1969) A Treatise on the Line Complex. Chelsea, New York
7. Livesley, R.K. (1964) Matrix Methods of Structural Analysis. Pergamon Press, Oxford
8. Kerr, D.R. (1988) Analysis, properties, and design of a Stewart-platform transducer. Trends and Developements in Mechanisms, Machines, and Robotics – 1988 ASME Design Technology Conferences, New York pp. 139–145
9. Merlet, J.-P. (1987) Parallel Manipulators. INRIA Rapports de Recherche N°646, Valbonne, France
10. Merlet, J.-P. (1988) Parallel Manipulator Part 2: Singular configurations and Grassmann Geometry. INRIA Research Report No. 791, Valbonne
11. Merlet, J.-P. (1990) Personal communication
12. Nielsen, J. (1935) Vorlesungen ber elementare Mechanik. Springer, Berlin
13. Phillips, J. (1984) (1990) Freedom in Machinery, Vol. 1-2, Cambridge Univ. Press, Cambridge
14. Rosen, C.A., Nitzan, D. (1977) Use of sensors in programmable automation. IEEE Computer, Vol. 10, Nr. 12, Reprinted in Tutorial on Robotics (eds. Lee, C.S.G., Gonzalez, R.C., Fu, K.S.) IEEE Press, New York 1986
15. Sattler, K. (1969) Lehrbuch der Statik, Vol. I/A. Springer, Berlin
16. Timerding, H.E. (1902) Geometrische Grundlegung der Mechanik eines starren Krpers. Enzyklopdie der Mathematischen Wissenschaften Vol. IV.2 pp. 125–189 (eds. Klein, F., Mller, C.) Teubner, Leipzig
17. Veblen, O., Young, J.W. (1910) Projective Geometry. 3 Vols. Ginn Boston
18. Waldron, K.J., Wang, S.L. (1987) A study of the singular configurations of serial manipulators. Transaction of the ASME, Journal of Mech., Transmiss., and Automation in Design Vol. 109, pp. 14–20
19. West, H.H. (1993) Fundamentals of Structural Analysis. Wiley, New York
20. Zindler, K. (1921) Algebraische Liniengeometrie. Enzyklopdie der Mathematischen Wissenschaften. (eds. Klein, F., Mller, C.) B.G. Teubner Leipzig Vol. III C8. pp. 972 – 1228

A CLOSED-FORM SOLUTION FOR THE DIRECT KINEMATICS OF A SPECIAL CLASS OF SPHERICAL THREE-DEGREE-OF-FREEDOM PARALLEL MANIPULATORS

C.M. GOSSELIN AND M. GAGNÉ
Département de Génie Mécanique
Université Laval
Québec, Québec, Canada, G1K 7P4

Abstract. It has been shown elsewhere that the solution of the direct kinematic problem of spherical three-degree-of-freedom parallel manipulators leads to a maximum of 8 solutions. Moreover, a polynomial of degree 8 can be obtained, whose roots will lead to all the solutions of the problem. In this paper, a particular geometry of spherical parallel manipulator is studied. This geometry arises from kinematic optimization which has been performed in previous work. The direct kinematic problem associated with this special architecture is studied here and it is shown that a simple closed-form solution can be obtained for this manipulator, which contrasts with the very complex polynomial solution obtained for the general case. This work is mainly motivated by the real-time trajectory planning and control of a prototype of parallel manipulator which is based on the simplified geometry studied here.

1. Introduction

Parallel manipulators were introduced in robotics almost two decades ago because of their inherent stiffness and their dynamic properties. Among other architectures which have attracted the attention of robotic engineers, spherical three-degree-of-freedom parallel manipulators have been investigated [1], [2], [3], [4], [5]. A spherical three-degree-of-freedom parallel manipulator could be used, for instance, as a stiff orientation wrist in robotics

231

J.-P. Merlet and B. Ravani (eds.), Computational Kinematics, 231–240.
© 1995 *Kluwer Academic Publishers.*

or as a mechanism for the orientation of machine tool beds. Prototypes of such manipulators are presented in [3], where a robotic shoulder module is described and in [6] where a high-performance camera-orienting device is introduced.

A general solution to the direct kinematic problem of spherical parallel manipulators has been presented in [7], [8], [9] and [10]. This solution leads to a polynomial of degree 8 whose roots lead to the feasible poses of the platform associated with given actuator coordinates. In general, this polynomial is very complex and cannot be used in real-time applications.

In this paper, the direct kinematic problem associated with a particular geometry of spherical three-degree-of-freedom parallel manipulator is studied. This geometry has been obtained through kinematic dexterity optimization [5] and has been used to build the prototype of the *agile eye* described in [6]. It is shown here that, for this particular geometry, the complexity of the direct kinematic problem reduces drastically and that a closed-form solution can be obtained. This result is of great interest in the context of real-time control of the prototype, which motivated the work presented in this paper.

2. Kinematics of spherical three-degree-of-freedom parallel manipulators

A spherical three-degree-of-freedom parallel manipulator is shown in Fig. 1. It consists of a platform connected to a fixed base via three kinematic chains. Each of the chains is composed of two intermediate links and three revolute joints. The structure of the manipulator is such that the axes of all nine revolute joints intersect at one common point which is referred to as the *center* of the mechanism. The three motors of the manipulator are fixed to the base and hence, only the revolute joints connecting each of the three chains to the base are actuated. Unit vectors pointing outwards from the center of the mechanism are defined along the axis of each of the revolute joints. The unit vectors along the axes of the actuators (fixed axes) are noted $\mathbf{u}_i$, $i = 1, 2, 3$ while the unit vectors along the axes attached to the platform are noted $\mathbf{v}_i$, $i = 1, 2, 3$. Finally, the unit vectors defined along the axes of the intermediate joints are noted $\mathbf{w}_i$, $i = 1, 2, 3$. The link angles are assumed to be identical on each of the kinematic chains connecting the base to the platform and are noted α_1 and α_2. Additionally, the platform and the base are assumed to be symmetric and the angle between the fixed axes (base) is noted γ_1 while the angle between the axes on the platform is noted γ_2. This is represented schematically on Fig. 1. Finally, the actuator angles are noted θ_i, $i = 1, 2, 3$ and are measured along the fixed axes — defined by vectors $\mathbf{u}_i$, $i = 1, 2, 3$ — with respect to a given reference.

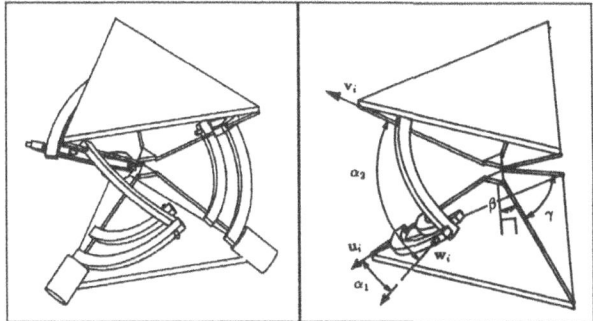

Figure 1. General architecture of a spherical three-degree-of-freedom parallel manipulator with revolute actuators.

The reference configuration of the platform is defined as the one for which vectors $\mathbf{v}_i$, $i = 1,2,3$ have a specific reference orientation noted $\mathbf{v}_{i0}$, $i = 1,2,3$. Moreover, the orientation of the platform is defined using a rotation matrix $\mathbf{Q}$ which relates the current orientation of vector $\mathbf{v}_i$ to the reference orientation $\mathbf{v}_{i0}$, i.e.,

$$\mathbf{v}_i = \mathbf{Q}\mathbf{v}_{i0}, \qquad i = 1,2,3 \tag{1}$$

Since the platform is in pure rotation about the center of the mechanism, its pose is completely specified by matrix $\mathbf{Q}$.

Using the above conventions, one can write the solution of the inverse and direct kinematic problems (see for instance [8], [9] and [10]). As mentioned above, the solution to the direct kinematic problem leads to very complex expressions and cannot be expressed in closed-form.

In order to reduce the number of equations involved in these derivations, Euler angles are used to describe the orientation of the platform. An Euler angle convention involving successive rotations about the Z, Y and X axes is chosen here. The rotation matrix describing the orientation of the platform can be written as

$$\mathbf{Q} = \begin{bmatrix} c_1c_2 & c_1s_2s_3 + s_1c_3 & -c_1s_2c_3 + s_1s_3 \\ -s_1c_2 & -s_1s_2s_3 + c_1c_3 & s_1s_2c_3 + c_1s_3 \\ s_2 & -c_2s_3 & c_2c_3 \end{bmatrix} \tag{2}$$

where c_i and s_i stand for $\cos\phi_i$ and $\sin\phi_i$ and where ϕ_1, ϕ_2 and ϕ_3 are the angles associated respectively with each of the successive rotations.

3. Special geometry of the spherical three-degree-of-freedom parallel manipulator

In [5], it was shown that entire families of isotropic spherical three-degree-of-freedom parallel mechanisms exist, i.e., families of mechanisms which lead

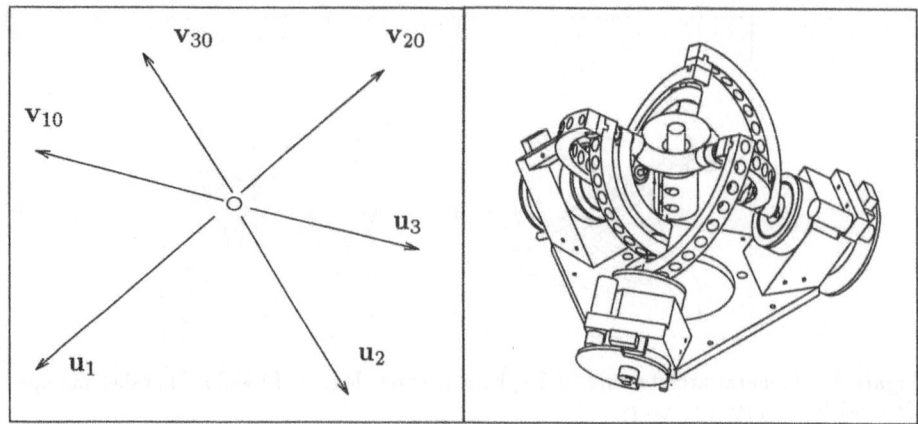

Figure 2. Unit vectors on the base and on the platform in the isotropic configuration for the special architecture and the camera-orienting device based on this architecture.

to the maximum kinematic accuracy for certain points in their workspace. Isotropic loci were shown in the space of the design parameters. These results have been generalized in [11]. In [6], one of the isotropic designs identified in the aforementioned work was used to built a high-performance camera orienting device. The motivation behind the work reported here arises from this special geometry of the spherical three-degree-of-freedom parallel manipulator.

Following the notation defined in the preceding section, one obtains the special geometry of the manipulator presented in [6] by setting

$$\alpha_1 = \alpha_2 = \gamma_1 = \gamma_2 = \pi/2 \tag{3}$$

In other words, the fixed unit vectors $\mathbf{u}_i$, $i = 1, 3, 2$ define an orthonormal reference frame on the base while unit vectors $\mathbf{v}_i$, $i = 1, 2, 3$ define an orthonormal reference frame on the platform. Moreover, the isotropic configuration is the one in which one has

$$\mathbf{v}_1 = -\mathbf{u}_3, \quad \mathbf{v}_2 = -\mathbf{u}_1, \quad \mathbf{v}_3 = -\mathbf{u}_2 \tag{4}$$

This is illustrated in Fig. 2. The resulting mechanism [6] is also shown in Fig. 2 in its isotropic configuration, which is used here as the reference configuration. Therefore, vectors $\mathbf{v}_i$, $i = 1, 2, 3$ are shown in Fig. 2 as the reference vectors $\mathbf{v}_{i0}$, $i = 1, 2, 3$.

Since the unit vectors on the base, vectors $\mathbf{u}_i$, $i = 1, 3, 2$ define an orthonormal reference frame, they will be used as the fixed reference frame. One then has

$$\mathbf{u}_1 = [1\,0\,0]^T, \quad \mathbf{u}_2 = [0\,0\,1]^T, \quad \mathbf{u}_3 = [0\,1\,0]^T \tag{5}$$

Moreover, using the reference configuration represented in Fig. 2, one can write

$$\mathbf{v}_{10} = -\mathbf{u}_3, \quad \mathbf{v}_{20} = -\mathbf{u}_1, \quad \mathbf{v}_{30} = -\mathbf{u}_2 \tag{6}$$

The actuated joint coordinates are then defined in the trigonometric direction around each of the $\mathbf{u}_i$ vectors. Since the proximal link angle, α_1, is equal to $\pi/2$, this leads to

$$\mathbf{w}_1 = \begin{bmatrix} 0 \\ -\sin\theta_1 \\ \cos\theta_1 \end{bmatrix}, \quad \mathbf{w}_2 = \begin{bmatrix} -\sin\theta_2 \\ \cos\theta_2 \\ 0 \end{bmatrix}, \quad \mathbf{w}_3 = \begin{bmatrix} \cos\theta_3 \\ 0 \\ -\sin\theta_3 \end{bmatrix} \tag{7}$$

For a given orientation of the platform, vectors $\mathbf{v}_i$, $i = 1, 2, 3$ are easily computed using eq.(1) since matrix $\mathbf{Q}$ is known. Moreover, since the distal link angle, α_2, is equal to $\pi/2$, one can write

$$\mathbf{w}_i \cdot \mathbf{v}_i = 0, \qquad i = 1, 2, 3 \tag{8}$$

The substitution of eq.(7) in eq.(8) then leads to simple equations in the sine and cosine of the actuated joint coordinates which are readily solved as

$$\tan\theta_1 = (v_{1z}/v_{1y}), \quad \tan\theta_2 = (v_{2y}/v_{2x}), \quad \tan\theta_3 = (v_{3x}/v_{3z}) \tag{9}$$

where v_{ix}, v_{iy} and v_{iz} are respectively the x, y and z components of vector $\mathbf{v}_i$.

The solution of the inverse kinematic problem for the special geometry of manipulator is simpler than for the general case since no quadratic equation is involved. As in the general case, 8 solutions are obtained because of the two branches of the inverse tangent function.

4. Direct kinematic problem for the special geometry

The solution of the direct kinematic problem for the special geometry will now be addressed. Using the expression given above for matrix $\mathbf{Q}$ — eq.(2) — and substituting in eq.(1) together with eqs.(5) and (6), expressions for vectors $\mathbf{v}_i$, $i = 1, 2, 3$ as functions of the Euler angles ϕ_2, ϕ_2 and ϕ_3 are obtained. Then, these expressions are substituted in eq.(8) together with eq.(7). This leads to three equations in the three unknown ϕ_1, ϕ_2 and ϕ_3 which can be written as

$$\begin{aligned} -\sin\theta_1(\sin\phi_1\sin\phi_2\sin\phi_3 - \cos\phi_1\cos\phi_3) & \\ + \cos\theta_1\cos\phi_2\sin\phi_3 &= 0 \end{aligned} \tag{10}$$

$$\sin\theta_2\cos\phi_1\cos\phi_2 + \cos\theta_2\sin\phi_1\cos\phi_2 = 0 \tag{11}$$

$$\begin{aligned} \cos\theta_3(\cos\phi_1\sin\phi_2\cos\phi_3 - \sin\phi_1\sin\phi_3) & \\ + \sin\theta_3\cos\phi_2\cos\phi_3 &= 0 \end{aligned} \tag{12}$$

The solution of these three equations for angles ϕ_1, ϕ_2 and ϕ_3 will give the solution of the direct kinematic problem. As can be easily noticed, the equations obtained here for the special geometry are much simpler than the equations obtained — see for instance [9] — for the general case. In fact, because of the fixed reference frame used here and because of a judicious choice of the Euler angle sequence, eq.(11) can be rewritten as

$$\cos\phi_2(\cos\theta_2\sin\phi_1 + \sin\theta_2\cos\phi_1) = \cos\phi_2\sin(\phi_1 + \theta_2) = 0 \qquad (13)$$

which leads to two distinct sets of solutions which have a different geometric interpretation. These sets are obtained with

$$\sin(\phi_1 + \theta_2) = 0, \quad \text{and} \quad \cos\phi_2 = 0 \qquad (14)$$

respectively. They will be discussed separately in the next subsections.

4.1. FIRST SET OF SOLUTIONS

The first set of solutions is obtained through the first condition of eq.(14), which leads directly to two solutions for angle ϕ_1, i.e.,

$$\phi_1 = -\theta_2, \quad \text{or} \quad \phi_1 = \pi - \theta_2 \qquad (15)$$

However, from the definition of the Euler angles used here — eq.(2) — and from eqs.(10) and (12), it is clear that these two solutions will lead to sets of Euler angles which will correspond to identical orientations. Therefore, only one of these solutions need to be considered. The first one will be used here.

Once angle ϕ_1 has been determined, eqs.(10) and (12) can be rewritten as follows

$$\begin{aligned}
A_1\cos\phi_3 + B_1\sin\phi_3 &= 0 & (16)\\
A_2\cos\phi_3 + B_2\sin\phi_3 &= 0 & (17)
\end{aligned}$$

with

$$\begin{aligned}
A_1 &= \cos\phi_1\sin\theta_1 & (18)\\
B_1 &= \cos\phi_2\cos\theta_1 - \sin\phi_1\sin\phi_2\sin\theta_1 & (19)\\
A_2 &= \cos\phi_1\sin\phi_2\cos\theta_3 + \cos\phi_2\sin\theta_3 & (20)\\
B_2 &= -\sin\phi_1\cos\phi_3 & (21)
\end{aligned}$$

Since $\cos\phi_3$ and $\sin\phi_3$ cannot vanish simultaneously, eqs.(16) and (17) lead to

$$A_1B_2 - A_2B_1 = 0 \qquad (22)$$

The substitution of eqs.(18) to (21) into eq.(22) then leads to, after rearranging

$$\cos \phi_2 (C_1 \cos \phi_2 + C_2 \sin \phi_2) = 0 \qquad (23)$$

with

$$\begin{aligned}
C_1 &= \cos \phi_1 \sin \phi_1 \sin \theta_1 \cos \theta_3 + \cos \theta_1 \sin \theta_3 & (24) \\
C_2 &= \cos \phi_1 \cos \theta_1 \cos \theta_3 - \sin \phi_1 \sin \theta_1 \sin \theta_3 & (25)
\end{aligned}$$

Since the first factor in eq.(23) corresponds to the second set of solution identified above — and which will be described in the next subsection —, it can be ignored in the present solution and one can then use the second factor to obtain

$$\phi_2 = \tan^{-1}(-C_1/C_2) \qquad (26)$$

which leads to two solutions for ϕ_2 since C_1 and C_2 are known. Once the values of ϕ_2 have been obtained, eqs.(16) or (17) can be used to compute the value of angle ϕ_3, i.e.,

$$\phi_3 = \tan^{-1}(-A_1/B_1) = \tan^{-1}(-A_2/B_2) \qquad (27)$$

which leads to two solutions for angle ϕ_3 for each value of angle ϕ_2.

Hence, 4 solutions to the direct kinematic problem will be obtained in this first set of solutions. They correspond to 4 distinct orientations of the moving platform since the sets of Euler angles duplicating these solutions have been eliminated from the outset. It is also noted that these solutions are obtained through very simple closed-form expressions which involve only two inverse tangent functions and a few multiplications. A numerical example illustrating this set of solutions will be presented in a later section.

4.2. SECOND SET OF SOLUTIONS

The second condition of eq.(14) will now be considered. This condition leads to

$$\phi_2 = \pi/2, \quad \text{or} \quad \phi_2 = -\pi/2 \qquad (28)$$

Considering the first solution — $\phi_2 = \pi/2$ — , one can simplify eqs.(10) and (12) and obtain

$$\cos(\phi_1 + \phi_3) = 0 \qquad (29)$$

Moreover, matrix $\mathbf{Q}$ can be rewritten, in this case, as

$$\mathbf{Q} = \begin{bmatrix} 0 & \sin(\phi_1 + \phi_3) & -\cos(\phi_1 + \phi_3) \\ 0 & \cos(\phi_1 + \phi_3) & \sin(\phi_1 + \phi_3) \\ 1 & 0 & 0 \end{bmatrix} \qquad (30)$$

which, together with eq.(29), leads to two possible solutions, i.e.,

$$\mathbf{Q} = \begin{bmatrix} 0 & 1 & 0 \\ 0 & 0 & 1 \\ 1 & 0 & 0 \end{bmatrix}, \text{ and } \mathbf{Q} = \begin{bmatrix} 0 & -1 & 0 \\ 0 & 0 & -1 \\ 1 & 0 & 0 \end{bmatrix} \tag{31}$$

Similarly, considering the second solution of eq.(28), one can simplify eqs.(10) and (12) and obtain

$$\cos(\phi_1 - \phi_3) = 0 \tag{32}$$

Moreover, matrix $\mathbf{Q}$ can be rewritten, in this case, as

$$\mathbf{Q} = \begin{bmatrix} 0 & \sin(\phi_1 - \phi_3) & \cos(\phi_1 - \phi_3) \\ 0 & \cos(\phi_1 - \phi_3) & -\sin(\phi_1 - \phi_3) \\ -1 & 0 & 0 \end{bmatrix} \tag{33}$$

which, together with eq.(32), leads to two possible solutions, i.e.,

$$\mathbf{Q} = \begin{bmatrix} 0 & 1 & 0 \\ 0 & 0 & -1 \\ -1 & 0 & 0 \end{bmatrix}, \text{ and } \mathbf{Q} = \begin{bmatrix} 0 & -1 & 0 \\ 0 & 0 & 1 \\ -1 & 0 & 0 \end{bmatrix} \tag{34}$$

which completes the second set of four solutions. The unit vectors of the moving platform associated with these 4 solutions are readily obtained from eq.(1).

It is pointed out that this second set of solutions, i.e., the four solutions identified above, is independent from the specified actuated joint coordinates. Indeed, these orientations of the platform can be attained with any actuated joint coordinates since they correspond to singularities of a higher order. When the platform is in one of these configurations, the actuators can be moved arbitrarily and will not affect the pose of the platform. This can also be seen from the inverse kinematics, i.e., eq.(9) leads to indeterminacy. In practice, such configurations should not be inside the workspace of the manipulator. Hence, since this set of solution can be readily eliminated, the direct kinematic problem for the special architecture of manipulator treated here can be limited to the first set of solutions presented in the preceding subsection. The latter set of solutions leads to simple closed-form expressions and requires only a few computations. Therefore, it can be implemented in real-time, which is another advantage of this particular geometry. With the prototype of parallel manipulator presented in [6], this procedure, together with the dynamic model of the manipulator — which is also simplified because of the particular geometry — can be performed at 500 Hz using a controller which is based on a single DSP chip.

5. Numerical example

A numerical example is now presented in order to illustrate the procedure described above. First, an orientation of the moving platform of the mechanism is specified. Then, joint angles are obtained through the solution of the inverse kinematic problem. These angles are then used for the solution of the direct kinematic problem and the initial orientation of the platform is reproduced.

The orientation of the platform used for the solution of the inverse kinematic problem is such that unit vectors $\mathbf{v}_i$, $i = 1, 2, 3$ are given by

$$\mathbf{v}_1 = \begin{bmatrix} -0.6493 \\ -0.7411 \\ 0.1708 \end{bmatrix}, \ \mathbf{v}_2 = \begin{bmatrix} -0.7571 \\ 0.6512 \\ -0.0530 \end{bmatrix}, \ \mathbf{v}_3 = \begin{bmatrix} -0.0720 \\ -0.1637 \\ -0.9839 \end{bmatrix} \tag{35}$$

This orientation is used for the solution of the inverse kinematic problem and one of the solutions obtained — the solution which can be attained by the mechanism shown in Fig. 2 — is given by

$$\theta_1 = -0.2266, \ \theta_2 = -0.7103, \ \theta_3 = 0.0730 \tag{36}$$

where the angles are given in radians.

¿From eq.(15), one then readily obtains

$$\phi_1 = -\theta_2 = 0.7103 \tag{37}$$

and from eq.(26), one finds

$$\phi_2 = 0.0530, \text{ and } \phi_2 = 3.1946 \tag{38}$$

Finally, eq.(27) leads to

$$\phi_3 = 0.1719, \ 3.3135, \ -0.1719, \ 2.9697 \tag{39}$$

when the two solutions of ϕ_2 are used. It can be readily verified, from eq.(35), that the first solution obtained corresponds to the original orientation of the platform chosen at the beginning of the example.

6. Conclusion

In this paper, a particular geometry of spherical parallel manipulator has been considered. This geometry arises from kinematic optimization which has been performed in previous work and involves symmetric link angles of 90 degrees in each of the subchains of the mechanism. For a general manipulator, it is well known that the direct kinematic problem leads to very

complicated expressions for the coefficients of a polynomial of degree 8. By
contrast, it has been shown here that a simple closed-form solution can be
obtained for the special geometry of the manipulator. This is due to the
decoupling of the solutions in two groups, one of which is associated with
higher order singularities and is easily factored out. The closed-form solution obtained requires very few computations and is therefore of practical
interest. It can be implemented in real-time in the controller of a prototype
of parallel manipulator which is based on the simplified geometry studied
here. Therefore, in addition to being kinematically optimal, this geometry
also provides significant advantages in terms of kinematic computational
simplicity.

Acknowledgements: The authors would like to acknowledge the financial support of
the Natural Sciences and Engineering Research Council of Canada (NSERC) and the
Fonds pour la Formation des Chercheurs et l'aide à la Recherche (FCAR) du Québec as
well as the help of Eric Lavoie and Jean-François Hamel for the figures.

References

1. H. Asada and J.A. Cro Granito, 'Kinematic and static characterization of wrist
   joints and their optimal design', Proceedings of the *IEEE International Conference
   on Robotics and Automation*, pp. 244–250, 1985.
2. D.J. Cox and D. Tesar, 'The dynamic model of a three-degree-of-freedom parallel
   robotic shoulder module', Proceedings of the *Fourth International Conference on
   Advanced Robotics*, Columbus, June 13–15, 1989.
3. W.M. Craver, 'Structural analysis and design of a three-degree-of-freedom robotic
   shoulder module', Master Thesis, The University of Texas at Austin, 1989.
4. C. Gosselin and J. Angeles, 'The optimum kinematic design of a spherical three-
   degree-of-freedom parallel manipulator', *ASME Journal of Mechanisms, Transmissions, and Automation in Design*, Vol. 111, No. 2, pp. 202–207, 1989.
5. C. Gosselin and E. Lavoie, 'On the kinematic design of spherical three-degree-of-
   freedom parallel manipulators', *The International Journal of Robotics Research*,
   Vol. 12, No. 4, pp. 394–402, 1993.
6. C. Gosselin and J.-F. Hamel, 'The agile eye: a high-performance three-degree-of-
   freedom camera-orienting device', Proceedings of the *IEEE International Conference
   on Robotics and Automation*, San Diego, May, pp. 781–786, 1994.
7. C. Innocenti and V. Parenti-Castelli, 'Echelon form solution of direct kinematics for
   the general fully-parallel spherical wrist', *Mechanism and Machine Theory*, Vol. 28,
   No. 4, pp. 553 561, 1993.
8. C. Gosselin, J. Sefrioui and M.J. Richard, 'On the direct kinematics of spherical three-degree-of-freedom parallel manipulators with a coplanar platform', *ASME
   Journal of Mechanical Design*, Vol. 116, No. 2, pp. 587–593, 1994.
9. C. Gosselin, J. Sefrioui and M.J. Richard, 'On the direct kinematics of spherical
   three-degree-of-freedom parallel manipulators of general architecture', *ASME Journal of Mechanical Design*, Vol. 116, No. 2, pp. 594–598, 1994.
10. K. Wohlhart, 'Displacement analysis of the general spherical Stewart platform',
    Mechanism and Machine Theory, Vol. 29, No. 4, pp. 581–589, 1994.
11. H.R.M. Daniali, P.J. Zsombor-Murray and J. Angeles, 'Isotropic design of spherical
    parallel manipulators', Proceedings of the *ASME Design Automation Conference*,
    Minneapolis, September 11–14, Volume 2, 1994.

ALGORITHMS FOR KINEMATIC CALIBRATION OF FULLY-PARALLEL MANIPULATORS

C INNOCENTI
Department of Mechanical Engineering - University of Bologna
Viale Risorgimento, 2 - 40136 Bologna - Italy

Abstract. The paper presents two new algorithms for kinematic calibration of fully-parallel manipulators with general geometry Both algorithms are insensitive to the magnitude of the kinematic parameter errors and unambiguously identify the solution of the calibration problem The first algorithm is aimed at determining the actual coordinates of the centers of the spherical pairs connecting base and platform to the six legs it requires seven different locations of the platform to be measured, along with the corresponding sets of leg lengths A generalization of the first one, the second algorithm is based on measurement of eight platform locations, as well as of the leg length variations with respect to unknown reference lengths Outputs of this algorithm are the values of the reference lengths of the legs, together with the coordinates of the attachment points on base and platform Finally, numerical examples are reported

1. Introduction

Robot calibration is the set of procedures aimed at enhancing the accuracy of a robot manipulator without affecting its hardware As neatly surveyed by Roth *et al* (1987) and Everett *at al* (1987), robot calibration can be carried out at different levels of sophistication Kinematic calibration, in particular, deals with the improvement of the kinematic model of a manipulator that is attainable by substituting the nominal values of the manipulator kinematic parameters with their actual values Thanks to kinematic calibration, the effects of machining and mounting errors on manipulator accuracy are neutralized or, at least, minimized

The kinematic parameters of a manipulator - defined as the kinematically relevant linear and angular dimensions of its links - are seldom measured directly More frequently, they are indirectly deduced after observation of the actual dependence of the end-effector rigid-body position (location) on the corresponding set of actuator displacements

For a fully-parallel manipulator (see Fig 1) the typical kinematic calibration procedure prescribes that the platform be moved through several locations At each location, the position of the platform reference frame, W_p, with respect to the base reference frame, W_b, is measured, and the length of each leg $A_j B_j$, $(j=1, ,6)$, is detected by the built-in linear transducer (Merlet, 1993) Based on the collected data, as well as on the kinematic model of the manipulator, the coordinates of the spherical pair centers A_j and B_j (referred to W_b and W_p respectively) can be estimated

Moreover, since a linear transducer might not provide the length of a leg, but only the leg length variations, a more ambitious kinematic calibration procedure could be

241

J P Merlet and B Ravani (eds) Computational Kinematics 241–250

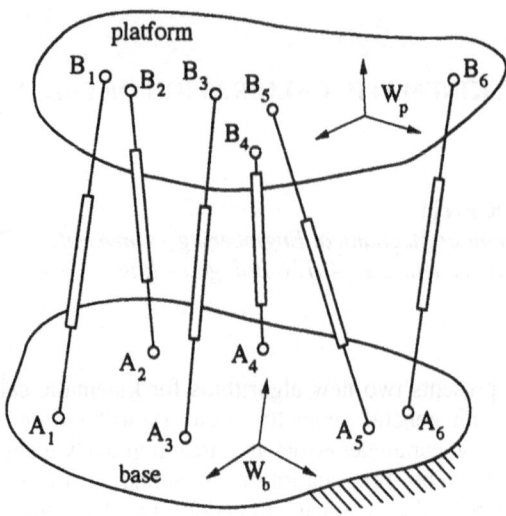

Figure 1 A fully-parallel manipulator with general geometry

aimed at evaluating - in addition to the positions of points A_j, B_j, (j=1,..,6) - the reference length of each leg (defined as the leg length that corresponds to the zero reading of the leg linear transducer).

Existing calibration algorithms for parallel manipulators, such as those proposed by Zhuang and Roth (1993) and Nahvi *et al.* (1994), resort to optimization techniques or to other numerical iterative procedures in order to solve a non-linear set of equations that contain the kinematic parameters as unknowns. Unfortunately, as has been pointed out by Merlet (1993), these algorithms might converge to several numerical solutions, whereas the answer to a calibration problem is physically only one.

Other calibration algorithms, like that presented by Kugiumtzis and Lillekjendlie (1994) for a serial-parallel manipulator, approximate to the first order the equations that provide the deviations of the actual kinematic parameter from their nominal values. Generally, this approach can be adopted only when an accurate estimation of the kinematic parameters is known beforehand.

Interestingly, all known approaches to calibration of serial and parallel manipulators accept computational approximation of one kind or another as unavoidable. A possible reason can be traced to the former need to find calibration algorithms for serial manipulators, whose structure imposes that any calibration equation contains all unknown kinematic parameters. Things are quite different for parallel manipulators, which allow sets of calibration equations to be laid down that contain only some of the unknown kinematic parameters. This distinction, although already pointed out by Zhuang and Roth (1993), has not been fully exploited yet.

In this paper, two original algorithms for determining the kinematic parameters of general-geometry 6-6 fully-parallel manipulators are presented. Both algorithms rely on the minimum number of measurement sets and are able to unambiguously identify the only one solution to a calibration problem. Moreover, since neither algorithm needs any preemptive estimate of the solution, they are not bound to cope with small construction errors only. The mathematical core of both algorithms rests on the procedure suggested by Innocenti (1994) for solving a special system of six second-order equations in six

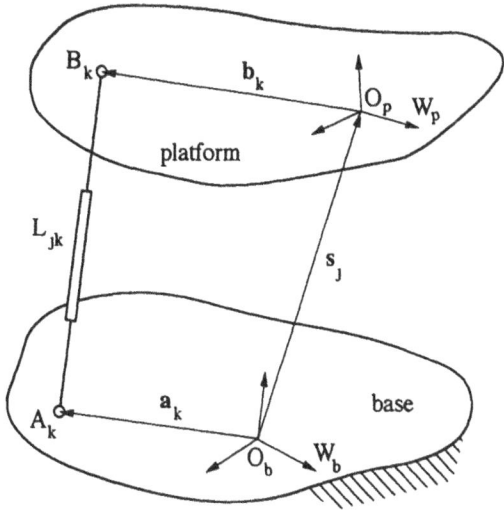

Figure 2 Kinematic entities associated with a calibration equation

unknowns.

The first of the proposed algorithms provides the coordinates of the leg attachment points on base and platform. Input data to the algorithm are seven data sets, each composed of one platform location and the corresponding leg lengths.

The second algorithm is based on eight platform position measurements in order to determine - in addition to the coordinates of the attachment points - the reference length of each leg. Among the data required by the second algorithm are the length variations of each leg with respect to the corresponding (unknown *a priori*) reference length. The second algorithm proves useful in case the linear transducers of the manipulator actuators need calibration themselves.

For each of the proposed kinematic calibration algorithms, a numerical example is reported.

2. The First Calibration Algorithm

In this section a kinematic calibration algorithm is presented that is able to determine, for each of the six legs of a fully-parallel manipulator, the coordinates of the centers of the extremity spherical pairs with respect to reference frames fixed to base and platform. The algorithm requires measuring seven sets of data. Each set comprises the platform rigid-body position (location) with respect to the base, and the length of six legs. Prior to explaining the algorithm in detail, some considerations are in order so as to justify the choice of the number (seven) of data sets the algorithm is based on.

Let us consider the sub-problem of determining the coordinates of the extremity attachment points of only one leg - the k-th one ($1 \leq k \leq 6$) - by relying on a generic number N of data sets, each set comprising the platform location and the corresponding length of leg k. Figure 2 represents base and platform of the parallel manipulator at the j-th measured location: W_b and W_p are arbitrarily-chosen cartesian reference frames fixed to base and platform respectively, whereas s_j represents the position vector of origin O_p of frame W_p with respect to frame W_b. Associated to vector s_j is the 3×3 orthogonal matrix R_j describing the orientation of frame W_p with respect to W_b. Leg k is also represented in Fig. 2: although the position vectors a_k and b_k of extremity points A_k and B_k are to be considered as unknown, the leg length L_{jk} of limb k at the j-th platform location is a problem datum.

Holding the above-introduced notation, the only constraint the j-th measurement set imposes on vectors a_k and b_k is represented by the following equation (Zhuang and Roth, 1993)

$$(R_j \, b_k + s_j - a_k)^T \, (R_j \, b_k + s_j - a_k) = L_{jk}^{\,2} \tag{1}$$

which, after rearrangement, becomes

$$(a_k^T \, a_k + b_k^T \, b_k + s_j^T \, s_j - L_{jk}^{\,2})/2 - a_k^T \, R_j \, b_k - s_j^T \, a_k + s_j^T \, R_j \, b_k = 0 \tag{2}$$

This is an algebraic second-order equation in six scalar unknowns, namely, the components of vectors a_k and b_k. In order to compute these components, the number N of data sets cannot be less than six.

Actually, even the choice N=6 is not satisfactory because the corresponding set of non-linear equations provides more than one solution in terms of vectors a_k and b_k. This circumstance, associated with the possible lack of a good estimate for the actual values of a_k and b_k, could hinder the solution of the calibration problem. Similar drawbacks would also emerge in case more than one solution of the equation set falls within the neighborhood of the (unknown) calibration solution.

To overcome the above-mentioned indeterminacy, more information must be collected from the manipulator, thus letting the number of measurement sets equal seven.

2.1. CALIBRATION EQUATIONS

With reference to the k-th leg only (1≤k≤6), the seven conditions involving the components of vectors a_k and b_k are represented by equation (2), where index j ranges from 1 to 7

$$(a_k^T \, a_k + b_k^T \, b_k + s_j^T \, s_j - L_{jk}^{\,2})/2 - a_k^T \, R_j \, b_k - s_j^T \, a_k + s_j^T \, R_j \, b_k = 0 \qquad (j=1,..,7) \tag{3}$$

These conditions are (non-linearly) dependent because they stem from compatible sets of data collected from the same manipulator.

In order to find the solution of equation set (3), the seventh equation is first subtracted from the previous six

$$(s_j^T \, s_j - s_7^T \, s_7 - L_{jk}^{\,2} + L_{7k}^{\,2})/2 + a_k^T \, (R_7 - R_j) \, b_k$$
$$+ (s_7 - s_j)^T \, a_k + (s_j^T \, R_j - s_7^T \, R_7) \, b_k = 0 \qquad (j=1,..,6) \tag{4}$$

Now the components of vector a_k in reference frame W_b and the components of vector b_k in reference frame W_p are explicitly introduced

$$a_k = (a_{1k}, a_{2k}, a_{3k})^T; \qquad\qquad b_k = (b_{1k}, b_{2k}, b_{3k})^T \tag{5}$$

Substitution of expressions (5) for a_k and b_k into equations (4) leads to

$$\sum_{\substack{u=0,3 \\ v=0,3}} e_{uvjk} \, a_{uk} \, b_{vk} = 0 \qquad (j=1,..,6) \qquad (6)$$

In equations (6), symbols a_{0k} and b_{0k} are defined by

$$a_{0k} = b_{0k} = 1 \qquad (7)$$

Moreover, coefficients e_{uvjk} ($u,v=0,..,3$; $j=1,..,6$) are constant terms, because they depend on the input data only. Their explicit expressions are omitted for the sake of conciseness.

2.2. SOLUTION PROCEDURE

Since all equations of set (6) lack terms depending on products like $a_{uk}a_{vk}$ and $b_{uk}b_{vk}$, ($u,v=1,2,3$), they can be solved by the procedure presented by Innocenti (1994). By this procedure, the solution of a set of six second-order equations with the same structure as equations (6) is reduced to finding the roots of a 20th order algebraic equation containing any one of the unknowns (for example, a_{3k} in the present case). Once the algebraic equation has been found, and all real roots $(a_{3k})_h$ ($h=1,..,2\alpha_k$; $1\le\alpha_k\le10$) computed, linear back-substitution provides all real solutions of equation set (4) in terms of vector pairs $(a_k, b_k)_h$. These vector pairs represent tentative solutions for the calibration problem.

Identification of the correct solution is performed by testing all $2\alpha_k$ vector pairs $(a_k, b_k)_h$, ($h=1,..,2\alpha_k$) and selecting the one that satisfies any equation of set (3) or, equivalently, only the seventh equation of set (3).

3. The Second Calibration Algorithm

The second of the proposed calibration algorithms is an extension of the first one, since it can be applied to those cases where the linear transducers equipping the legs of the parallel manipulator need calibration too. Precisely, it is assumed that the reading ΔL_{jk} provided by the transducer of leg k, and concurring to define the j-th calibration data set, does not exactly correspond to the leg length, L_{jk}, but differ from it by an offset value, L_{0k}, according to the following relation

$$L_{jk} = L_{0k} + \Delta L_{jk} \qquad (8)$$

Quantity L_{0k}, here referred to as the reference length of leg k, is one of the kinematic parameters the calibration algorithm has to determine. The remaining unknown parameters are - like the previous algorithm - the coordinates of attachment points on base and platform with respect to reference frames W_b and W_p.

Since there are as many as seven unknown kinematic parameters for each leg, considerations similar to those reported at the beginning of the previous section lead to fixing at eight the minimum number of data sets that are required to remove indeterminacy.

3.1. CALIBRATION EQUATIONS

Based on eight data sets, eight equations in seven unknowns can be written for the generic leg k ($1 \leq k \leq 6$). Each equation can be derived from condition (2) after replacing quantity L_{jk} with the expression provided by position (8)

$$(a_k^T a_k + b_k^T b_k + s_j^T s_j - L_{0k}^2 - 2 L_{0k} \Delta L_{jk} - \Delta L_{jk}^2) / 2$$

$$- a_k^T R_j b_k - s_j^T a_k + s_j^T R_j b_k = 0 \qquad (j=1,..,8) \qquad (9)$$

Although the number of the equations exceeds the number of the unknowns by one, equation set (9) is not overconstrained because it derives from measurements on the same manipulator. Indeed, equation set (9) generally admits only one solution for parameters a_k, b_k, and L_{0k}.

3.2. SOLUTION PROCEDURE

In order to find the solution of equation set (9), the last equation is subtracted from the previous ones, thus obtaining

$$(\Delta L_{8k} - \Delta L_{jk}) L_{0k} + (s_j^T s_j - s_8^T s_8 - \Delta L_{jk}^2 + \Delta L_{8k}^2) / 2 + a_k^T (R_8 - R_j) b_k$$

$$+ (s_8 - s_j)^T a_k + (s_j^T R_j - s_8^T R_8) b_k = 0 \qquad (j=1,..,7) \qquad (10)$$

Supposing that, at least for one value i of index j ($1 \leq i \leq 7$), the following inequality is satisfied

$$(\Delta L_{8k} - \Delta L_{ik}) \neq 0 \qquad (11)$$

unknown L_{0k} can be linearly obtained from the i-th equation of set (10) and substituted in the remaining six equations. Consequently, the following set of six equations in six unknowns (the components of vectors a_k and b_k, see relations (5)) can be obtained

$$(\Delta L_{8k} - \Delta L_{ik}) [(s_j^T s_j - s_8^T s_8 - \Delta L_{jk}^2 + \Delta L_{8k}^2) / 2 + a_k^T (R_8 - R_j) b_k + (s_8 - s_j)^T a_k$$

$$+ (s_j^T R_j - s_8^T R_8) b_k] - (\Delta L_{8k} - \Delta L_{jk}) [(s_i^T s_i - s_8^T s_8 - \Delta L_{ik}^2 + \Delta L_{8k}^2) / 2$$

$$+ a_k^T (R_8 - R_i) b_k + (s_8 - s_i)^T a_k + (s_i^T R_i - s_8^T R_8) b_k] = 0 \qquad (j=1,...,7; j \neq i) \qquad (12)$$

By adopting positions (5) and (7), equations (12) can be rewritten in the following form

$$\sum_{\substack{u=0,3 \\ v=0,3}} f_{uvnk} a_{uk} b_{vk} = 0 \qquad (n=1,..,6) \qquad (13)$$

where coefficients f_{uvnk} (u,v=0,1,2,3; n=1,..,6) are known quantities because they depend only on the measured data.

Equations (13) have the same structure as equations (6). Once again, the solving procedure reported by Innocenti (1994) can be adopted in order to compute all real solutions in terms of vector pairs $(a_k, b_k)_h$, (h=1,..,2β_k; $1 \leq \beta_k \leq 10$). For every vector pair, the i-th equation of set (10) linearly provides the corresponding value, $(L_{0k})_h$, for the reference length of leg k.

The selection of the proper solution out of the just determined $2\beta_k$ tentative solutions $(a_k, b_k, L_k)_h$ $(h=1,..,2\beta_k)$ can be performed by singling out the specific tentative solution that satisfies each equation of set (9) or, equivalently, only the eighth equation of set (9).

4. Remarks

The procedures explained in the previous two sections can be extended to include the possibility that the length of some legs do not vary during calibration, either because such legs are intentionally kept locked or because they are constant-length connecting rods (as for some legs of parallel manipulators with less than six degrees of freedom).

For these special cases, the first of the proposed calibration algorithms does not formally change. With reference to any constant-length leg, the algorithm can now be interpreted as the resolution of the problem addressed by Innocenti (1994), with the additional selection of the only solution that corresponds to the known leg length.

Also the second calibration algorithm reduces, for any constant-length leg, to the procedure reported by Innocenti (1994). Based on seven of the eight data sets, a maximum of twenty pairs of attachment points on base and platform can be found that are placed at the same mutual distance for each of the seven locations of the platform. The sought-for two points are those that, for the eighth platform location, are set apart by the same amount (the sought-for leg reference length) as for the previous seven locations.

5. Numerical Examples

This section shows application of the proposed algorithms to calibration of the same six-degree-of-freedom fully-parallel manipulator. All input data, reported in Table 1, are hypothetical. Nevertheless, the limited accuracy of experimental measurements is simulated by providing all values for lengths (expressed in arbitrary length unit) and direction cosines with six meaningful digits at most.

Each column of Table 1 refers to a different data set. Within the same data set, the location of the platform is defined by the components of vector s_j $(j=1,..,8)$ in reference frame W_b (see also Fig. 2), and by the elements r_{uvj} $(u,v=1,2,3; j=1,..,8)$ of orthogonal matrix R_j. Moreover, for each location of the platform the set of leg lengths, L_{jk}, $(k=1,..,6)$, together with the consistent set of sensor readings, ΔL_{jk}, are also provided.

5.1. APPLICATION OF THE FIRST CALIBRATION ALGORITHM

The first of the proposed calibration algorithms is based on seven data sets, each comprising a platform location and the corresponding leg lengths. Accordingly, the eighth data set of Table 1 is ignored, together with all ΔL_{jk} values of the considered seven data sets.

The procedure explained in subsection 2.2 is applied six times, once for each leg. By referring, for example, to the first leg (k=1), a 20th order algebraic equation containing parameter a_{31} as unknown is obtained. The real roots of this equation are reported in the second column of Table 2. Based on them, twelve tentative solutions in terms of vector pairs $(a_1, b_1)_h$ $(h=1,..,12)$ can be computed. The third column of Table 2 lists, for every

TABLE 1 Input data sets for the calibration algorithms

	J = 1	2	3	4	5	6	7	8
	−0 93944	0 07317	0 89079	0 63796	0 26825	−0 70053	0 38101	0 11651
s_j	0 56375	0 69121	0 29502	−0 69927	−0 70566	0 22848	0 07461	−0 03347
	1 12486	2 50209	1 28736	1 59766	2 91310	2 12017	2 82749	1 89539
r_{11j}	−0 014362	−0 516680	0 131792	−0 620199	−0 277261	−0 708385	0 356581	−0 772641
r_{12j}	−0 216350	0 825490	−0 951924	−0 487992	0 584649	−0 239258	−0 631356	0 633417
r_{13j}	0 976210	−0 227175	0 276535	−0 614180	0 762438	0 664038	0 688651	0 042528
r_{21j}	−0 276196	−0 646198	0 744296	0 293657	0 695280	−0 677336	−0 835698	0 342979
r_{22j}	−0 937468	−0 201921	−0 089226	0 581588	−0 425591	0 495028	0 113991	0 360117
r_{23j}	−0 211827	0 735973	−0 661863	−0 758631	0 579188	−0 544209	0 537229	0 867572
r_{31j}	0 960994	0 561666	0 654717	0 727405	0 663108	−0 198511	−0 417683	0 534220
r_{32j}	−0 272667	0 527062	0 293052	−0 650860	0 690694	−0 835286	−0 767069	0 684908
r_{33j}	−0 046291	0 637759	0 696753	−0 217399	−0 288494	−0 512728	−0 486976	−0 495490
L_{j1}	3 71462	4 72007	2 57456	3 98720	4 03617	4 28984	2 89696	4 05329
L_{j2}	6 46724	6 46205	5 42955	5 63069	6 94127	5 94020	2 79995	6 82776
L_{j3}	6 14994	6 32322	4 43030	4 37290	6 60437	5 13941	4 66062	5 09096
L_{j4}	4 60004	5 91219	3 32287	2 17104	7 49362	1 16258	2 85582	5 52735
L_{j5}	4 02568	5 05487	4 19615	5 34127	3 44035	5 58947	5 85121	4 33236
L_{j6}	5 38154	5 11180	3 47977	2 71115	2 57773	4 58100	5 93354	2 27480
ΔL_{j1}	3 27081	4 27626	2 13075	3 54339	3 59236	3 84603	2 45315	3 60948
ΔL_{j2}	6 52927	6 52408	5 49158	5 69272	7 00330	6 00223	2 86198	6 88979
ΔL_{j3}	7 07055	7 24383	5 35091	5 29351	7 52498	6 06002	5 58123	6 01157
ΔL_{j4}	4 02225	5 33440	2 74508	1 59325	6 91583	0 58479	2 27803	4 94956
ΔL_{j5}	3 66426	4 69345	3 83473	4 97985	3 07893	5 22805	5 48979	3 97094
ΔL_{j6}	5 42078	5 15104	3 51901	2 75039	2 61697	4 62024	5 97278	2 31404

value of $(a_{31})_h$, the corresponding deviation in terms of the absolute value of the left-hand side of the seventh equation (3). It can be easily verified that the correct solution is the fourth one (h=4), because it generates the smallest deviation value.

The correct solution for all legs is reported in Table 3 in terms of position vectors a_k and b_k (k=1,..,6).

5.2. APPLICATION OF THE SECOND CALIBRATION ALGORITHM

This time all eight data sets of Table 1 are taken into account. Consistently with the features of the second calibration algorithm, all L_{jk} values, (j=1,..,8; k=1,..,6), are ignored.

According to the procedure outlined in subsection 3.2, a twentieth-order algebraic equation is found for each leg and subsequently solved. By considering only the real roots, back-substitution provides a number of tentative solutions for the generic leg k in term of quantities a_k, b_k, and L_{0k}. Table 4 reports the eight tentative solutions for the first leg (k=1) - synthetically identified by kinematic parameter $(a_{31})_h$, (h=1,..,8) - together with the corresponding values of the left-hand side of the eighth equation of set (9); clearly, the correct solution is the third one (h=3).

The whole solution to the calibration problem is reported in Table 5 in terms of position vectors a_k and b_k, and reference lengths L_{0k} (k=1,..,6). The components of vectors a_k and b_k are very close to those obtained by the first algorithm (see Table 3). The slight differences are a consequence of the intentionally low accuracy affecting the input data of Table 1.

TABLE 2   Tentative a_{31} values and deviations
for the first calibration algorithm

h	$(a_{31})_h$	deviation
1	2 69181	1 72e+00
2	14 86363	1 89e+02
3	−0 22341	1 49e+00
4	−0 00922	6 36e−05
5	2 03123	7 08e+02
6	4 13684	5 12e+01
7	6 12723	2 75e+00
8	3 62540	4 33e+00
9	1 32048	6 65e+01
10	−7 57180	1 73e+02
11	−0 14296	4 99e+00
12	0 12579	1 50e+00

TABLE 3   The output of the first calibration algorithm   coordinates of attachment points
on base and platform, for all legs

k	a_k			b_k		
1	1 98742	0 11186	−0 00922	1 50011	−0 84126	0 43115
2	3 16121	−2 54102	0 22715	2 52291	−1 27395	−0 05006
3	1 63043	3 05991	−0 44017	1 78477	1 76638	0 11348
4	−1 31197	2 86717	−0 20033	−0 18937	2 52552	−0 32491
5	−2 78639	−1 19524	0 42210	−2 44736	−0 75484	0 02262
6	−0 03905	−3 01485	0 16102	−1 06603	−1 65458	0 28982

Finally, the correctness of reference length values L_{0k}, (k=1,..,6), reported in Table 5 can be checked through equation (8) by considering, for any data set of Table 1, the L_{jk} values and the corresponding ΔL_{jk} values.

6. Conclusions

Two original algorithms for solving the kinematic calibration of fully-parallel manipulators with general geometry have been presented. Neither algorithm requires preemptive estimation of the unknown kinematic parameters.

The first of the proposed calibration algorithms is based on seven data sets, each composed of one platform location and the corresponding leg lengths. The algorithm unambiguously provides the coordinates of all spherical pair centers with respect to arbitrary reference frames fixed to base and platform.

The features of the second algorithm are similar to those of the first one, apart from the additional ability to determine the offset values of the manipulator linear transducers. The algorithm relies upon eight data sets, each composed of one platform location and the corresponding transducer readings.

Numerical examples showing application of the proposed algorithms to calibration of a fully-parallel manipulator have been reported.

TABLE 4 Tentative a_{31} values and deviations
for the second calibration algorithm

h	$(a_{31})_h$	deviation
1	3 23920	1 03e+01
2	1 86303	9 44e−01
3	−0 00922	7 60e−05
4	1 29953	1 59e+02
5	3 04544	4 46e+01
6	1 95530	2 40e+00
7	−0 31454	3 21e+00
8	3 26065	3 31e+00

TABLE 5 The output of the second calibration algorithm coordinates of attachment points on base and platform plus reference length, for all legs

k	a_k			b_k			L_{0k}
1	1 98742	0 11186	−0 00922	1 50011	−0 84126	0 43115	0 44380
2	3 16097	−2 54079	0 22727	2 52301	−1 27410	−0 05007	−0 06222
3	1 63045	3 05992	−0 44020	1 78478	1 76639	0 11348	−0 92056
4	−1 31195	2 86714	−0 20031	−0 18938	2 52552	−0 32492	0 57776
5	−2 78639	−1 19528	0 42208	−2 44730	−0 75483	0 02257	0 36144
6	−0 03906	−3 01486	0 16102	−1 06603	−1 65458	0 28982	−0 03922

Acknowledgments

The financial support of Italy's MURST is greatly appreciated

References

Everett, L J , Driels, M , and Mooring, B W , 1987, "Kinematic Modelling for Robot Calibration", *Proc of the 1987 IEEE International Conference on Robotics and Automation* pp 183-189

Innocenti, C , 1995, "Polynomial Solution of the Spatial Burmester Problem", *ASME Journal of Mechanical Design* Vol 117, No 1, pp 64-68

Kugiumtzis, D , and Lillekjendlie, B , 1994, "Estimation Model for Kinematic Calibration of Manipulators with a Parallel Structure", *Journal of Robotic Systems* Vol 11, No 5, pp 399-410

Merlet, J P , 1993, "Calibration of Parallel Manipulators", in J Angeles, G Hommel, and P Kovacs, "Computational Kinematics", *Dagstuhl Seminar Report 75* p 18

Nahvi, A , Hollerbach, J M , and Hayward, V , 1994, "Calibration of a Parallel Robot Using Multiple Kinematic Closed Loops", *Proc of the IEEE Int Conf on Robotics and Automation* pp 407-412

Roth, Z S , Mooring, B W , and Ravani, B , 1987, "An Overview of Robot Calibration", *IEEE Journal of Robotics and Automation* Vol RA-3, No 5, pp 377-385

Zhuang, H , and Roth, Z S , 1993, "Method for Kinematic Calibration of Stewart Platforms", *Journal of Robotic Systems* Vol 10, No 3, pp 391-405

PARALLEL REDUNDANT MANIPULATORS BASED ON OPEN AND CLOSED NORMAL ASSUR CHAINS

S. LÖSCH
Institute of Mechanics
Graz, University of Technology
Kopernikusgasse 24, A – 8010 Graz
email: loesch@mech.tu-graz.ac.at

Abstract. This paper deals with the position analysis of two manipulators based on open normal Assur chains (A(3.5),A(3.6)) and one manipulator based on a closed normal Assur chain (A(4.4)). The manipulators consist of movable rigid bodies interconnected with rotary joints, which are either lined up in a row (open normal chain) or form a ring (closed normal chain). Legs, whose lengths can be varied using either linear actuators (driven P - joints) or rotary actuators (driven R - joints), establish the connection with the ground. For all three mechanisms a lexicographic Gröbner Basis was computed from polynomials that describe the position of the manipulator. The univariate polynomial of the basis for case A(3.5) has degree 54, for case A(3.6) it has degree 162 and for A(4.4) degree 56, which leads to 54, 162 and 56 possible complex positions respectively.

1. Introduction

Some tasks in robotics demand accurate positioning and large load capacities. Since they sum up the backshlashes to a smaller extent, parallel manipulators solve such problems more precisely than serial manipulators. However, their disadvantage is their drastically reduced workspace. Because of the redundancy of the parallel manipulators discussed here there are many different ways to carry out a motion of the manipulator hand and one can choose among them in such a way that the power input is minimized. In this article we analyze the forward kinematics of the manipulators described above. This means calculating all possible positions of the end ef-

251

J.-P. Merlet and B. Ravani (eds.), Computational Kinematics, 251–260.
© 1995 *Kluwer Academic Publishers.*

fector for a given set of input leg lengths. Our tool for solving the nonlinear constraint equations is a Gröbner Basis algorithm modified in such a way that we do not use integer arithmetics, but floating point arithmetics with arbitrary precision floating point coefficients. To compute a lexicographic Gröbner Basis directly, from which it is easy to deduce all solutions, long computational times and incredible amounts of storage are often required. To overcome these problems, we first compute a Gröbner Basis in a finite coefficient field with degree reverse ordering and, in a second run, a basis using arbitrary floating point coefficients following the trace produced by the first run, thus excluding all unnecessary S - polynomial reductions for the second run [11]. The transformation into a lexicographic basis is done with Faugère's algorithm [5]. All the following results are computed from polynomials taking arbitrary values for the parameters. Therefore the solutions in this paper can only be seen as nearly generic.

2. Floating point Gröbner basis

Because of the fact that the Buchberger Algorithm has become increasingly important in recent years we assume the reader to be familiar with *Gröbner Bases*. For an introduction see for instance [4] or [2].

2.1. CHECKING FOR ZEROS

The main part of the polynomial arithmetic during the Gröbner Basis Algorithm consists in the subtraction of two polynomials. Starting with the highest monomial with respect to the chosen ordering we take one monomial from each polynomial and compare them. If they differ we append the higher one to the new S - polynomial and take the next monomial of the corresponding polynomial and compare again. In case of two monomials with the same size their coefficients have to be subtracted. To detect whether a monomial vanishes we have to make a statement like the following which asks for exact zeros:

if $(\mathrm{coeff}_1 - \mathrm{coeff}_2 = \mathrm{integercoeff}) = 0$ **then** monomial is zero

It has to be formulated in the following way when using floats:

if $\dfrac{\|\mathrm{floatcoeff}_1 - \mathrm{floatcoeff}_2\|}{\|\mathrm{floatcoeff}_1 + \mathrm{floatcoeff}_2\|} \leq \mathrm{limit}$ **then** monomial is zero

This relative formulation enables us to compare the fraction with a constant limit. But this test is not very stable because of the large influence of roundoff errors which often drastically decrease the accuracy of the floats. The crucial point here is that a limit greater than the machine accuracy has to be determined before the calculation. If we take the machine accuracy as the least limit, we get a result depending on hardware floating point arithmetic. Even when we detect zero monomials with this statement, we can never be sure if the monomial is exactly zero. That is why we have to look for a certain kind of arithmetic which supports this test. Taking coefficients as a tuple consisting of an integer and an arbitrary precision float it is possible to make the zero test in the following manner:

$$\textbf{if } \text{modularcoeff}_1 - \text{modularcoeff}_2 = 0 \land \frac{\|\text{floatcoeff}_1 - \text{floatcoeff}_2\|}{\|\text{floatcoeff}_1 + \text{floatcoeff}_2\|} \leq \text{limit}$$
$$\textbf{then } \text{monomial is zero}$$

A coefficient asserted to be zero by the modular value must have a small floating point value, but if the modular value is zero and the floating point value is greater than the limit two cases might have occurred:

1. The result of the modular subtraction is a multiple of the modulus. This can be avoided by either using two moduli which have to be zero concurrently or by making a second modular run with another prime as the modulus. Our experience shows that the occurrence of a multiple of the modulus is very rare, and to keep the algorithm fast we have chosen the latter option.

2. The roundoff errors have been too big and the second run of the algorithm must be computed once again taking floats with higher precision.

Thus the zero check has to be done in the following manner:

$$\textbf{if } \text{modularcoeff}_1 - \text{modularcoeff}_2 = 0$$
$$\qquad \textbf{if } \frac{\|\text{floatcoeff}_1 - \text{floatcoeff}_2\|}{\|\text{floatcoeff}_1 + \text{floatcoeff}_2\|} \leq \text{limit}$$
$$\qquad\qquad \textbf{then } \text{monomial is zero}$$
$$\qquad \textbf{else}$$
$$\qquad\qquad \text{raise precision and restart second run}$$
$$\textbf{else}$$
$$\qquad \textbf{if } \frac{\|\text{floatcoeff}_1 - \text{floatcoeff}_2\|}{\|\text{floatcoeff}_1 + \text{floatcoeff}_2\|} \leq \text{limit}$$
$$\qquad\qquad \textbf{then } \text{a multiple of the modulus occurred. Take}$$
$$\qquad\qquad \text{another prime as modulus and restart first run}$$

This representation of coefficients by a pair consisting of a floating point approximation and a modular value is due to D. Lazard [9].

2.2. ARBITRARY PRECISION FLOATS

Our software is modeled after a preliminary version of a multiprecision floating point and interval arithmetics package by Krandick and Johnson [8] to be used within the *saclib* environment. The *saclib* floats are represented by a pair consisting of an integer of arbitrary length, the mantissa, and a second integer of fixed length called the exponent. To represent the mantissa as a binary number of 2 - length, a positive power of 2 (ζ) is used for the basis (β): $\beta = 2^\varsigma$. A non – zero floating point number is thus characterized by the sign s, a certain number of β – digits a_i and the exponent e: $F = \left(s, a_0\beta^0 + a_1\beta^1 + \ldots + a_n\beta^n, e\right)$. The number of β – digits n is called the precision of F and the mantissa a is normalized, which means

$$L_\beta\left(a\right) = \frac{L_2\left(a\right)}{\varsigma} \qquad \begin{array}{ll} L_\beta & \ldots \quad \beta - \text{length of } a \text{ (precision)} \\ L_2 & \ldots \quad 2 - \text{length of } a \end{array}$$

In order to get a floating point Gröbner Basis algorithm with high performance we wrote our own routines, which work with fixed size arrays. The main idea was to use floating point numbers which all have the same precision from the beginning, thus simulating the use of hardware floating point arithmetics with a much higher machine accuracy. Well knowing that arithmetic is not exact with numbers in floating point representations like the one mentioned above, we have to interpret our results carefully. With an increasing amount of calculation the roundoff errors accumulate and we cannot be sure whether these roundoffs were magnified in such a way that the true answer is swamped. To overcome this problem we did our calculation twice – the second time with increased precision – and compared the coefficients of the resulting polynomials. Corresponding digits of the mantissa were declared to be valid. But this was an expensive and rather empirical way to determine the *accuracy* of the floats. A better way would be to observe the accumulation of the roundoff errors during the algorithm.

2.3. LOSS OF ACCURACY

To express the influence of roundoff errors we use the basic concept of *relative errors* mentioned in [6], a rough but reasonably useful way to determine the significant digits of the numbers. Let f be an exact real number which has to be expressed by the the floating point approximation $\hat{f}$. Then the fraction $(\hat{f} - f)/f$ is called the *relative error* of approximation. The magnification of the relative error by the operations of multiplication and division is not assumed to be significant. On the other hand, a subtraction of nearly equal values causes a substantiel loss of accuracy and greatly increases the relative error. To observe the loss of accuracy during the calculation in this way we first implemented some instructions in the routines *fpsum* and *fpdif* which detect *bad* additions and subtractions and compute the number of digits which have been lost. Taking this number as the scale for the relative error we are able to quote the accuracy of each float at any stage. Furthermore, an additional integer is appended to the internal representation of an arbitrary precision float which characterizes the accuracy: $F = \left(s, a_0\beta^0 + a_1\beta^1 + \ldots + a_n\beta^n, e, acc\right)$. The zero check described in 2.1 can now be done in a more elegant way. Since we know the accuracy we are able to compare the difference between two floating point numbers with the value zero. In this case *zero* is the current machine accuracy, which is determined by the present number of β digits. In spite of this attendant calculation of accuracy we do not leave our modular support because of the remaining uncertainty which is caused by the fact that we do not alter the accuracy when calling the functions *fpdiv* and *fpprod*.

3. Assur groups

The basic mechanisms of all the manipulators investigated in the following are Assur groups which are "the smallest kinematic chain, which, when added to or subtracted from a mechanism, results in mechanisms that have the same mobility as the original" [10]. In order to get all possible positions of the end – effector for a given set of leg lengths, we have to carry out a position analysis of the respective Assur group.

4. Parallel redundant manipulators based on open normal chains

An open normal Assur chain consists of triangles which are lined up in a row interconnected by rotary joints. At the remaining corners, legs are connected with the triangles via rotary joints: the first and the last triangles have 2 legs each and the in – between triangles have one. According to a classification of Assur groups proposed by [1] the open normal Assur chains always have class 3 and the order equals the number of legs.

4.1. THE ASSUR GROUP (3.5)

Without loss of generality we choose a coordinate system whose origin is identical with point A_1 and whose x – axis passes through point A_2. To define the geometry of the three triangles we take the lengths of two sides and the angle included by them $(a_i, b_i, \phi_i;\ i = 1 \div 3)$ (see Fig. 4.1). The position of the group for a given n – tuple $(l_1, l_2, l_3, l_4, l_5)$ is described taking the four angles $\alpha, \beta, \gamma, \delta$. We should remark here that it was our purpose to choose one angle as the third triangle parameter to justify our floating point arithmetics. Anyone computing a Gröbner Basis using integer arithmetics has difficulties in making the same input polynomials algebraic because all angles are arguments of trigonometric functions. One possible solution would be to evaluate the sines and cosines to floats with a certain number of digits and then to convert these floats to rationals. In doing so, however, the input polynomials do not define the problem exactly but only approximately, and it remains to investigate whether this fact causes a discrepancy between the integer results and the floating point results. Furthermore, we have to ask how precise the input parameters have to be given if we want to have a correct result. For the numeric examples given

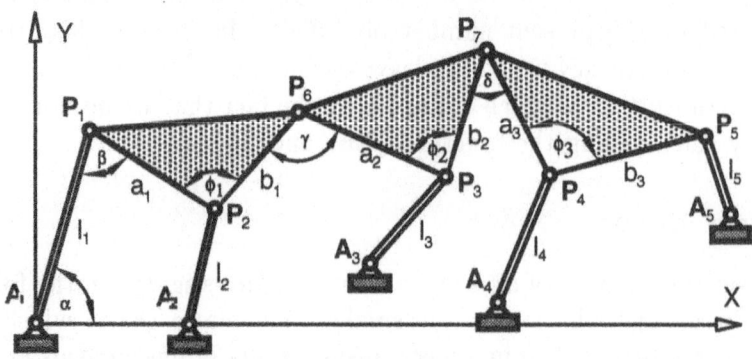

Figure 1. Assur group (3.5)

in this paper we did the computations several times, continually increasing the precision of the floats and thus checking our results. All roots were backsubstituted into the input polynomials to confirm them, so that we could be sure that our solutions were correct. With the following vector definitions ($\mathbf{v}_1 = \vec{A_1P_1}, \mathbf{v}_2 = \vec{P_1P_2}, \mathbf{v}_3 = \vec{A_1A_2}, \mathbf{v}_4 = \vec{P_2P_6}, \mathbf{v}_5 = \vec{P_6P_3}, \mathbf{v}_6 = \vec{A_1A_6}, \mathbf{v}_7 = \vec{P_3P_7}, \mathbf{v}_8 = \vec{P_7P_4}, \mathbf{v}_9 = \vec{A_1A_4}, \mathbf{v}_{10} = \vec{P_4P_5}, \mathbf{v}_{11} = \vec{A_1A_5}$):

$$\mathbf{v}_1 = \begin{bmatrix} l_1 \cos\alpha \\ l_1 \sin\alpha \end{bmatrix}, \mathbf{v}_2 = \begin{bmatrix} -a_1 \cos(\alpha+\beta) \\ -a_1 \sin(\alpha+\beta) \end{bmatrix}, \mathbf{v}_3 = \begin{bmatrix} A_{2x} \\ 0 \end{bmatrix},$$

$$\mathbf{v}_4 = \begin{bmatrix} b_1 \cos(\alpha+\beta-\phi_1) \\ b_1 \sin(\alpha+\beta-\phi_1) \end{bmatrix}, \mathbf{v}_5 = \begin{bmatrix} -a_2 \cos(\alpha+\beta+\gamma-\phi_1) \\ -a_2 \sin(\alpha+\beta+\gamma-\phi_1) \end{bmatrix}, \mathbf{v}_6 = \begin{bmatrix} A_{3x} \\ A_{3y} \end{bmatrix}$$

$$\mathbf{v}_7 = \begin{bmatrix} b_2 \cos(\alpha+\beta+\gamma-\phi_1-\phi_2) \\ b_2 \sin(\alpha+\beta+\gamma-\phi_1-\phi_2) \end{bmatrix}, \mathbf{v}_8 = \begin{bmatrix} -a_3 \cos(\alpha+\beta+\gamma+\delta-\phi_1-\phi_2) \\ -a_3 \sin(\alpha+\beta+\gamma+\delta-\phi_1-\phi_2) \end{bmatrix}$$

$$\mathbf{v}_9 = \begin{bmatrix} A_{4x} \\ A_{4y} \end{bmatrix}, \mathbf{v}_{10} = \begin{bmatrix} b_3 \cos(\alpha+\beta+\gamma+\delta-\phi_1-\phi_2-\phi_3) \\ b_3 \sin(\alpha+\beta+\gamma+\delta-\phi_1-\phi_2-\phi_3) \end{bmatrix}, \mathbf{v}_{11} = \begin{bmatrix} A_{5x} \\ A_{5y} \end{bmatrix}$$

it is possible to write the four constraint equations:

$$(\mathbf{v}_1 + \mathbf{v}_2 - \mathbf{v}_3)^2 - l_2^2 = 0 \quad (1), (\mathbf{v}_1 + \mathbf{v}_2 + \mathbf{v}_4 + \mathbf{v}_5 - \mathbf{v}_6)^2 - l_3^2 = 0 \quad (2)$$

$$(\mathbf{v}_1 + \mathbf{v}_2 + \mathbf{v}_4 + \mathbf{v}_5 + \mathbf{v}_7 + \mathbf{v}_8 - \mathbf{v}_9)^2 - l_4^2 = 0 \quad (3)$$

$$(\mathbf{v}_1 + \mathbf{v}_2 + \mathbf{v}_4 + \mathbf{v}_5 + \mathbf{v}_7 + \mathbf{v}_8 + \mathbf{v}_{10} - \mathbf{v}_{11})^2 - l_5^2 = 0 \quad (4)$$

The left – hand sides of equations 1 to 4 together with the trigonometric identities $\sin^2\alpha + \cos^2\alpha - 1$, $\sin^2\beta + \cos^2\beta - 1$, $\sin^2\gamma + \cos^2\gamma - 1$ and $\sin^2\delta + \cos^2\delta - 1$ form eight polynomials in the variables $\sin\alpha, \cos\alpha, \sin\beta, \cos\beta, \sin\gamma, \cos\gamma, \sin\delta, \cos\delta$, which generate the ideal for which we determine a lexicographic Gröbner Basis using $a_1 = 5, b_1 = 9/2, \phi_1 = 8/5, a_2 = 48/10, b_2 = 9/2, \phi_2 = 37/25, a_3 = 6, b_3 = 53/10, \phi_3 = 3/2, l_1 = 15/2, l_2 = 4, l_3 = 7/2, l_4 = 3, l_5 = 5/2, A_{2x} = 6, A_{3x} = 27/2, A_{3y} = 3/2, A_{4x} = 17, A_{4y} = 2, A_{5x} =$

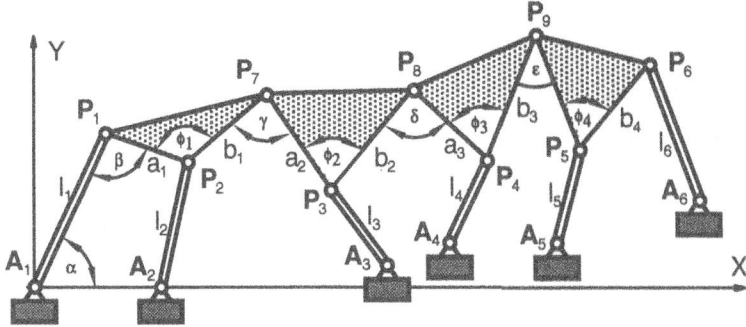

Figure 2. Assur group (3.6)

$49/2$, $A_{5y} = 5$ as numerical values for the input parameters. We use *La-guerre's Method* to find all roots of the univariate polynomial. According to the degree of the univariate polynomial in the basis, which is 54, there are 54 complex positions of the Assur group (3.5). Because of the size of the different bases (modular, degree reverse, lexicographic) we do not include them in this paper, but they can be accessed at our ftp server [1].

4.2. THE ASSUR GROUP (3.6)

This group is similar to the Assur group (3.5) and is built by adding one triangle and one further leg. It has class 3 and order 6. The parameter space is given by $(a_i, b_i, \phi_i); i = 1 \div 4, l_i; i = 1 \div 6)$ and $A_{2x}, A_{3i}, A_{4i}, A_{5i}, A_{6i}; i = x, y$. In addition to the vectors defined in 4.1 we need to redefine $\mathbf{v}_{11}$ and to define the following vectors $(\mathbf{v}_{11}=\overrightarrow{P_9P_5}, \mathbf{v}_{12}=\overrightarrow{A_1A_5}, \mathbf{v}_{13}=\overrightarrow{P_5P_6}, \mathbf{v}_{14}=\overrightarrow{A_1A_6})$:

$$\mathbf{v}_{11} = \begin{bmatrix} -a_4 \cos(\alpha + \beta + \gamma + \delta + \varepsilon - \phi_1 - \phi_2 - \phi_3) \\ -a_4 \sin(\alpha + \beta + \gamma + \delta + \varepsilon - \phi_1 - \phi_2 - \phi_3) \end{bmatrix}, \mathbf{v}_{12} = \begin{bmatrix} A_{5x} \\ A_{5y} \end{bmatrix}$$

$$\mathbf{v}_{13} = \begin{bmatrix} b_4 \cos(\alpha + \beta + \gamma + \delta + \varepsilon - \phi_1 - \phi_2 - \phi_3 - \phi_4) \\ b_4 \sin(\alpha + \beta + \gamma + \delta + \varepsilon - \phi_1 - \phi_2 - \phi_3 - \phi_4) \end{bmatrix}, \mathbf{v}_{14} = \begin{bmatrix} A_{6x} \\ A_{6y} \end{bmatrix}$$

In this case we have five constraint equations whose left – hand sides together with five trigonometric identities as polynomials in $\sin\alpha, \cos\alpha, \sin\beta, \cos\beta, \sin\gamma, \cos\gamma, \sin\delta, \cos\delta, \sin\varepsilon, \cos\varepsilon$ generate the ideal the variety of which gives all possible positions of the Assur group (3.6).

$$(\mathbf{v}_1 + \mathbf{v}_2 - \mathbf{v}_3)^2 - l_2^2 = 0 \quad (5), (\mathbf{v}_1 + \mathbf{v}_2 + \mathbf{v}_4 + \mathbf{v}_5 - \mathbf{v}_6)^2 - l_3^2 = 0 \quad (6)$$

$$(\mathbf{v}_1 + \mathbf{v}_2 + \mathbf{v}_4 + \mathbf{v}_5 + \mathbf{v}_7 + \mathbf{v}_8 - \mathbf{v}_9)^2 - l_4^2 = 0 \quad (7)$$

$$(\mathbf{v}_1 + \mathbf{v}_2 + \mathbf{v}_4 + \mathbf{v}_5 + \mathbf{v}_7 + \mathbf{v}_8 + \mathbf{v}_{10} + \mathbf{v}_{11} - \mathbf{v}_{12})^2 - l_5^2 = 0 \quad (8)$$

$$(\mathbf{v}_1 + \mathbf{v}_2 + \mathbf{v}_4 + \mathbf{v}_5 + \mathbf{v}_7 + \mathbf{v}_8 + \mathbf{v}_{10} + \mathbf{v}_{11} + \mathbf{v}_{13} - \mathbf{v}_{14})^2 - l_6^2 = 0 \quad (9)$$

[1] ftp.tu–graz.ac.at, /pub/papers/mechanic

$$\sin^2\alpha + \cos^2\alpha = \sin^2\beta + \cos^2\beta = \sin^2\gamma + \cos^2\gamma = \sin^2\delta + \cos^2\delta = \sin^2\varepsilon + \cos^2\varepsilon = 1$$

The final lexicographic Gröbner Basis was computed for the parameters: $a_1=5, b_1=5, \phi_1=8/5, a_2=5, b_2=4, \phi_2=37/25, a_3=6, b_3=6, \phi_3=3/2, a_4=2, b_4=5, \phi_4=7/5, l_1=8, l_2=4, l_3=3, l_4=2, l_5=5, l_6=7, A_{2x}=6, A_{3x}=13, A_{3y}=1, A_{4x}=17, A_{4y}=2, A_{5x}=25, A_{5y}=5, A_{6x}=31, A_{6y}=2$. The degree of the univariate polynomial is rather high (162), indicating 162 complex solutions [2].

4.3. CONCLUSION

The position analysis of the general planar parallel manipulator based on the open normal Assur chain A(3.3) gives 6 solutions [12]. The Assur group A(3.4) and corresponding manipulators have 18 possible positions [13]. Here we found 54 solutions for the Assur group A(3.5) and 162 solutions for the manipulator based on the Assur group A(3.6). This induces 2×3^n as the formula to get all possible positions of manipulators based on open normal Assur chains when n is the number of mobile bodies. The same result was detected much earlier by *Wunderlich* while investigating higher coupler curves [15].

5. Parallel redundant manipulators based on closed normal chains

Taking the first and last triangles of an open normal Assur chain in order to connect them with a rotary joint and removing the legs at this joint, we get a closed normal Assur chain. This assembly can be classified by counting the number of corners of the inner loop, which is always identical with the number of legs that connect the bodies with the base. Thus we get the class and order of the group. The simplest configuration which can be used as a parallel manipulator is the Assur group (4.4) (see Fig. 5.1).

5.1. THE ASSUR GROUP (4.4)

In order to get polynomial equations whose Gröbner Basis is simpler to compute, we defined the geometry of the four triangles by taking their lengths ($a_i, b_i, c_i; i = 1 \div 4$). The following vector definitions are needed to express three constraint equations using the leg lengths ($l_i; i = 2 \div 4$) and one equation which represents the four – bar linkage of the inner loop analytically ($\mathbf{v}_1 = \overrightarrow{A_1 P_1}, \mathbf{v}_2 = \overrightarrow{P_1 P_5}, \mathbf{v}_3 = \overrightarrow{P_5 P_2}, \mathbf{v}_4 = \overrightarrow{A_1 A_2}, \mathbf{v}_5 = \overrightarrow{P_1 P_8}, \mathbf{v}_6 = \overrightarrow{P_8 P_3}, \mathbf{v}_7 = \overrightarrow{A_1 A_3}, \mathbf{v}_8 = \overrightarrow{P_5 P_6}, \mathbf{v}_9 = \overrightarrow{P_6 P_4}, \mathbf{v}_{10} = \overrightarrow{A_1 A_4}$).

$$\mathbf{v}_1 = \begin{bmatrix} l_1\cos\phi_1 \\ l_1\sin\phi_1 \end{bmatrix}, \mathbf{v}_2 = \begin{bmatrix} -c_1\sin\phi_2 - a_1\cos\phi_2 \\ c_1\cos\phi_2 - a_1\sin\phi_2 \end{bmatrix}, \mathbf{v}_7 = \begin{bmatrix} A_{3x} \\ A_{3y} \end{bmatrix}$$

[2]ftp.tu-graz.ac.at, /pub/papers/mechanic

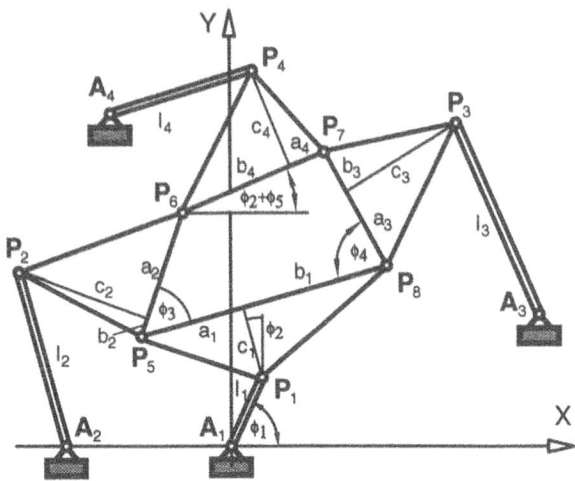

Figure 3. Assur group (4.4)

$$\mathbf{v}_3 = \begin{bmatrix} b_2 \cos(\phi_2+\phi_3) - c_2 \sin(\phi_2+\phi_3) \\ b_2 \sin(\phi_2+\phi_3) + c_2 \cos(\phi_2+\phi_3) \end{bmatrix}, \mathbf{v}_5 = \begin{bmatrix} -c_1 \sin(\phi_2) + b_1 \cos(\phi_2) \\ c_1 \cos(\phi_2) + b_1 \sin(\phi_2) \end{bmatrix}$$

$$\mathbf{v}_6 = \begin{bmatrix} -a_3 \cos(-\phi_4+\phi_2) - c_3 \sin(-\phi_4+\phi_2) \\ -a_3 \sin(-\phi_4+\phi_2) + c_3 \cos(-\phi_4+\phi_2) \end{bmatrix}, \mathbf{v}_8 = \begin{bmatrix} (a_2+b_2) \sin(\phi_2+\phi_3) \\ (a_2+b_2) \cos(\phi_2+\phi_3) \end{bmatrix}$$

$$\mathbf{v}_4 = \begin{bmatrix} A_{2x} \\ 0 \end{bmatrix}, \mathbf{v}_9 = \begin{bmatrix} b_4 \cos(\phi_2+\phi_5) - c_4 \sin(\phi_2+\phi_5) \\ b_4 \sin(\phi_2+\phi_5) + c_4 \cos(\phi_2+\phi_5) \end{bmatrix}, \mathbf{v}_{10} = \begin{bmatrix} A_{4x} \\ A_{4y} \end{bmatrix}$$

The angle ϕ_5 can be represented by $\sin\phi_5 = [(a_3+b_3)\sin\phi_4 - (a_2+b_2)\sin\phi_3]$ $/(a_4+b_4)$, $\cos\phi_5 = [(a_1+b_1) - (a_2+b_2)\cos\phi_3 - (a_3+b_3)\cos\phi_4]/(a_4+b_4)$. The vectorial constraint equations

$$(\mathbf{v}_1 + \mathbf{v}_2 + \mathbf{v}_3 - \mathbf{v}_4)^2 - l_2^2 = 0 \quad (9), \quad (\mathbf{v}_1 + \mathbf{v}_5 + \mathbf{v}_6 - \mathbf{v}_7)^2 - l_3^2 = 0 \quad (10)$$

$$(\mathbf{v}_1 + \mathbf{v}_2 + \mathbf{v}_8 + \mathbf{v}_9 - \mathbf{v}_{10})^2 - l_4^2 = 0 \quad (11)$$

together with the analytical representation of the four – bar linkage

$$[(a_3+b_3)\sin\phi_4 - (a_2+b_2)\sin\phi_3]^2 +$$

$$[(a_1+b_1) - (a_2+b_2)\cos\phi_3 - (a_3+b_3)\cos\phi_4]^2 - (a_4+b_4)^2 = 0 \quad (12)$$

and the trigonometric identities $\sin^2\phi_1 + \cos^2\phi_1 - 1$, $\sin^2\phi_2 + \cos^2\phi_2 - 1$, $\sin^2\phi_3 + \cos^2\phi_3 - 1$, $\sin^2\phi_4 + \cos^2\phi_4 - 1$ generate the ideal for which we computed a lexicographic Gröbner Basis. The input parameters were: $a_1 = 4, b_1 = 2, c_1 = 3, a_2 = 3, b_2 = 2, c_2 = 2, a_3 = 2, b_3 = 5, c_3 = 4, a_4 = 1, b_4 = 5/2, c_4 = 5$ $l_1 = 2, l_2 = 5, l_3 = 11/2, l_4 = 3$ $A_{2x} = -6, A_{3x} = 7, A_{3y} = 15/2, A_{4x} = -4, A_{4x} = 9$. Again all results are accessible at our ftp server [3]. 56 complex solutions

[3] ftp.tu–graz.ac.at, /pub/papers/mechanic

were found according to the degree of the univariate polynomial. The same result was found by *Wohlhart* for a special case of the A(4.4) group, the rhombic Assur group [14].

6. Future work

We want to extend our algorithm to the computation of Gröbner Bases of ideals of higher dimensions to be able to investigate coupler curves and further input – output relations of mechanisms of mobility one and higher. The loss of accuracy of the arbitrary precision floats should be observed in a preciser way. So far we only have a lower limit which gives us a minimum accuracy by taking the *worst case*.

References

1. Artobolevski, I. I. (1977) Theorie des Mechanismes et des Machines,Edition Mir,p. 50
2. Becker, T., Weispfenning V. (1993) Gröbner Bases,*A Computational Approach to Commutative Algebra*,Graduate Texts in Mathematics,Volume 141,Springer – Verlag.
3. Collins, G., E. et al., (1993) `saclib 1.1` users guide. Research Institute for Symbolic Computation, RISC-Linz Report Series Number 93-19, RISC-Linz, Johannes Kepler University, A – 4040 Linz, Austria.
4. Cox, D., Little, J., O'Shea, D., (1992) Ideals,Varieties,and Algorithms, *An Introduction to Computional Algebraic Geometry and Commutative Algebra*.Springer – Verlag.
5. Faugère, J., C., Gianni, P., Lazard, D., Mora, T., (1988) Efficient computation of zero – dimensional Gröbner bases by change of ordering. Technical Report LITP 89 - 52. *J. Symbolic Computation* submitted.
6. Knuth, D., E., (1981) The Art of Computer Programing,Volume 2,*Seminumerical Algorithms*.Addison – Wesley Puplishing Company,Reading Massachusetts,2nd edition.
7. Krandick, W., Johnson, J., R., (1994) A Multiprecision Floating Point and Interval Arithmetic Package for Symbolic Computation. RISC – Linz Report Series. 1994. RISC – Linz, Johannes Kepler University, A – 4040 Linz, Austria, to appear.
8. Krandick, W. (1993) personal communication.
9. Lazard, D., (1992), (a) Stewart platforms and Gröbner basis. *Proceedings of Advance in Robot Kinematics*. Ferrara, Italy.
10. Leinonen, T., ed., (1991) Terminology for the theory of machines and mechanisms. Mechanism and Machine Theory, Vol. 26, No. 5, p. 450.
11. Traverso, C., (1988) Gröbner trace algorithms. *ISSAC 88, Lecture Notes in Computer Science*. Springer Verlag.
12. Wohlhart, K., (1992) Direct Kinematic Solution of the General Planar Stewart Platform.International Conference on *Computer Integrated Manufacturing*,Zakopane,Poland.
13. Wohlhart, K., (1994), (a) A Parallel Redundant Manipulator Based on the Assur Group (3.4), International Conference on *Computer Integrated Manufactoring*, CIM 94, Zakopane, Poland.
14. Wohlhart, K., (1994), (b) Position Analysis of the Rhombic Assur Group. Tenth CISM – IFToMM symposium on *Theory and Practice of Robots and Manipulators*, Ro.Man.Sy.'94, Gdańsk, Poland.
15. Wunderlich, W., (1963) Höhere Koppelkurven, Österreichisches Ingenieur Archiv, XVII 3, pp. 162 – 165.

FORWARD KINEMATICS OF A PARALLEL MANIPULATOR WITH ADDITIONAL ROTARY SENSORS MEASURING THE POSITION OF PLATFORM JOINTS

L. TANCREDI, M. TEILLAUD AND J-P. MERLET

INRIA

B.P. 93, 06902 Sophia-Antipolis Cedex, France

Abstract. The measurement of the link lengths of a general six degrees of freedom parallel manipulator is not sufficient to determine its actual posture. In fact, there are 40 complex solutions to this problem. We show that adding additional rotary sensors allows us to know the position of some points on the platform and reduces the number of solutions. We give conditions under which the manipulator has a unique posture.

1. Introduction

A general six degrees of freedom parallel manipulator is made of two rigid bodies connected to each other by six links. One of the bodies is fixed, the *base*; the other body is a movable *platform*. Each link is connected to the base by a universal joint and to the platform by a ball-and-socket joint. Linear actuators enable the links to change their length. For each actuator a sensor measures the length of the leg (see Figure 1(a)).

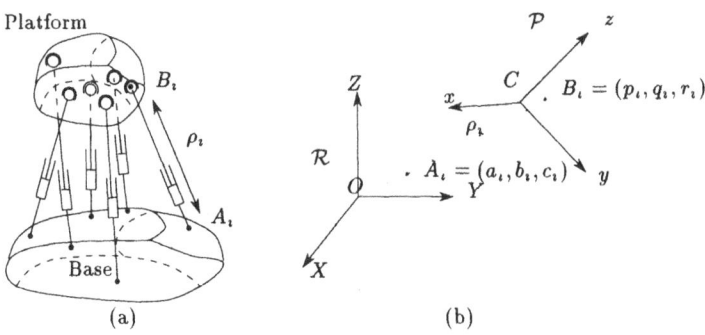

Figure 1. (a) General 6 degrees of freedom parallel manipulator - (b) Notations

261

J.-P. Merlet and B. Ravani (eds.), Computational Kinematics, 261–270.
© *1995 Kluwer Academic Publishers.*

We consider the forward (or direct) kinematics problem (FKP): for a given set of link lengths, as measured by the actuators' sensors, determine the posture of the platform in the reference frame. In this paper *posture* means position *and* orientation of the platform. The FKP does not have a unique solution. The number of real solutions to this problem for a general parallel manipulator is not known yet, but an upper bound on the number of complex solutions is 40. This result has been obtained by continuation methods *[13, 17]*. Algebraic proofs are also available *[11]*, and recently an algorithm has been presented which gives the forty solutions of the FKP *[5]*. At present examples with 16 real solutions have been found *[1]*. Even for a special manipulator architecture where the platform is a triangle, the bound is still 16, and this bound has been reached by real solutions *[2, 6]*.

Computing all the solutions is very slow *[9]*. Moreover, it still remains an open and difficult problem to extract the unique actual solution from the set of computed postures. It appears necessary to have more data on the posture. One way to do this is to add extra sensors. There are two main types of sensors: rotary sensors and displacement sensors. NAIR *[12]*, CHEOK et al. *[3]* used displacement sensors. Their studies solve the FKP, but only in the case of a manipulator with a planar platform. Adding displacement sensors involves adding passive legs and therefore increases the risk of intersections between legs. STOUGHTON et al. *[15]*, MERLET *[10]* used rotary sensors located on the base joints.

Outline. In Section 2 we introduce the method we have used. In Sections *3, 4* and *5*, respectively, we then consider the cases when the extra sensors enable us to know the position of three, two or one platform joints. In the discussion we will always suppose that the manipulator is *well* built, *i.e.* the problem admits at least one solution. We determine the number of solutions of the FKP for each case and the conditions under which a unique solution can be found. This allows deduction of some design rules that guarantee a unique solution according to the number of extra sensors. The results are applied to special architectures in Section *6*.

Notations. The main conventions are illustrated in Figure 1(b). Let $\mathcal{R} = (O, X, Y, Z)$ be the reference frame fixed to the base. Let $\mathcal{P} = (C, x, y, z)$ be the mobile frame fixed to the movable platform. The coordinates of C in $\mathcal{R}$ are (C_X, C_Y, C_Z). A_i denotes the centre of the base joint of the ith leg. Its coordinates are (a_i, b_i, c_i) in frame $\mathcal{R}$. B_i denotes the centre of the platform joint of the ith leg. Its coordinates are (p_i, q_i, r_i) in $\mathcal{P}$. ρ_i is the measured length of $[A_i B_i]$. We denote by $\mathcal{S}_i$ the sphere centered at A_i with

2. Method

Our method is based on two main points: the first is the type of the sensors we use and their location on the manipulator, the second is the way the problem is stated.

2.1. MEASURE OF THE EXTRA SENSORS

Let us recall that a universal joint can be equipped with two rotary sensors: one per rotation axis of the joint.

We define a joint frame (A_i, u, v, w) attached to each joint i, for i in $\{1, \ldots, 6\}$, near the base where $(A_i u)$ is coincident with the first rotation axis of the joint. Let $(A_i w'')$ be the supporting axis of the link $[A_i B_i]$. This axis is obtained from $(A_i w)$ by applying a rotation of angle α around $(A_i u)$ (rotation matrix R_α) followed by a rotation around $(A_i v') = R_\alpha(A_i v)$ with the angle β (rotation matrix R_β). The joint frame coordinates of $A_i B_i$ are given by $A_i B_i = \|A_i B_i\| R_\alpha R_\beta \begin{bmatrix} 0 & 0 & 1 \end{bmatrix}^T$. As $\|A_i B_i\| = \rho_i$ is given by the prismatic sensor, two extra rotary sensors measuring α and β are enough to determine the position of B_i. The measurement of only one angle, α, enables us to determine a plane containing B_i and A_i.

2.2. ALGEBRAIC FORMULATION OF THE PROBLEM

A way to express the forward kinematics problem algebraically is to write down expressions for the lengths of the legs, which are known by measurement. Since the coordinates of the platform joints are only known in the platform frame $\mathcal{P}$, we have to express their coordinates in the base frame $\mathcal{R}$. This can be done by applying a displacement to the platform frame from an initial posture, which is the posture of the base frame, to its current posture. This displacement from $\mathcal{R}$ to $\mathcal{P}$ is a composition of a translation from O to C with a rotation. We have for i in $\{1, \ldots, 6\}$:

$$A_i B_i = C B_i + O C - O A_i,$$

which yields in $\mathcal{R}$

$$\rho_i^2 = \left(R \begin{bmatrix} p_i & q_i & r_i \end{bmatrix}^T + \begin{bmatrix} C_X & C_Y & C_Z \end{bmatrix}^T - \begin{bmatrix} a_i & b_i & c_i \end{bmatrix}^T \right)^2 \quad (1)$$

where R is a rotation matrix, i.e. it satisfies the usual orthonormality equations. The unknowns of the problem are the components of the matrix R and those of the vector OC. When a rotary sensor is added to a leg i, we get a linear equation, because the measure of the sensor defines a plane $\mathcal{E}_i$ containing A_i and B_i (see Section 2.1). Let us suppose the normal vector $N_i = (N_{iX}, N_{iY}, N_{iZ})$ to this plane to be known. Then we have

$A_i B_i \cdot N_i = 0$. This system of equations, which involves quadratic equations in twelve unknowns, is very difficult to solve. In the following, we consider cases where at least one joint position is known using extra sensors. Then the coordinates of C are known, thus the quadratic equations (1) become linear and three unknowns vanish.

3. Three known joints

With two rotary sensors per leg on each of three legs, the positions of three points B_i are determined (see Section 2.1). The knowledge of three non-collinear points is sufficient to deduce the posture of the manipulator (see [14, 16]). Thus, this case admits a unique solution to the FKP. The case when the joints equipped with extra sensors are collinear is degenerate.

4. Two known joints

When the positions of two different points of the platform are known by means of four rotary sensors placed on two segments, and when the lengths of only two other segments are given, there are at most two possible postures for the robot and usually one, as shown in [10]. We systematically study the cases when there are exactly one or two postures (or an infinity for degenerate cases) when the joints of two legs are known, and the lengths of the four other legs are measured.

By renumbering the joints, we can assume that these two known joints are B_1 and B_2. They determine a rotation axis for the platform. Each point of the platform is thus moving on a circle (possibly reduced to a point) centered at line $(B_1 B_2)$. We will denote by C_i the circle associated with B_i, for $i = 3, \ldots, 6$. From the constraints given by the lengths of the segments, B_i also lies on sphere S_i centered on A_i. In the general case when C_i is not contained in S_i, B_i can have at most two positions.

We can choose C and O to be equal to B_1, and both first axes (OX) of $\mathcal{R}$ and (Ox) of $\mathcal{P}$ to be equal to the line $(B_1 B_2)$. The rotation matrix from $\mathcal{R}$ to $\mathcal{P}$ is thus:

$$R = \begin{bmatrix} 1 & 0 & 0 \\ 0 & m_y & -m_z \\ 0 & m_z & m_y \end{bmatrix} \quad \text{where } m_y^2 + m_z^2 = 1 \tag{2}$$

If there exists $i \in \{3, \ldots, 6\}$, say $i = 3$ without loss of generality, such that A_3 does not lie on line $(B_1 B_2)$, we can choose the axis (OY) of $\mathcal{R}$ such that A_3 defines the plane (XOY). Similarly, if B_3 does not lie on $(B_1 B_2)$, we choose (Oy) such that B_3 defines the plane (xOy).

Let us consider the following condition:

There exists i, say $i = 3$, such that neither A_i nor B_i lies on $(B_1 B_2)$. $\tag{3}$

In what follows, whenever condition (3) holds, we will choose (OY) and (Oy) as above; hence $c_3 = r_3 = 0$, and Equations (1) become:

$$\rho_3^2 = a_3^2 + b_3^2 + p_3^2 + q_3^2 - 2a_3 p_3 - 2b_3 q_3 m_y \tag{4}$$

$$\rho_i^2 = 2(-c_i r_i - b_i q_i)m_y - 2(c_i q_i - b_i r_i)m_z + \|OA_i\|^2 + \|CB_i\|^2 - 2a_i p_i, \ i \in \{4,5,6\} \tag{5}$$

m_y is the only unknown appearing in Equation (4).

If Condition (3) does not hold, we choose axes (OY) and (Oy) arbitrarily. For each i in $\{3,\dots,6\}$, A_i or B_i lies on $(B_1 B_2)$, so the coefficients of the unknowns m_y and m_z in Equations (1) vanish. Any rotation will be a solution of the system, which implies that the robot is in a posture where a singularity appears, corresponding to a Grassman variety of rank 5 [8].

Property 4.1 *If, for each i in $\{3,\dots,6\}$, A_i or B_i lies on $(B_1 B_2)$, then the robot is in a posture corresponding to a singularity.*

From now on in Section 4, we will always assume that Condition (3) holds.

As $b_3 q_3 \neq 0$ (by Condition (3)), we can compute a unique value for m_y from Equation (4). Then one equation (5) will allow us to deduce m_z, provided that there exists i in $\{4,5,6\}$ such that the coefficient $-c_i q_i + b_i r_i$ of m_z is not equal to zero.

Let us look at the geometric meaning of the following equation, for some $i \in \{4,5,6\}$:

$$-c_i q_i + b_i r_i = 0. \tag{6}$$

For two points P and Q not lying on $(B_1 B_2)$, let us denote by $\widehat{PQ}$ the oriented angle between the two planes $(B_1 B_2 P)$ and $(B_1 B_2 Q)$. Equation (6) clearly holds when A_i or B_i lies on $(B_1 B_2)$, because in this case $b_i = c_i = 0$ or $q_i = r_i = 0$. Let us now assume that neither A_i nor B_i lies on $(B_1 B_2)$. Then (6) can be rewritten as $\frac{c_i}{b_i} = \frac{r_i}{q_i}$, where $\frac{c_i}{b_i} = \tan \widehat{A_3 A_i}$ et $\frac{r_i}{q_i} = \tan \widehat{B_3 B_i}$ (see Figure 2). Note that these angles are well defined whenever none of the points A_i, B_i, A_3, B_3 lies on $(B_1 B_2)$. Thus Condition (6) holds if and only if the following condition is fulfilled:

$$\left(\begin{array}{c} \text{either } A_i \text{ or } B_i \text{ lies on } (B_1 B_2) \\ \text{or else } \widehat{A_3 A_i} = \widehat{B_3 B_i} \end{array} \right) \tag{7}$$

Suppose that (7) holds for each i in $\{4,5,6\}$; then none of the equations (5) allows us to compute m_z. In this case, another equation will be necessary. We will use Equation (2): $m_y^2 + m_z^2 = 1$. This equation generally has two solutions. It admits one double solution $m_z = 0$ if and only if $m_y = \pm 1$, corresponding to a rotation of angle $\widehat{A_3 B_3} = 0 \pmod{\pi}$, which means that B_1, B_2, A_3, B_3 are coplanar.

Theorem 4.1 *Let the positions of two joints, say B_1 and B_2, be known. Then the FKP admits a unique solution, if and only if*

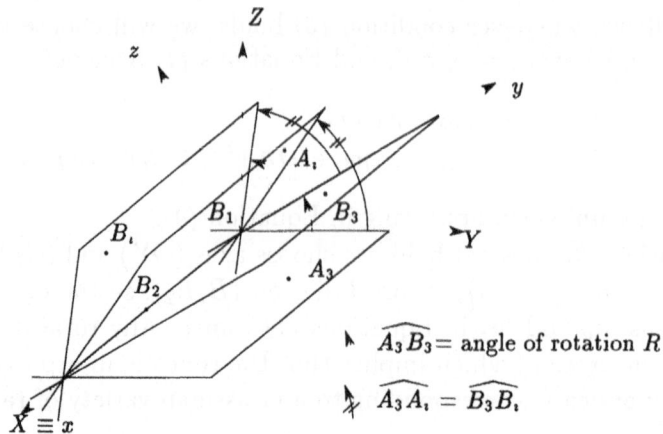

$$\widehat{A_3B_3} = \text{angle of rotation } R$$

$$\widehat{A_3A_\imath} = \widehat{B_3B_\imath}$$

Figure 2. Geometric interpretation of Condition (6)

- there exists i, say $i = 3$, such that neither $A_\imath$ nor $B_\imath$ lies on (B_1B_2)
- and $\exists\, i \in \{4,5,6\}$ such that $\left(\begin{array}{c} \text{neither } A_\imath \text{ nor } B_\imath \text{ lies on } (B_1B_2) \\ \text{and } \widehat{A_3A_\imath} \neq \widehat{B_3B_\imath} \end{array} \right)$
  or B_1, B_2, A_3, B_3 are coplanar

Theorem 4.2 *Let the positions of two joints, say B_1 and B_2, be known. Then the FKP admits two distinct solutions, if and only if*

- there exists i, say $i = 3$, such that neither $A_\imath$ nor $B_\imath$ lies on (B_1B_2)
- and $\forall\, i \in \{4,5,6\}$, $\left(\begin{array}{c} \text{either } A_\imath \text{ or } B_\imath \text{ lies on } (B_1B_2) \\ \text{or else } \widehat{A_3A_\imath} = \widehat{B_3B_\imath} \end{array} \right)$
- and B_1, B_2, A_3, B_3 are not coplanar.

These conditions are rather restrictive, which shows that, in most cases, there will be only one solution.

4.1. ADDING MORE ROTARY SENSORS

We can prove the following theorems, which assume that the conditions given in Theorem 4.2 hold:

Theorem 4.3 *A fifth sensor, placed on a segment $[A_\imath B_\imath]$, for any choice of $i \in \{3,\ldots,6\}$ will allow the FKP to have only one solution if and only if*

- $B_\imath$ is not aligned with B_1 and B_2,
- and the angle between the plane $(B_1B_2A_3)$ and the vector normal to $\mathcal{E}_\imath$ is different from $\widehat{B_3B_\imath}$.

Theorem 4.4 *Two extra sensors, respectively placed on segments $[A_\imath B_\imath]$ and $[A_\jmath B_\jmath]$, for $i,j \in \{3,\ldots,6\}, i \neq j$, still do not give a unique solution to the FKP if and only if*

- $B_i = B_j$, $A_i \neq A_j$, and the vectors N_i, N_j and $A_i A_j$ are dependent, where N_i (resp. N_j) denotes the normal vector to the plane $\mathcal{E}_i$ (resp. $\mathcal{E}_j$) given by the sensor,
- or $B_i \neq B_j$ and, for $k = i, j$, B_k is aligned with B_1 and B_2 or the angle between the plane $(B_1 B_2 A_3)$ and the vector N_k is equal to $\widehat{B_3 B_k}$.

In the other cases, six sensors will always give a unique solution.

5. One known joint

In this section we equip the manipulator with two additional rotary sensors, both located on leg 1. As seen in Section 2.1, it allows complete determination of the coordinates of B_1. Unless all joints on the platform are identical, we can renumber the segments so that B_2 is different from B_1. We choose O and C to be superimposed with B_1, the vector $B_1 A_2$ to define the axis (OX) (since $A_2 \neq B_1$), and $B_1 B_2$ to define (Ox). Thus Equations (1) yield:

$$\rho_2{}^2 = -2 a_2 p_2 l_x + a_2{}^2 + p_2{}^2 \tag{8}$$

$$\rho_i{}^2 = -2 a_i p_i l_x - 2 b_i p_i l_y - 2 c_i p_i l_z - 2 a_i q_i m_x - 2 b_i q_i m_y - 2 c_i q_i m_z$$
$$-2(a_i r_i n_x + b_i r_i n_y + c_i r_i n_z) + \|OA_i\|^2 + \|CB_i\|^2 \quad i \in \{3, \ldots, 6\} \tag{9}$$

5.1. GENERAL PLATFORM

In [7, 4] the authors consider the case of a manipulator with one of the platform joints positions determined by three legs connected to this joint, and only three legs connected to other joints. This case is shown to have at most eight solutions (in a non-singular posture). So, in our case, more constrained, the FKP cannot have more than eight solutions.

5.2. PLANAR PLATFORM

For this particular robot we choose the plane of the platform to be the plane (xOy). This case leads to a system of equations (8), (9) which turns out to be a system of equations with only six unknowns (since $r_i = 0$, for each i). Let us give some sufficient conditions for this system to become degenerate, i.e. for the determinant to vanish.

- there exists i such that $A_i = B_1$ (not allowed by design)
- there exists i such that $B_i = B_1$
- $B_3, \ldots, B_6$ are collinear with B_1 (ZHANG and SONG's design [18])
- $A_2, \ldots, A_6, B_1$ are coplanar

If the system is linearly independent, then we have five linear equations that enable us to solve for five unknowns. The last unknown is found by one of the orthonormality equations of the rotation matrix. Since this last equation is of degree two, there may be up to two solutions. The conditions that determine whether there are one or two solutions are complex. When the rank of the system is not full and decreases by one, we need to add one more quadratic equation to solve our problem; hence the number of solutions doubles. Depending on the rank of the linear system, the number of solutions for the FKP can thus be 2 (full rank), 4 (rank $= 4$), 8 (rank $= 3$) or infinity (rank < 3).

5.3. ADDING MORE ROTARY SENSORS

Now we consider one leg equipped with two sensors (leg 1) and another one equipped with only one rotary sensor (leg 2).

5.3.1. *General platform*
In contrast to Section 4, we use a geometric approach to solve this case. First, we determine the position of B_2. Then, using the results of Section 4, we consider the postures for the whole platform. The joint B_2 is located at the intersection of the sphere S_2 centered at A_2 with radius ρ_2, the sphere centered at B_1 with radius $\|B_1 B_2\|$, and the plane $\mathcal{E}_2$ given by the third sensor. When B_1, A_2, B_2 are collinear, the position of B_2 is uniquely determined. Otherwise, there are two solutions for the joint B_2. We know that there are at most two solutions for the FKP when the positions of B_1 and B_2 are known and that the most general case has one solution (see Section 4). Thus there may be up to four solutions to the FKP, and it has two solutions in the most general case. It is complicated to determine exactly the conditions yielding three or four solutions.

It will be necessary to add six sensors to obtain a linear system. That design leads to a unique solution, except in the degenerate cases.

5.3.2. *Planar platform*
Let us use again the method described in Section 2. Now the third sensor adds a sixth linear equation to a system with six unknowns. When this system is of full rank, the FKP has a unique solution. This system is degenerate when the determinant of the subsystem *(8)*, *(9)* based on the five legs is zero (see 5.2), or when the third sensor gives redundant information. From Section *5.2* we know that there are up to two solutions for the posture when the subsystem *(8)*, *(9)* is non-degenerate. The third sensor gives no useful information when its measured plane contains the two possible locations for B_2 previously found in this case. This occurs for example when the normal N_2 is parallel to the axis $(B_1 A_2)$, i.e. A_2 is also the center of the circle formed by the intersection of S_2 and the sphere centered at B_1 with radius $\|B_1 B_2\|$. In general there will be one solution for this design. Adding more sensors reduces the risk of having a degenerate case.

6. Applications

It would be interesting to study some special parallel manipulator architectures for which the number of solutions for the FKP is almost independent of the posture. Let us consider the case where the position of two joints are known and B_1 and B_2 denote these joints.

• Let us first consider an architecture in which the platform has five collinear joints, i.e. B_4, B_5, B_6 lie on $(B_1 B_2)$, as in [18]. From Theorems 4.1 and 4.2, we deduce that, in general, we will obtain two postures for this robot, the two positions of B_3 being symmetric with respect to the plane $(B_1 B_2 A_3)$, and that these two solutions reduce to one double solution when B_1, B_2, A_3, B_3 are coplanar, but this yields a singular posture.

• Consider a special parallel manipulator where the platform is planar and none of its joints are lined up with $(B_1 B_2)$. There are two solutions for the FKP if and only if A_3 is not in the plane of the platform and if all points A_4, A_5, A_6 are either coplanar with A_3, B_1, B_2 or aligned with B_1 and B_2. In the special case where the platform is planar but the base is not, there is usually only one solution.

• Let us now study a robot with planar platform and base. It can easily be seen from Theorem 4.1 that, in most cases, there will be only one solution for this robot (we still assume that Condition (3) holds). By Theorem 4.2, there are two solutions if and only if B_1 and B_2 lie in the plane of the base. The case of a double solution is obtained when the base and the platform lie in the same plane; it corresponds to a classical singular configuration [8].

• Let us consider a special manipulator which has a triangular platform. The joints are superimposed two by two ($B_1 \equiv B_4$, $B_2 \equiv B_5$, $B_3 \equiv B_6$). The forward kinematics problem admits two solutions if and only if B_1, B_2, A_3 and A_6 are coplanar. In fact, when this condition holds and in addition, B_3 is coplanar with B_1, B_2, A_3 and A_6, the FKP admits a double solution. But this latter case is the well-known *Hunt's singular configuration* [8].

Let us now consider the case where the position of one joint is known. In order to have the smallest number of solutions for the FKP, we should design the manipulator with a planar platform such that the sufficient conditions of Section 5.2 are not fulfilled. With a planar platform and a *nonplanar* base we minimize the risk of having a degenerate case. We should be careful not to choose collinear joints or superimposed joints equipped with extra sensors. There will be one solution for a general platform with one joint known and four additional sensors; for a planar platform, one more joint suffices, except in the degenerate cases.

7. Conclusion

We showed under which conditions the forward kinematics problem admits various numbers of solutions when the position of at least one point of the mobile platform is determined by additional sensory data. These conditions are geometric as well as analytic. They make it possible to compute the solutions and their associated postures.

References

1. M. Ait-Ahmed. *Contribution à la modélisation géométrique et dynamique des robots parallèles*. PhD thesis, Paul Sabatier University, Toulouse, France, February 1993.
2. S. Charentus and M. Renaud. Calcul du modèle géométrique direct de la plate-forme de Stewart. Technical Report 89260, LAAS, Toulouse France, July 1989.
3. K.C. Cheok, J.L. Overholt, and R.R. Beck. Exact methods for determining the kinematics of a Stewart platform using additional displacement sensors. *J. of Robotics Systems*, 10(5):689–707, 1993.
4. K.H. Hunt and E.J.F. Primrose. Assembly configurations of some in-parallel actuated manipulators. *Mechanism and Machine Theory*, 28(1):31–42, January 1993.
5. M.L. Husty. An algorithm for solving the direct kinematics of Stewart-Gough-type platforms. Technical Report TR-CIM-94-7, McGill University, Montréal Canada, June 1994.
6. C. Innocenti and V. Parenti-Castelli. Direct position analysis of the Stewart platform mechanism. *Mechanism and Machine Theory*, 25(6):611–621, 1990.
7. C. Innocenti and V. Parenti-Castelli. Direct kinematics of the 6-4 fully parallel manipulator with position and orientation uncoupled. In *European Robotics and Intelligent Systems Conf.*, Corfu, June 23-28, 1991.
8. J-P. Merlet. Parallel manipulators, Part 2, Singular Configurations and Grassmann geometry. Technical Report 791, INRIA, February 1988.
9. J-P. Merlet. An algorithm for the forward kinematics of general 6 d.o.f. parallel manipulators. Technical Report 1331, INRIA, November 1990.
10. J-P. Merlet. Closed-form resolution of the direct kinematics of parallel manipulators using extra sensors data. In *IEEE Int. Conf. on Robotics and Automation*, pages 200–304, Atlanta, May 2-7, 1993.
11. B. Mourrain. The 40 generic positions of a parallel robot. In M. Bronstein, editor, *ISSAC'93*, ACM press, pages 173–182, Kiev (Ukraine), July 1993.
12. R. Nair. On the kinematics geometry of parallel robot manipulators. Master's thesis, University of Maryland, College Park, 1992.
13. M. Raghavan. The Stewart platform of general geometry has 40 configurations. In *ASME Design and Automation Conf.*, volume 32-2, pages 397–402, Chicago, September 22-25, 1991.
14. X. Shi and R.G. Fenton. A complete and general solution to the forward kinematics problem of a platform-type robotic manipulator. In *IEEE Int. Conf. on Robotics and Automation*, pages 3055–3062, San Diego CA USA, 1994.
15. R. Stoughton and T. Arai. Optimal sensor placement for forward kinematics evaluation of a 6-dof parallel link manipulator. In *IEEE Int. Workshop on Intelligent Robots and Systems, (IROS)*, pages 785–790, Osaka, November 3-5, 1991.
16. L. Tancredi and J-P. Merlet. Evaluation of the errors when solving the direct kinematics of parallel manipulators with extra sensors. In J. Lenarcic and B. Ravani, editors, *Advances in Robot Kinematics and Computational Geometry*, pages 439–448, Ljubljana, Slovenia, 1994. Springer-Verlag.
17. C.W. Wampler. Forward displacement analysis of general six-in-parallel (Stewart) platform manipulators using soma coordinates. Technical Report 8179, General Motors, May 1994.
18. C.D Zhang and S.M. Song. Forward kinematics of a class of parallel (Stewart) platforms with closed-form solutions. In *IEEE Int. Conf. on Robotics and Automation*, pages 2676–2681, Sacramento, April 11-14,1991.

ON THE MODELING OF GRASPS WITH A MULTI-FINGERED HAND

V BRODSKY AND M SHOHAM
Technion - Israel Institute of Technology
Department of Mechanical Engineering
Technion City, Haifa 32000, Israel

Abstract

This research investigates the stability of planar grasps with a multi-fingered robotic hand, using energy approach and geometric interpretation A more general non-linear finger model was adopted, which reveals that the conditions for stability, obtained by traditional linearized model, are too relaxed

Geometrically, the critical conditions for the linearized planar model constitute a hyperplane in the space of grasping forces, whereas the non-linearized model constitutes a third order surface contained within the permissible region of the former Hence, allowable grasping forces calculated by a linearized model may practically lead to instability

The linearized finger model analysis shows that at critical force there is one and only one instantaneous instability center fixed in the plane (which coincides with the compliance center during loading), infinitesimal rotation about which causes instability When a non-linearized finger model is considered, the compliance center position depends on the applied forces and it moves in the plane during loading Furthermore, in some cases, there appear a set of instantaneous instability centers as the critical level of forces is reached

1. Introduction

Within the wide scope of artificial hand design and grasp analysis, this paper concentrates on the subject of grasp stability, which by itself has already been dealt with before by many investigators [Hanafusa and Asada, 1977, Cutkosky, 1985, Mason and Salisbury, 1985 Kerr and Roth, 1986, Li and Sastry, 1987, Nguyen, 1988, Grupen, Henderson, and McCammon, 1989] In our analysis we use the grasp stability definition of the first approach, given qualitatively by Cutkosky and Howe [in Venkataraman, 1990] Will the grasp return to its initial configuration after being disturbed by an external force or moment?

We deal in this investigation with a quasi static case which provides the necessary conditions for a stable grasp in the above mentioned sense (An analysis of the Liapunov stability of a grasp that takes into consideration also dynamic and control effects, can be found in e g [Jen, Shoham, and Longman, 1994])

The present work compares the model of a spring-like finger as introduced in previous investigations by Hanafusa and Asada [1977], Nguyen [1985b], Cutkosky [1985] and others, with a more comprehensive one which takes into consideration infinitesimal change of finger orientation due to a disturbance, and hence involves second-order terms This rigorous analysis shows that the linearized finger model cannot be used to determine practically the region of stable grasping force and the location of the grasp compliance center

271

J -P Merlet and B Ravani (eds), Computational Kinematics, 271–280
© 1995 *Kluwer Academic Publishers*

To formulate the problem we made the following assumptions:
- rigid object and fingers,
- finger-object contact is a point one with friction,
- the fingers act as 'spring-like' fingers, namely, force applied at the fingertips is the sum
    of a given initial force and a disturbance force which is negatively proportional to
    the fingertip displacement.
- the object is initially in equilibrium.

The goal of this investigation is to calculate, under these assumptions, the boundaries of
the grasping forces within which the grasp is stable.

2. Finger Models

Consider a robot hand with four-jointed fingers grasping an object as shown in Figs. 1 and
2. If fingertip motions and forces applied by the finger to the object are in the same plane
(the shaded area in Fig. 1), one can equivalently describe the finger behavior in this plane
with only a two-jointed finger. This simplification reduces the problem to a planar one, and
it is used throughout the paper.

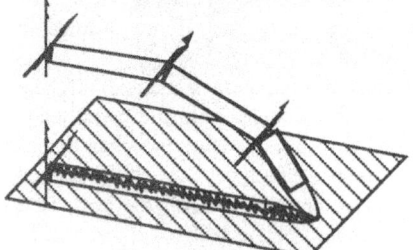

Fig. 1. Projection of a spatial finger
on the plane

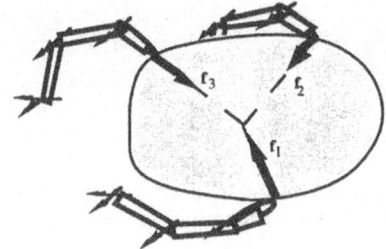

Fig. 2. Body grasped by a three-fingered hand

The 'spring-like' finger model contains at each joint a spring which models the flexibility
of the controller (proportional terms) and transmission elasticity. The four-hinge jointed
finger model contains four torsional springs, whereas its planar simplification contains
only two - one torsional and one linear - as shown in Fig. 3.

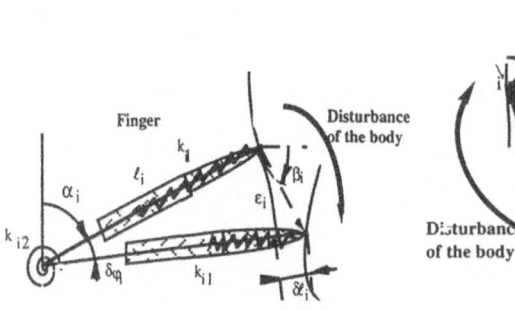

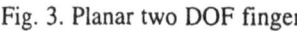

Fig. 3. Planar two DOF finger

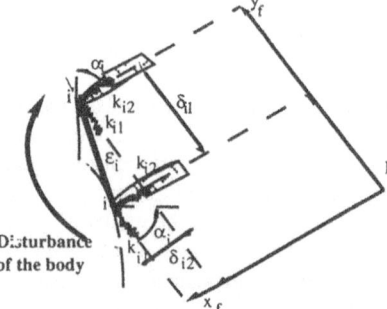

Fig. 4. Linearized finger tip displacement

Previous investigations assumed that infinitesimal motion of the fingertip is negligible
compared to the finite length of the finger, namely, following a disturbance the finger still
maintains its orientation. Noting that stability analysis requires investigation of the second

derivative of energy, it implies that infinitesimal terms must be kept up to the second order. As a result, infinitesimal orientational changes in hinged-finger due to infinitesimal motion of the object cannot be neglected.

2.1. LINEARIZED FINGER MODEL

Starting with the linearized model, Fig. 4, we attach a coordinate system x_f, y_f to the fingertip along which spring deformations are measured. As depicted in Fig. 5, linear translation of the object along a line directed to β_i, causes the fingertip to move from position i to i', which, in turn, cause spring deflections, d_{i1}, δ_{i2} given by:

$$\delta_{i1} = \varepsilon_i \cos(\beta_i - \alpha_i), \quad \delta_{i2} = \varepsilon_i \sin(\beta_i - \alpha), \tag{1}$$

where ε_i is the i-th fingertip motion (also object motion at i-th fingertip contact point), α_i is the initial finger orientation, and compression is assumed to be positive.

At this point we represent a general displacement of a rigid body in a plane by its two linear and one rotational components. In Section 3, where the concept of instantaneous instability center is introduced, this motion is represented as only a rotation about some point. Hence, object disturbance is written as: $\mathbf{p} = \begin{bmatrix} \delta a & \delta b & \delta\theta \end{bmatrix}^T$.

Fingertip displacements, α_i, due to small disturbance, $\mathbf{p}$, are given in x-y system by :

$$\varepsilon_{ix} = r_i[\cos(\gamma_i + \delta\theta) - \cos\gamma_i] + \delta a, \quad \varepsilon_{iy} = r_i[\sin(\gamma_i + \delta\theta) - \sin\gamma_i] + \delta b. \tag{2}$$

where r_i is the length of a radius-vector from the center of rotation to fingertip i, and γ is the angle of this radius-vector with respect to the hand coordinate system. We use these relations and Eq. (1) to calculate the grasp energy Hessian matrix of the linearized model, U_i, the vanishing of its determinant implies instable grasp.

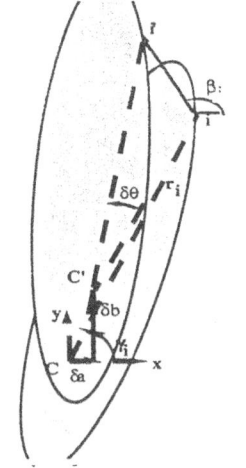

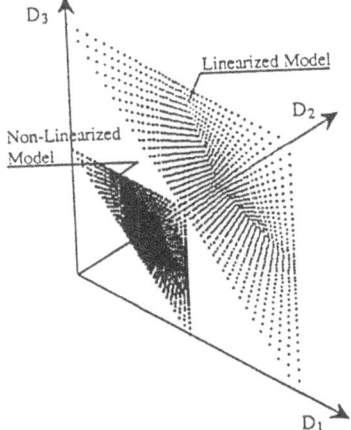

Fig. 5. Object Displacement   Fig. 6. The instability planes

Hence, the grasp becomes unstable when: $\det\left[\dfrac{\partial^2 U_i}{\partial\delta_n \partial\delta_m}\right] = 0 \tag{3}$

where $\partial\delta_n$ and $\partial\delta_m$ are two out of three components of planar disturbance.

Investigating the components of the Hessian matrix, one can observe that out of nine elements only one, $U_{i,33}$, depends on the initial deflections (initial grasping forces), and this dependency is expressed in a linear manner. In addition, equilibrium equations contribute three other linear equations, resulting in a linear equation in 2N-3 parameters, which geometrically describes a hyper-plane. In the case of a three-fingered hand (N=3), this set of equations is a function of only three components, which implies that instability occurs on a plane in the three-dimensional space of independent deflections.

2.2 NON-LINEARIZED FINGER MODEL

A more accurate non-linearized finger model of a hinge-jointed finger is considered next where each finger contains schematically, one linear and one torsional spring (rather than two linear springs) as shown in Fig. 3. In this case the orientation of the finger due to a small disturbance changes. Springs deflections due to small displacement of the fingertip are:

$$\delta\varphi_i = \tan^{-1}\frac{\varepsilon_i \sin\beta_i}{\ell_i + \varepsilon_i \cos\beta_i}, \quad \delta\ell_i = \left\{\left[\ell_i + \varepsilon_i \cos\beta_i\right]^2 + \left[\varepsilon_i \sin\beta_i\right]^2\right\}^{\frac{1}{2}} - \ell_i. \tag{4}$$

where $\delta\varphi_i$ is the change in i-th finger orientation, ℓ_i is its length, and $\delta\ell_i$ is the change in its length. This leads to the following energy Hessian matrix of the non-linearized finger model:

$$U_{j,11} = \sum_{i=1}^{N}\{k_{i1}\cos^2\alpha_i + k_{i2}\frac{\sin^2\alpha_i}{\ell_i^2} - k_{i1}\Delta_i\frac{\sin^2\alpha_i}{\ell_i} - 2k_{i2}\Phi_i\frac{\sin\alpha_i\cos\alpha_i}{\ell_i}\}, \tag{5a}$$

$$U_{j,12} = U_{j,21} = \sum_{i=1}^{N}\{\tfrac{1}{2}\sin 2\alpha_i\left(k_{i1} - \frac{k_{i2}}{\ell_i^2}\right) + k_{i1}\Delta_i\frac{\sin\alpha_i\cos\alpha_i}{\ell_i} + k_{i2}\Phi_i\frac{\cos 2\alpha_i}{\ell_i^2}\}, \tag{5b}$$

$$U_{j,22} = \sum_{i=1}^{N}\{k_{i1}\sin^2\alpha_i + k_{i2}\frac{\cos^2\alpha_i}{\ell_i^2} - k_{i1}\Delta_i\frac{\cos^2\alpha_i}{\ell_i} + 2k_{i2}\Phi_i\frac{\sin\alpha_i\cos\alpha_i}{\ell_i}\}, \tag{5c}$$

$$U_{j,13} = U_{j,31} = \sum_{i=1}^{N}\{-r_i[k_{i1}\cos\alpha_i\sin(\gamma_i - \alpha_i) + k_{i2}\frac{\sin\alpha_i\cos(\gamma_i - \alpha_i)}{\ell_i^2} +$$
$$+k_{i1}\Delta_i\frac{\sin\alpha_i\cos(\gamma_i - \alpha_i)}{\ell_i} + k_{i2}\Phi_i\frac{\cos(\gamma_i - 2\alpha_i)}{\ell_i^2}]\}, \tag{5d}$$

$$U_{j,13} = U_{j,31} = \sum_{i=1}^{N}\{-r_i[k_{i1}\cos\alpha_i\sin(\gamma_i - \alpha_i) + k_{i2}\frac{\sin\alpha_i\cos(\gamma_i - \alpha_i)}{\ell_i^2} +$$
$$+k_{i1}\Delta_i\frac{\sin\alpha_i\cos(\gamma_i - \alpha_i)}{\ell_i} + k_{i2}\Phi_i\frac{\cos(\gamma_i - 2\alpha_i)}{\ell_i^2}]\}, \tag{5e}$$

$$U_{j,13} = U_{j,31} = \sum_{i=1}^{N}\{-r_i[k_{i1}\cos\alpha_i\sin(\gamma_i - \alpha_i) + k_{i2}\frac{\sin\alpha_i\cos(\gamma_i - \alpha_i)}{\ell_i^2} +$$
$$+k_{i1}\Delta_i\frac{\sin\alpha_i\cos(\gamma_i - \alpha_i)}{\ell_i} + k_{i2}\Phi_i\frac{\cos(\gamma_i - 2\alpha_i)}{\ell_i^2}]\}, \tag{5f}$$

As is expected, all the components of the Hessian matrix depend on the grasping forces (the initial springs deflections) which make the solution of stable grasping forces boundaries more difficult. Actually, after substituting the linear conditions for equilibrium, the equation of critical forces is a third order algebraic equation in 2N-3 independent deflections (forces).

3. Instantaneous Instability Center (IIC)

The definiteness of the Hessian matrix (and, correspondingly, the stability of grasping) depends on the principal minors of the Hessian. It can be shown that both the first and the second principle minors of the linearized model are always positive (except where all springs' stiffness are zero, which obviously has no physical sense). Observing that all terms composing the first and second principal minors of the Hessian matrix are invariant to coordinate system origin location, one can choose this origin to coincide with the instantaneous center, $x_c\, y_c$, the coordinates of which are obtained, in the linearized case, by

$$x_c \tfrac{1}{2}\sum_i \sin 2\alpha_i \left(k_{i1} - k_{i2}\right) - y_c \sum_i \left(k_{i1}\cos^2\alpha_i + k_{i2}\sin^2\alpha_i\right) =$$

$$= \sum_i \tfrac{1}{2}x_i\left(k_{i1} - k_{i2}\right)\sin 2\alpha_i - \sum_i y_i\left(k_{i1}\cos^2\alpha_i + k_{i2}\sin^2\alpha_i\right)$$

$$x_c \sum_i \left(k_{i1}\sin^2\alpha_i + k_{i2}\cos^2\alpha_i\right) - y_c \tfrac{1}{2}\sum_i \sin 2\alpha_i\left(k_{i1} - k_{i2}\right) =$$

$$= \sum_i x_i\left(k_{i1}\sin^2\alpha_i + k_{i2}\cos^2\alpha_i\right) - \sum_i \tfrac{1}{2}y_i\left(k_{i1} - k_{i2}\right)\sin 2\alpha_i \qquad (6)$$

It is helpful to note that the determinant of the above system is the same second order minor of the Hessian matrix and it is always positive. It follows that system (6) has a solution for all grasps, and this solution is unique. In the case of a linearized finger model, such a solution is also independent of the initial deflections of the springs, or equivalently, of initial grasping forces. Consequently, since the determinant of (6) is always positive (and cannot vanish), the instantaneous instability center (IIC), i.e., the point about which the object rotate once instability occur, does not lie at infinity. It means that instability of grasp with linearized finger model is caused by a rotation, and not by pure translation (rotation about point at infinity).

When grasp compliance is considered, the same point coincides with the compliance center as was derived by Nguyen [1985] and Shimoga and Goldenberg [1992]. We will demonstrate next that its constant position and uniqueness is, however, not guaranteed for the non-linearized finger model.

4. Force Boundaries for Non-Linearized Finger Model Grasps

Similarly to the linearized model, we also redefine the problem by considering a general disturbance in the plane to be a rotation about some point. One can investigate the occurrence of instability by looking for a center of rotational instability. Hence:

$$\sum_i \left(k_{i1}B_i^2 + \frac{k_{2i}}{\ell_i^2}A_i^2 + k_{i1}\Delta_i\left(-A_i - \frac{A_i^2}{\ell_i}\right) + k_{2i}\Phi_i\left(-\frac{B_i}{\ell_i} - 2\frac{A_iB_i}{\ell_i^2}\right)\right) = 0 \qquad (7)$$

where A_i and B_i are:

$$A_i = \left(x_i - x_c\right)\cos\alpha_i + \left(y_i - y_c\right)\sin\alpha_i, \quad B_i = -\left(x_i - x_c\right)\sin\alpha_i + \left(y_i - y_c\right)\cos\alpha_i.$$

It is worth noting that the introduced parameters A_i and B_i are functions of the grasp geometry and unknown IIC coordinates only, and are independent of the disturbance of the object. It enables us to write the derivatives of springs deflection with respect to small rotation of the object, in terms of those parameters.

Eq. (7) delineates the limits of grasp stability in the space of applied forces. It cannot, however, be directly solved since the coefficients of A_i and B_i contain the coordinates of IIC which are unknown. In order to solve this, we utilize the fact that a set of forces in equilibrium remains in equilibrium after being multiplied by some scalar. This fact enables us to divide the problem into two - one being the geometric property of the grasp while the other, denoted by Force Intensity Level (FIL), is its intensity, .

The coefficients A_i and B_i, are linear functions of IIC's coordinates. Hence, the second derivative of the energy is a quadratic form of x_c and y_c. It is also possible to separate the coefficients into two expressions - a_{ij}, that contains only geometrical properties of the grasp and stiffness of the springs; and - b_{ij}, that contains the initial grasping forces (springs' deflections), which we assume to be normalized and in equilibrium. Writing the second derivative of energy in such a form yields:

$$\left.\frac{d^2U}{d(\delta\theta)^2}\right|_{\delta\theta=0} = (a_{11} - t \cdot b_{11})x_c^2 + 2(a_{12} - t \cdot b_{12})x_c y_c + (a_{22} - t \cdot b_{22})y_c^2 +$$

$$2(a_{13} - t \cdot b_{13})x_c + 2(a_{23} - t \cdot b_{23})y_c + (a_{33} - t \cdot b_{33}) = 0 \tag{8}$$

where t is FIL. The coefficients of this equation given explicitly in Appendix A.

4.1 GEOMETRIC INTERPRETATION

Consider the second derivative of the energy given in (8), as being a spatial quadratic surface. It can be proven that for small FIL this surface whole lies above the xy plane which physically means a stable grasp. Increasing FIL, t causes stretching of the surface along the z axis until, at a critical force it touches the xy plane. At this point there appears an IIC in the plane the rotation about which cause instability. It is possible, however, that the intersection of the surface with the xy plane occurs along a curve which causes the appearance of a set of IICs the rotation about each causes instability. This differs from the linearized model where only one IIC might exist.

Mathematically, one can consider (8) as a planar quadratic form the behavior of which is fully described by its matrix:

$$A = \begin{bmatrix} (a_{11} - t \cdot b_{11}) & (a_{12} - t \cdot b_{12}) & (a_{13} - t \cdot b_{13}) \\ (a_{12} - t \cdot b_{12}) & (a_{22} - t \cdot b_{22}) & (a_{23} - t \cdot b_{23}) \\ (a_{13} - t \cdot b_{13}) & (a_{23} - t \cdot b_{23}) & (a_{33} - t \cdot b_{33}) \end{bmatrix}, \tag{9}$$

and its invariants:

$$A(t) = \begin{vmatrix} (a_{11} - t \cdot b_{11}) & (a_{12} - t \cdot b_{12}) & (a_{13} - t \cdot b_{13}) \\ (a_{12} - t \cdot b_{12}) & (a_{22} - t \cdot b_{22}) & (a_{23} - t \cdot b_{23}) \\ (a_{13} - t \cdot b_{13}) & (a_{23} - t \cdot b_{23}) & (a_{33} - t \cdot b_{33}) \end{vmatrix}, \tag{9a}$$

$$D(t) = \begin{vmatrix} (a_{11} - t \cdot b_{11}) & (a_{12} - t \cdot b_{12}) \\ (a_{12} - t \cdot b_{12}) & (a_{22} - t \cdot b_{22}) \end{vmatrix}, \tag{9b} \qquad I(t) = (a_{11} - t \cdot b_{11}) + (a_{22} - t \cdot b_{22}). \tag{9c}$$

Note that for t=0, all these invariants are positive, which can easily be proved as follows.

All components of $I(0)$ are positive (see Appendix A). The expression of $D(0)$ coincides with the second principal minor of the linearized finger model Hessian matrix, M_2, and as was noted earlier, is positive for all grasping configurations. Since the determinant $A(0)$ is independent of the choice of the coordinate system origin, one may choose such a system that terms a_{13} and a_{23} vanish. In this case $A(0)$ becomes: $A(0) = M_2 \cdot a'_{33}$, which is positive. Note that positive definiteness of a'_{33} is obtained for non-zero springs stiffness and for finite object size. The above discussion implies that <u>positive values of invariants (12a-c) assure a stable grasp.</u>

Note that invariants (12a-c) are all continuous functions of FIL, t, since $A(t)$, $D(t)$ and $I(t)$ are, respectively, polynomials of the third, second and first order in t, with constant coefficients. Since the invariants at t=0 are all positive and continuous, there exists a region of FIL $0 \le t \le t_1$, in which these invariants remain positive. As mentioned above, such positive definiteness of these invariants means a stable grasp. Thus, we have proved the next Theorem:

A planar grasp in equilibrium with one torsional and one linear 'spring-like' fingers with friction, is stable for small grasping forces.

Geometrically, the vanishing of A(t) while D(t) remains positive means that the surface (8) touches the plane in a single point. Further increasing t leads to transformation of this point into an ellipse. Simultaneous vanishing of A(t) and D(t) leads to more complicated cases, such as the appearance of a line or a hyperbola at a critical point. The various cases of instability are illustrated in the next section.

5. Illustrative Example

Two examples, that illustrate the spatial behavior of the energy second derivative of a grasp with non-linearized finger model, are given next. The first example is a symmetrical grasp of a round object. A drawing of the grasped object and corresponding finger data are given in Fig. 7.a . As was proven earlier, A(t) starts from a positive value at t=0, and after reaching zero at a critical point, remains negative. In this case the shape is maintained as elliptic paraboloid while crossing the xy plane, which means IIC is a point in the plane. From the view of our geometric analysis this is a typical case, and it is confirmed by Fig. 7c.

The second case is illustrated by Figs. 8a-c. The geometry of the grasp and its parameters are identical to the first case. The only difference is that one normal force is a unit, the two others equal zero. Obviously, considering a circular object, slippage may occur; but we can easily choose an appropriate shape (rectangle, for example), so that the given system of forces will not cause finger slippage. This distribution of forces leads to a different behavior of the invariant A(t). It vanishes at the critical point, but does not change its sign. Still, the invariant I(t) is sufficiently positive, and only the second invariant D(t) defines the form of the intersection curve after the critical point. The negative value of D(t) in addition to the positive A(t) and I(t) means that the curve is a hyperbola.

Geometrically, at the critical point, the intersection of the surface with the xy plane gives two coincided lines. Further increasing FIL leads to transformation of those lines into two branches of hyperbola. The surface of energy second derivative in this case gains infinity curvature in one direction at the critical point, and this curvature becomes negative immediately after the force intensity level exceeds this point.

6. Numerical Solution of the Grasping Force Boundaries

In this section, the algorithm for obtaining the instability surface is presented. For a planar object grasped by three point fingers with friction, there are six interacting forces of which three are dependent through equilibrium equations. Derivation of the surface that bounds stable grasp, takes place in the space of three independent forces where each point in this space describes an equilibrium state.

Increasing the FIL produces a ray in the first octant (for squeezing forces) extending from the origin until at a point t it pierces the stable grasp boundary surface. Each ray in the space of independent forces can be obtained as a combination of some equilibrium system of forces and intensity, t. There is a one-to-one correspondence between the rays starting at the origin of independent forces' coordinate system and all possible states of equilibrium with the specifically determined metrics - force intensity level.

With this representation, one can solve the equation of instability condition for a given system of forces, or in another words, for a specific ray. The minimal positive root of this equation is the critical force intensity. Note that this is a cubic equation and can be rewritten as follows:

$$A(t) = e_0 t^3 + e_1 t^2 + e_2 t + e_3 = 0, \tag{10}$$

where e_0, e_1, e_2, and e_3 are given in Appendix A.

The critical surfaces in the space of three independent deflections, for linearized and non-linearized finger models, are given in Fig. 6. This is for a symmetrical grasp of a circular object with symmetrical distribution of forces and the same size of object and fingers. The closer-to-origin surface corresponds to the non-linearized finger model. The further one is a plane obtained by the linearized finger model. The comparison of those surfaces leads to the conclusion that the non-linearized model shows instability at about one half of the forces allowed by the linearized model, which is obviously unacceptable for practical applications.

This indicates that only when the dimensions of the fingers are sufficiently larger compared to the object, one can use the formulae for the linearized model to calculate the critical forces. Otherwise, the linearized model conditions allow too high a load and ultimately can lead to instability.

7. Conclusions

The boundaries of grasping forces that maintain stable grasp with 'spring-like' fingers are derived in this paper. Both linearized and more accurate non-linearized finger model are discussed.

It has been shown that critical forces describe a hyper-plane in the space of applied forces when a linearized finger model is used, and a third-order surface when a non-linearized model is used. The critical surface corresponding to the non-linearized finger model lies closer to the origin, and hence permits smaller grasping forces than the surface corresponding to the linearized model. In a common case, when the object and the fingers are of similar sizes, the allowed forces can be as low as half of that calculated by the linearized model.

The linearized finger model leads to a unique point in the plane, the rotation about which has a minimal stiffness (compliance center): and for the critical level of forces the infinitesimal rotation about it leads to instability. In our work where instability is concerned it is termed the instantaneous instability center. An important feature of the compliance center is that its position is function of the geometry of grasp and stiffness of fingers and is independent on the grasping forces.

When the non-linearized model is used, this observation no longer holds. First, the location of the compliance center depends on the forces system applied to the object and can move in the plane during loading. Secondly, in some cases, as the critical force intensity level is reached, a set of instantaneous instability centers may appear.

The use of geometric interpretation of the grasp stability problem, simplifies the calculation of the critical forces. This algorithm requires the solution of a set of cubic equations instead of the investigation of a third-order surface as obtained by evaluating the Hessian matrix of the grasp energy function.

References

Cutkosky, M. R., *Robotic Grasping and Fine Manipulation*, Kluwer Academic Publishers, Hingham, Massachusetts, 1985.

Grupen, R. A., Henderson, T. C. and McCammon, I. D., "A Survey of General Purpose Manipulation," *The Int. Journal of Robotics Research*, Vol. 8, No. 1, pp. 38-62, 1989.

Hanafusa, H. and Asada, H., "Stable Prehension by a Robot Hand with Elastic Fingers," Proc. of the *7th Int. Symposium on Industrial Robots*, Tokyo, pp. 361-368, 1977.

Jen. F., Shoham, M. and Longman, R. W., "Liapunov Stability of Force-Controlled Grasps with a Multi-Fingered Hand," to be published in *The Int. Journal of Robotics Research*.

Kerr, J. and Roth, B., "Analysis of Multifingered Hands," *The Int. Journal of Robotics Research*, Vol. 4, No. 4, pp. 3-17, 1986.

Li, Z. and Sastry, S., "Task Oriented Optimal Grasping by Multifingered Robot Hands," Proc. of the *IEEE Int. Conference on Robotics and Automation*, Raleigh, pp. 389-394, 1987.

Mason, M. T. and Salisbury, J. K. Jr., *Robot Hands and the Mechanics of Manipulation*, MIT Press, Cambridge, Massachusetts, 1985.

Nguyen, V.-D., "The Synthesis of Force-Closure Grasps in the Plane," MIT AI Memo 861, MIT Artificial Intelligence Lab., September, 1985a.

Nguyen, V.-D., "The Synthesis of Stable Grasps in the Plane," MIT AI Memo 862, MIT Artificial Intelligence Lab, October, 1985b.

Nguyen, V.-D., "Constructing Force-Closure Grasps," *The International Journal of Robotics Research*, Vol. 7, No. 3, pp. 3-16, 1988.

Shimoga, K. B. and Goldenberg, A. A., "Constructing Multifingered Grasps to Achieve Admittance Center," Proc. of the *IEEE Int. Conference on Robotics and Automation*, Raleigh, pp. 2296-2301, 1992.

Venkataraman, S. T. and Iberall, T., *Dexstrous Robot Hands*, Springer-Verlag Inc., New York, 1990.

Appendix A.

The coefficients of the second derivative of the grasp energy function in a quadratic form are:

$$a_{11} = \sum_i \left(k_{i1} \sin^2 \alpha_i + k_{i2} \frac{\cos^2 \alpha_i}{\ell_i^2} \right), \quad a_{12} = \sum_i \tfrac{1}{2} \sin 2\alpha_i \left(-k_{i1} + \frac{k_{i2}}{\ell_i^2} \right),$$

$$a_{22} = \sum_i \left(k_{i1} \cos^2 \alpha_i + k_{i2} \frac{\sin^2 \alpha_i}{\ell_i^2} \right), \quad a_{13} = \sum_i \left(k_{i1} M_i \sin \alpha_i - k_{i2} N_i \frac{\cos \alpha_i}{\ell_i^2} \right),$$

$$a_{23} = \sum_i \left(-k_{i1} M_i \cos \alpha_i - k_{i2} N_i \frac{\sin \alpha_i}{\ell_i^2} \right), \quad a_{33} = \sum_i \left(k_{i1} M_i^2 + k_{i2} N_i^2 \right),$$

where: $\quad M_i = -x_i \sin \alpha_i + y_i \cos \alpha_i, \qquad N_i = x_i \cos \alpha_i + y_i \sin \alpha_i.$

The coefficients that are depended on the initial deflections of the springs (initial grasping forces) are as follows:

$$b_{11} = \sum_i \left[\left(k_{i1} \Delta_i C_i - 2k_{i2} \Phi_i \frac{S_i}{\ell_i} \right) \frac{C_i}{\ell_i} \right], \qquad b_{12} = \sum_i \left[\frac{k_{i1} \Delta_i}{\ell_i} S_i C_i + \frac{k_{i2} \Phi_i}{\ell_i^2} \left(C_i^2 - S_i^2 \right) \right],$$

$$b_{22} = \sum_i \left[\frac{k_{i1} \Delta_i}{\ell_i} S_i^2 + 2 \frac{k_{i2} \Phi_i}{\ell_i^2} S_i C_i \right], \qquad b_{13} = \sum_i \left[-\frac{k_{i1} \Delta_i}{\ell_i} N_i C_i + \frac{k_{i2} \Phi_i}{\ell_i^2} \left(N_i S_i - M_i C_i \right) \right],$$

$$b_{23} = \sum_i \left[-\frac{k_{i1} \Delta_i}{\ell_i} N_i S_i - \frac{k_{i2} \Phi_i}{\ell_i^2} \left(N_i C_i + M_i S_i \right) \right],$$

$$b_{33} = \sum_i \left[k_{i1} \Delta_i N_i \left(1 + \frac{N_i}{\ell_i} \right) + \frac{k_{i2} \Phi_i M_i}{\ell_i} \left(1 + 2 \frac{N_i}{\ell_i} \right) \right].$$

The above expressions are used to derive Eq.(14):

$$e_0 = - \begin{vmatrix} b_{11} & b_{12} & b_{13} \\ b_{12} & b_{22} & b_{23} \\ b_{13} & b_{23} & b_{33} \end{vmatrix}, \quad e_1 = \begin{vmatrix} a_{11} & a_{12} & a_{13} \\ b_{12} & b_{22} & b_{23} \\ b_{13} & b_{23} & b_{33} \end{vmatrix} + \begin{vmatrix} b_{11} & b_{12} & b_{13} \\ b_{12} & b_{22} & b_{23} \\ b_{13} & b_{23} & b_{33} \end{vmatrix} + \begin{vmatrix} b_{11} & b_{12} & b_{13} \\ b_{12} & b_{22} & b_{23} \\ a_{13} & a_{23} & a_{33} \end{vmatrix},$$

$$e_2 = - \begin{vmatrix} a_{11} & a_{12} & a_{13} \\ a_{12} & a_{22} & a_{23} \\ b_{13} & b_{23} & b_{33} \end{vmatrix} - \begin{vmatrix} b_{11} & b_{12} & b_{13} \\ a_{12} & a_{22} & a_{23} \\ a_{13} & a_{23} & a_{33} \end{vmatrix} - \begin{vmatrix} a_{11} & a_{12} & a_{13} \\ b_{12} & b_{22} & b_{23} \\ a_{13} & a_{23} & a_{33} \end{vmatrix}, \quad e_3 = \begin{vmatrix} a_{11} & a_{12} & a_{13} \\ a_{12} & a_{22} & a_{23} \\ a_{13} & a_{23} & a_{33} \end{vmatrix}.$$

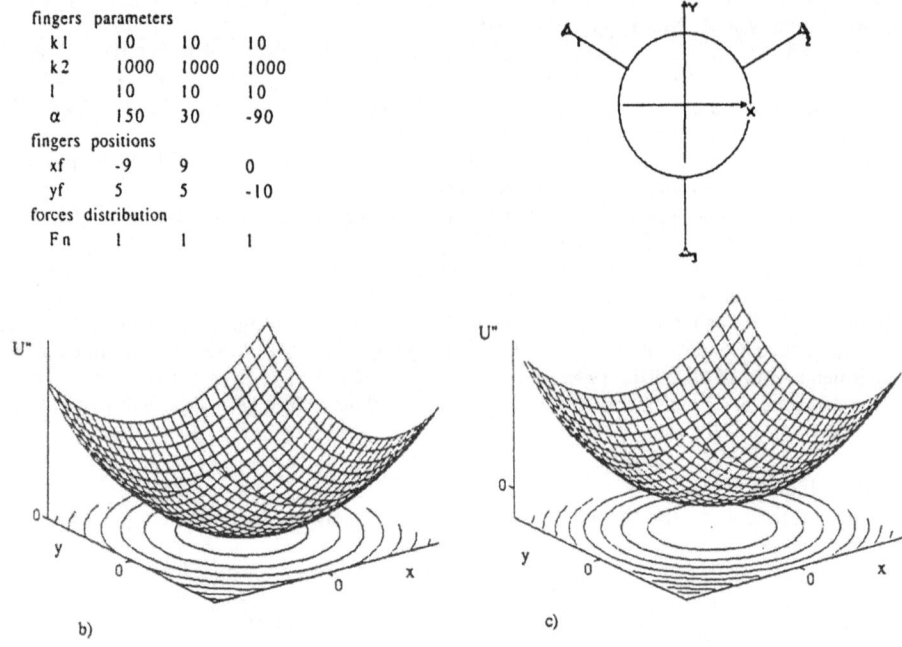

Fig. 7. Symmetrical Grasp. Instability as a function of FIL

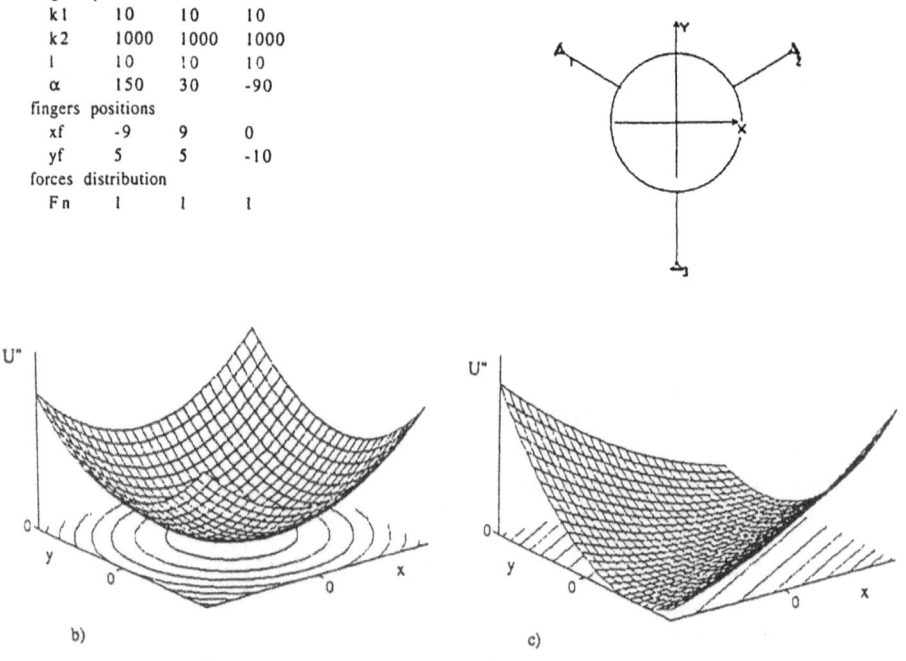

Fig. 8. Non-symmetrical Grasp. Instability as a function of FIL

A SPECIAL CLASS OF C^3 RATIONAL QUARTIC SPLINE CURVES FOR TWO-HARMONIC TRAJECTORY SYNTHESIS

Q. J. GE AND J. RASTEGAR
Department of Mechanical Engineering
State University of New York at Stony Brook
Stony Brook, NY 11794-2300, USA

Abstract

This paper develops an algorithm for constructing C^3 continuous rational quartic B-spline curves such that each rational Bézier curve segment corresponds to a two-harmonic trajectory pattern. Such smooth low-harmonic spline curves can be used for synthesizing robot joint trajectories that are least susceptible to vibrational excitation.

1. Introduction

Polynomial and spline curves have been widely used for robot trajectory synthesis (see, for example, Lin et al (1983), Fu et al (1987)). The resulting trajectories, when expressed in Fourier series, generally indicate a considerable number of harmonics with high frequencies in the joint (actuator) coordinate system. The problem becomes even greater for robot manipulators constructed with revolute joints for the nonlinearity of their dynamics will increase the harmonic content of the actuating torques or forces (Tu and Rastegar, 1993; Tu et al., 1994). For higher operating speeds, the higher harmonics present in the actuating torques required for accurate tracking of the trajectory may be well above the dynamic response limitations ("bandwidth") of the currently available (even high performance and direct drive) actuators.

Recently, the concept of Trajectory Patterns (Fardanesh and Rastegar, 1992; Rastegar et al., 1994) was introduced for robot trajectory synthesis with dynamics limitations. The class of the trajectory patterns that has been shown to be most appropriate for low harmonic motion synthesis are

281

J.-P. Merlet and B. Ravani (eds.), Computational Kinematics, 281–290.
© 1995 *Kluwer Academic Publishers.*

those formed by a number of sinusoidal time functions and an appropriate (small) number of their harmonics. Selection of the coefficients (or trajectory parameters) of the sinusoidal time functions allows for specification of a particular trajectory. Low harmonic trajectory patterns have been used to synthesize robot joint trajectories with minimal actuator high harmonic content (amplitudes) (Tu and Rastegar, 1993; Tu et al 1994).

More recently, Ge and Rastegar (1995a) have shown that a trajectory pattern with maximal harmonics $(n\omega)$, where ω is the fundamental frequency and n is an integer, corresponds to a class of rational Bézier curves of degree $2n$ with a set of specially selected weights. Ge and Rastegar (1995b) presented an algorithm for generating piecewise two-harmonic rational Bézier curves with C^2 continuity. The purpose of the present paper is to develop an algorithm for generating piecewise two-harmonic rational Bézier curves with C^3 continuity. Such smooth low-harmonic spline curves can be used for synthesizing robot joint trajectories that are least susceptible to vibrational excitation.

The paper is organized as follows. Section 2 briefly reviews the two-harmonic rational Bézier (2HRB) curves. Section 3 derives conditions for piecing two 2HRB curves with C^3 continuity. Section 4 presents a geometric algorithm for constructing a two-harmonic C^3 rational B-spline curves.

2. Two-Harmonic Rational Bézier Curves

Two-harmonic rational Bézier (2HRB) curves are special class of rational quartic Bézier curves given by (Ge and Rastegar, 1995a):

$$\mathbf{q}(u) = \sum_{i=0}^{4} \mathbf{b}_i \bar{B}_i^4(u), \tag{1}$$

where the $\mathbf{b}_i$ are the *Bézier points* and $\bar{B}_i^4(u)$ are rational polynomials given by

$$\bar{B}_i^4(u) = w_i B_i^4(u)/(1 + u^2)^2, \tag{2}$$

with the weights $w_0 = w_1 = 1, w_2 = 4/3, w_3 = 2, w_4 = 4$ and $B_i^4(u)$ being quartic Bernstein polynomials. Let $u = \tan(\omega t/2)$, then $\bar{B}_i^4(u)$ are expressible as linear combinations of the sinusoidal functions:

$$
\begin{bmatrix}
\bar{B}_0^4(u) \\
\bar{B}_1^4(u) \\
\bar{B}_2^4(u) \\
\bar{B}_3^4(u) \\
\bar{B}_4^4(u)
\end{bmatrix}
=
\begin{bmatrix}
3/2 & -2 & 0 & 0 & -1/2 \\
-3 & 4 & 2 & -1 & 1 \\
4 & -4 & -4 & 2 & 0 \\
-3 & 2 & 4 & -1 & -1 \\
3/2 & 0 & -2 & 0 & 1/2
\end{bmatrix}
\begin{bmatrix}
1 \\
\sin(\omega t) \\
\cos(\omega t) \\
\sin(2\omega t) \\
\cos(2\omega t)
\end{bmatrix}
$$

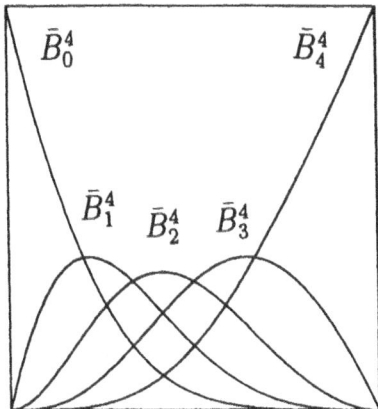

Figure 1. 2-Harmonic rational Bernstein polynomials

and are therefore referred to as *2-harmonic rational Bernstein polynomials*. The plots of these polynomials are shown in Figure 1. The 2HRB curves may be viewed as 2-harmonic trajectory patterns in rational Bézier form.

3. C^3 Continuity of 2HRB Curve Segments

In the field of Computer Aided Geometric Design, C^3 continuity conditions for two rational Bézier curve segments are typically expressed in homogeneous coordinates (Farin, 1993; Hoschek and Lasser, 1993). This projective approach, however, is not directly applicable to piecewise 2HRB curves due to the special choice for the weights of these 2HRB curves. In this section, we present conditions for piecing two 2HRB curves with C^3 continuity. These conditions will be used in the next section to develop 2-Harmonic Rational Spline (2HRS) curves.

Consider two 2HRB curve segments, $q_i(u)$ and $q_{i+1}(v)$, joined together at b_{4i+2}, where u and v are local parameters defined over the interval $[0, 1]$. The curve segment on the left, $q_i(u)$, has $b_{4i-2}, b_{4i-1}, b_{4i}, b_{4i+1}, b_{4i+2}$ as its control points, and $1, 1, 4/3, 2, 4$ as its weights; the curve segment on the right, $q_{i+1}(v)$, has $b_{4i+2}, b_{4i+3}, b_{4i+4}, b_{4i+5}, b_{4i+6}$, as its control points, and $4, 4, 16/3, 8, 16$ as its weights (Figure 2).

The first three derivatives of the right segment $q_{i+1}(u)$ with respect to u at the junction point b_{4i+2} are given by

$$\dot{q}_{i+1}(0) = 4(b_{4i+3} - b_{4i+2}), \tag{3}$$

$$\ddot{q}_{i+1}(0) = 8(2b_{4i+4} - 3b_{4i+3} + b_{4i+2}), \tag{4}$$

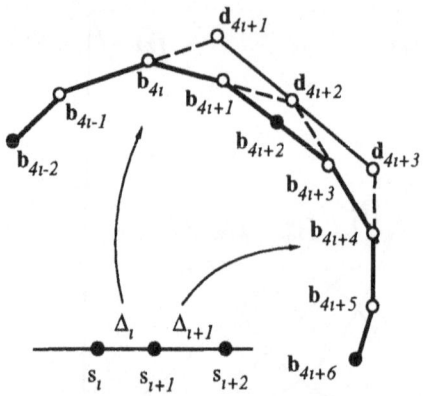

Figure 2. C^3 continuity of 2HRB curves.

$$\dddot{\mathbf{q}}_{i+1}(0) = 24(2\mathbf{b}_{i+5} - 4\mathbf{b}_{4i+4} + \mathbf{b}_{4i+3} + \mathbf{b}_{4i+2}). \tag{5}$$

The first three derivatives of the left segment are given by

$$\dot{\mathbf{q}}_i(1) = 2(\mathbf{b}_{4i+2} - \mathbf{b}_{4i+1}), \tag{6}$$

$$\ddot{\mathbf{q}}_i(1) = -4(\mathbf{b}_{4i+1} - \mathbf{b}_{4i}), \tag{7}$$

$$\dddot{\mathbf{q}}_i(1) = -6(\mathbf{b}_{4i+2} - 2\mathbf{b}_{4i+1} + \mathbf{b}_{4i-1}). \tag{8}$$

The two curves $\mathbf{q}_i$ and $\mathbf{q}_{i+1}$ may also be considered as two segments of one composite curve with a global parameter s, defined over the interval $[s_i, s_{i+2}]$. The left segment $\mathbf{q}_i$ is defined over the interval $[s_i, s_{i+1}]$; while the right segment $\mathbf{q}_{i+1}$ is defined over $[s_{i+1}, s_{i+2}]$. Let $\Delta_i = s_{i+1} - s_i$ and $\Delta_{i+1} = s_{i+2} - s_{i+1}$. Then the two curve segments are C^3 continuous at $\mathbf{b}_{4i+2}$ if their derivatives satisfy the relation (Farin, 1993):

$$\Delta_i \dot{\mathbf{q}}_{i+1}(0) = \Delta_{i+1} \dot{\mathbf{q}}_i(1), \tag{9}$$

$$\Delta_i^2 \ddot{\mathbf{q}}_{i+1}(0) = \Delta_{i+1}^2 \ddot{\mathbf{q}}_i(1), \tag{10}$$

$$\Delta_i^3 \dddot{\mathbf{q}}_{i+1}(0) = \Delta_{i+1}^3 \dddot{\mathbf{q}}_i(1). \tag{11}$$

The C^1 condition (9) can be simplified to yield, after the substitution of (3) and (6) into (9):

$$\mathbf{b}_{4i+2} = \frac{\Delta_{i+1}\mathbf{b}_{4i+1} + 2\Delta_i\mathbf{b}_{4i+3}}{2\Delta_i + \Delta_{i+1}}. \tag{12}$$

Equation (12) indicates that the junction point $\mathbf{b}_{4i+2}$ divides $\mathbf{b}_{4i+1}$ and $\mathbf{b}_{4i+3}$ into the ratio $2\Delta_i/\Delta_{i+1}$. Ge and Rastegar (1995b) showed that the

C^2 condition (10) is the existence of a point, $\mathbf{d}_{4i+2}$, such that (Figure 2)·

$$\mathbf{b}_{4i+1} = \frac{(\Delta_{i+1}^2 + 2\Delta_{i+1}\Delta_i)\mathbf{b}_{4i} + 2\Delta_i^2\mathbf{d}_{4i+2}}{2\Delta_i^2 + \Delta_{i+1}^2 + 2\Delta_{i+1}\Delta_i}, \tag{13}$$

$$\mathbf{b}_{4i+3} = \frac{\Delta_{i+1}\mathbf{d}_{4i+2} + (4\Delta_i + 2\Delta_{i+1})\mathbf{b}_{4i+4}}{4\Delta_i + 3\Delta_{i+1}}. \tag{14}$$

To express C^3 condition in terms of Bézier control points, we first substitute (5) and (8) into (11) and then combine the result with C^1 condition (12) and C^2 conditions (13), (14). After some algebra, we conclude that the C^3 condition for a C^2 curve is the existence of two points $\mathbf{d}_{4i+1}$ and $\mathbf{d}_{4i+3}$ such that they are collinear with $\mathbf{d}_{4i+2}$·

$$\mathbf{d}_{4i+2} = \frac{\gamma_{2,i}\mathbf{d}_{4i+1} + \gamma_{1,i}\mathbf{d}_{4i+3}}{\gamma_{1,i} + \gamma_{2,i}}, \tag{15}$$

and that they are on the following linear combinations:

$$\mathbf{d}_{4i+1} = (1 + \frac{\alpha_{2,i}}{\alpha_{1,i}})\mathbf{b}_{4i} - \frac{\alpha_{2,i}}{\alpha_{1,i}}\mathbf{b}_{4i-1}, \tag{16}$$

$$\mathbf{d}_{4i+3} = (1 + \frac{\alpha_{3,i+1}}{\alpha_{4,i+1}})\mathbf{b}_{4i+4} - \frac{\alpha_{3,i+1}}{\alpha_{4,i+1}}\mathbf{b}_{4i+5}, \tag{17}$$

where

$$\begin{aligned}
\gamma_{1,i} &= 2(2\Delta_i + \Delta_{i+1})(2\Delta_i^2 + \Delta_{i+1}^2 + 2\Delta_{i+1}\Delta_i), \\
\gamma_{2,i} &= -\Delta_{i+1}(\Delta_i + \Delta_{i+1})(4\Delta_i + 3\Delta_{i+1}), \\
\alpha_{1,i} &= 2\Delta_i(2\Delta_i - \Delta_{i+1})(\Delta_i + \Delta_{i+1}), \\
\alpha_{2,i} &= -\Delta_{i+1}(2\Delta_i^2 + \Delta_{i+1}^2 + 2\Delta_{i+1}\Delta_i), \\
\alpha_{3,i+1} &= 2\Delta_i^2(4\Delta_i + 3\Delta_{i+1}), \\
\alpha_{4,i+1} &= \Delta_{i+1}(4\Delta_i^2 - \Delta_{i+1}^2).
\end{aligned} \tag{18}$$

Figure 2 illustrates the arrangement of Bézier points and the points $\mathbf{d}_{4i+1}$, $\mathbf{d}_{4i+2}$, and $\mathbf{d}_{4i+3}$ that satisfy C^3 conditions. Note that, the purpose of Figure 2 is to show only collinearity points such as $\mathbf{d}_{4i+1}$, $\mathbf{d}_{4i+2}$, and $\mathbf{d}_{4i+3}$ Although the point $\mathbf{d}_{4i+2}$ is drawn between $\mathbf{d}_{4i+1}$ and $\mathbf{d}_{4i+3}$, for clarity in graphical illustration, the actual location of the point $\mathbf{d}_{4i+2}$ is on linear extrapolation of $\mathbf{d}_{4i+1}$ and $\mathbf{d}_{4i+3}$. Figure 3 should be read in the same spirit

4. An Algorithm for Constructing 2HRS Curves

This section develops a geometric algorithm for constructing a quartic rational spline (or 2HRS) curve made up of 2HRB curve-segments with C^3 continuity.

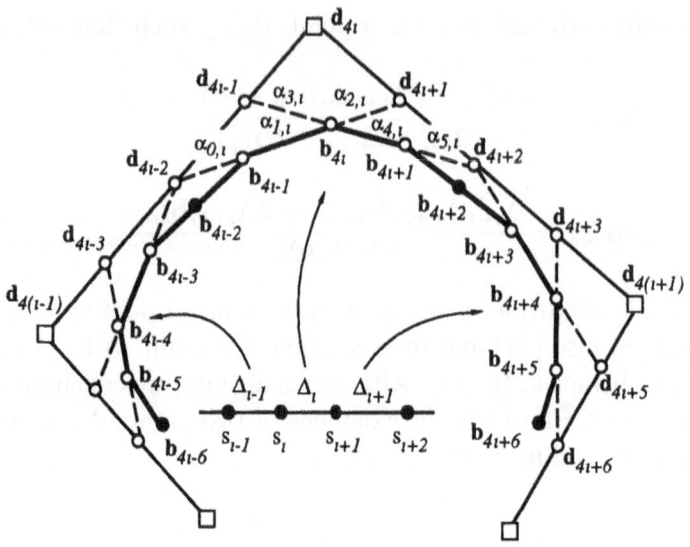

Figure 3 C^3 2HRS curve the auxiliary points $\mathbf{d}_{4i}$ define its control polygon

4 1 C^3 CONTINUITY REVISITED

Consider three 2HRB curve-segments, $\mathbf{q}_{i-1}, \mathbf{q}_i$, and $\mathbf{q}_{i+1}$ joined together with C^3 continuity at $\mathbf{b}_{4i-2}$ and $\mathbf{b}_{4i+2}$ (Figure 3). From C^3 conditions, we obtain two pairs of three collinear points, $\mathbf{d}_{4i-3}, \mathbf{d}_{4i-2}, \mathbf{d}_{4i-1}$ and $\mathbf{d}_{4i+1}$, $\mathbf{d}_{4i+2}, \mathbf{d}_{4i+3}$. We now show that the two lines defined by these points intersect at a point $\mathbf{d}_{4i}$ and that points such as $\mathbf{d}_{4i}$ can be used as control points for designing 2HRS curves.

Applying C^2 condition (14) to the junction point $\mathbf{b}_{4i-2}$, we obtain

$$\mathbf{b}_{4i-1} = \frac{\Delta_i \mathbf{d}_{4i-2} + (4\Delta_{i-1} + 2\Delta_i)\mathbf{b}_{4i}}{4\Delta_{i-1} + 3\Delta_i} \tag{19}$$

Equations (19) and (16) can be used to obtain

$$\mathbf{b}_{4i-1} = \frac{(\alpha_{1,i} + \alpha_{2,i})\mathbf{d}_{4i-2} + \alpha_{0,i}\mathbf{d}_{4i+1}}{\alpha_{0,i} + \alpha_{1,i} + \alpha_{2,i}}, \tag{20}$$

$$\mathbf{b}_{4i} = \frac{\alpha_{2,i}\mathbf{d}_{4i-2} + (\alpha_{0,i} + \alpha_{1,i})\mathbf{d}_{4i+1}}{\alpha_{0,i} + \alpha_{1,i} + \alpha_{2,i}}, \tag{21}$$

where

$$\alpha_{0,i} = 4(2\Delta_i - \Delta_{i+1})(\Delta_{i+1} + \Delta_i)(\Delta_i + 2\Delta_{i-1}), \tag{22}$$

and $\alpha_{1,i}, \alpha_{2,i}$ are given by (18).

Applying the C^3 condition (17) to the junction point $\mathbf{b}_{4\imath-2}$, we obtain

$$\mathbf{d}_{4\imath-1} = (1 + \frac{\alpha_{3,\imath}}{\alpha_{4,\imath}})\mathbf{b}_{4\imath} - \frac{\alpha_{3,\imath}}{\alpha_{4,\imath}}\mathbf{b}_{4\imath+1}. \tag{23}$$

Equations (13) and (23) can be used to obtain

$$\mathbf{b}_{4\imath} = \frac{(\alpha_{4,\imath} + \alpha_{5,\imath})\mathbf{d}_{4\imath-1} + \alpha_{3,\imath}\mathbf{d}_{4\imath+2}}{\alpha_{3,\imath} + \alpha_{4,\imath} + \alpha_{5,\imath}}, \tag{24}$$

$$\mathbf{b}_{4\imath+1} = \frac{\alpha_{5,\imath}\mathbf{d}_{4\imath-1} + (\alpha_{3,\imath} + \alpha_{4,\imath})\mathbf{d}_{4\imath+2}}{\alpha_{3,\imath} + \alpha_{4,\imath} + \alpha_{5,\imath}}, \tag{25}$$

where

$$\alpha_{5,\imath} = \Delta_{\imath+1}(2\Delta_{\imath} + \Delta_{\imath+1})(4\Delta_{\imath-1}^2 - \Delta_{\imath}^2)/(2\Delta_{\imath}). \tag{26}$$

Since equations (24) and (21) must define the same point $\mathbf{b}_{4\imath}$, we equate the right-hand sides of these equations to obtain, after some algebra,

$$-\frac{\gamma_{5,\imath}}{\gamma_{4,\imath}}\mathbf{d}_{4\imath-2} + (1 + \frac{\gamma_{5,\imath}}{\gamma_{4,\imath}})\mathbf{d}_{4\imath-1} = (1 + \frac{\gamma_{3,\imath}}{\gamma_{4,\imath}})\mathbf{d}_{4\imath+1} - \frac{\gamma_{3,\imath}}{\gamma_{4,\imath}}\mathbf{d}_{4\imath+2}, \tag{27}$$

where

$$\begin{aligned}
\gamma_{3,\imath} &= \alpha_{3,\imath}(\alpha_{0,\imath} + \alpha_{1,\imath} + \alpha_{2,\imath}), \\
\gamma_{4,\imath} &= (\alpha_{0,\imath} + \alpha_{1,\imath})(\alpha_{4,\imath} + \alpha_{5,\imath}) - \alpha_{2,\imath}\alpha_{3,\imath}, \\
\gamma_{5,\imath} &= \alpha_{2,\imath}(\alpha_{3,\imath} + \alpha_{4,\imath} + \alpha_{5,\imath}).
\end{aligned} \tag{28}$$

Eq. (27) suggests that the line-segments $\mathbf{d}_{4\imath-3}\mathbf{d}_{4\imath-1}$ and $\mathbf{d}_{4\imath+3}\mathbf{d}_{4\imath+1}$ intersect at a point, which is denoted by $\mathbf{d}_{4\imath}$. It follows that Eq. (27) can be broken into two:

$$\mathbf{d}_{4\imath} = (1 + \frac{\gamma_{3,\imath}}{\gamma_{4,\imath}})\mathbf{d}_{4\imath+1} - \frac{\gamma_{3,\imath}}{\gamma_{4,\imath}}\mathbf{d}_{4\imath+2}, \tag{29}$$

$$\mathbf{d}_{4(\imath+1)} = -\frac{\gamma_{5,\imath+1}}{\gamma_{4,\imath+1}}\mathbf{d}_{4\imath+2} + (1 + \frac{\gamma_{5,\imath+1}}{\gamma_{4,\imath+1}})\mathbf{d}_{4\imath+3} \tag{30}$$

Note that Eq.(30) is obtained from (27) by replacing $\imath$ with $\imath+1$. Eqs. (29) and (30) can be rearranged to yield

$$\mathbf{d}_{4\imath+1} = \frac{\gamma_{4,\imath}\mathbf{d}_{4\imath} + \gamma_{3,\imath}\mathbf{d}_{4\imath+2}}{\gamma_{3,\imath} + \gamma_{4,\imath}}, \tag{31}$$

$$\mathbf{d}_{4\imath+3} = \frac{\gamma_{5,\imath+1}\mathbf{d}_{4\imath+2} + \gamma_{4,\imath+1}\mathbf{d}_{4\imath+4}}{\gamma_{4,\imath+1} + \gamma_{5,\imath+1}}. \tag{32}$$

From Eqs. (15), (31), and (32), we can solve the inner points $\mathbf{d}_{4\imath+1}$, $\mathbf{d}_{4\imath+2}$, and $\mathbf{d}_{4\imath+3}$ in terms of the end points $\mathbf{d}_{4\imath}$ and $\mathbf{d}_{4(\imath+1)}$:

$$\begin{aligned}
\mathbf{d}_{4\imath+1} &= ((\eta_{\imath} - \eta_{1,\imath})/\eta_{\imath})\mathbf{d}_{4\imath} + (\eta_{1,\imath}/\eta_{\imath})\mathbf{d}_{4(\imath+1)}, &\tag{33} \\
\mathbf{d}_{4\imath+2} &= ((\eta_{3,\imath} + \eta_{4,\imath})/\eta_{\imath})\mathbf{d}_{4\imath} + ((\eta_{1,\imath} + \eta_{2,\imath})\eta_{\imath})\mathbf{d}_{4(\imath+1)}, &\tag{34} \\
\mathbf{d}_{4\imath+3} &= (\eta_{4,\imath}/\eta_{\imath})\mathbf{d}_{4\imath} + ((\eta_{\imath} - \eta_{4,\imath})/\eta_{\imath})\mathbf{d}_{4(\imath+1)}, &\tag{35}
\end{aligned}$$

where $\eta_i = \eta_{1,i} + \eta_{2,i} + \eta_{3,i} + \eta_{4,i}$ and

$$\begin{aligned}
\eta_{1,i} &= \gamma_{1,i}\gamma_{3,i}\gamma_{4,i+1}, & \eta_{2,i} &= \gamma_{1,i}\gamma_{4,i}\gamma_{4,i+1}, \\
\eta_{3,i} &= \gamma_{2,i}\gamma_{4,i}\gamma_{4,i+1}, & \eta_{4,i} &= \gamma_{2,i}\gamma_{4,i}\gamma_{5,i+1}.
\end{aligned} \tag{36}$$

Equations (24), (25), (21), (20), (33), (34), and (35) indicate that the end points $\mathbf{d}_{4i}$ and $\mathbf{d}_{4(i+1)}$ can be used as control points for generating the piecewise Bézier polygon for a 2HR spline curve.

4.2. THE ALGORITHM FOR 2HRS CURVES

Given a set of L intervals defined over $s_0 < \ldots < s_L$ (called a *knot sequence*) and a set of $L + 1$ control points, $\mathbf{D}_0, \mathbf{D}_1, \ldots, \mathbf{D}_L$ (where $\mathbf{D}_0 = \mathbf{D}_{4L}$) that form a control polygon for a closed 2HRS curve, the algorithm proceeds as follows:

1. Let $\Delta_i = s_{i+1} - s_i$ $(i = 0, 1, \ldots, L - 1)$ and $\Delta_L = \Delta_0, \Delta_{L+1} = \Delta_1$. Compute the parameters $\alpha_{j,i}$, $\gamma_{j,i}$, and $\eta_{j,i}$ using (18), (22), (26), (28), and (36).
2. Let $\mathbf{d}_{4i} = \mathbf{D}_i$ and then compute the inner points $\mathbf{d}_{4i+1}, \mathbf{d}_{4i+2}, \mathbf{d}_{4i+3}$ on the polygon leg $\mathbf{d}_{4i}\mathbf{d}_{4(i+1)}$ using (33), (34), (35).
3. Compute the Bézier points $\mathbf{b}_{4i}$, $\mathbf{b}_{4i+1}$ on the line-segment $\mathbf{d}_{4i-1}\mathbf{d}_{4i+2}$ and the Bézier point $\mathbf{b}_{4i-1}$ on the line-segment $\mathbf{d}_{4i-2}\mathbf{d}_{4i+1}$ using (24), (25), and (20).
4. Compute Bézier junction point $\mathbf{b}_{4i+2}$ on the line-segment $\mathbf{b}_{4i+1}\mathbf{b}_{4i+3}$ using (12).
5. Each 2HRB curve segment can then be generated from (1) with $u = (s - s_i)/\Delta_i$.

For an open 2HRS curve, we are given a knot sequence $s_0 < \ldots < s_L$ and a set of $L + 4$ control points, $\mathbf{D}_{-2}, \mathbf{D}_{-1}, \ldots, \mathbf{D}_{L+1}$. In this case, things are a bit complicated near the ends:

$$\begin{aligned}
\Delta_{-1} &= \Delta_L = 0, \quad \mathbf{d}_{-2} = \mathbf{b}_{-2} = \mathbf{D}_{-2}, \quad \mathbf{d}_{-1} = \mathbf{b}_{-1} = \mathbf{D}_{-1}, \quad \mathbf{d}_1 = \mathbf{D}_0, \\
\mathbf{d}_0 &= \mathbf{b}_0 = (\alpha_{2,0}\mathbf{b}_{-1} + \alpha_{1,0}\mathbf{d}_1)/(\alpha_{1,0} + \alpha_{2,0}), \\
\mathbf{d}_2 &= ((\eta_{3,0} + \eta_{4,0})\mathbf{d}_1 + \eta_{2,0}\mathbf{d}_4)/(\eta_{2,0} + \eta_{3,0} + \eta_{4,0}), \\
\mathbf{d}_3 &= (\eta_{4,0}\mathbf{d}_1 + (\eta_{2,0} + \eta_{3,0})\mathbf{d}_4)/(\eta_{2,0} + \eta_{3,0} + \eta_{4,0}), \\
\mathbf{d}_{4(L-2)+2} &= (\eta_{3,L-2}\mathbf{d}_{4(L-2)} + (\eta_{1,L-2} + \eta_{2,L-2})\mathbf{d}_{4(L-2)+3})/(\eta_{1,L-2} + \eta_{2,L-2} + \eta_{3,L-2}), \\
\mathbf{d}_{4(L-2)+1} &= ((\eta_{2,L-2} + \eta_{3,L-2})\mathbf{d}_{4(L-2)} + \eta_{1,L-2}\mathbf{d}_{4(L-2)+3})/(\eta_{1,L-2} + \eta_{2,L-2} + \eta_{3,L-2}), \\
\mathbf{d}_{4(L-1)} &= \mathbf{b}_{4(L-1)} = (\alpha_{4,L-1}\mathbf{d}_{4(L-1)-1} + \alpha_{3,L-1}\mathbf{d}_{4(L-1)+1})/(\alpha_{3,L-1} + \alpha_{4,L-1}), \\
\mathbf{d}_{4(L-1)-1} &= \mathbf{D}_{L-1}, \quad \mathbf{d}_{4L-3} = \mathbf{D}_L, \quad \mathbf{d}_{4L-2} = \mathbf{b}_{4L-2} = \mathbf{D}_{L+1}
\end{aligned}$$

Figure 4 shows a closed 2HRS curve with its deBoor polygon (the square) and its piecewise Bézier polygon (between the spline curve and the square). Figure 5 shows an open 2HRS curve with its deBoor polygon. Note that although 2HRB curves have convex-hull property, C^3 2HRS curves in general do not have convex-hull property

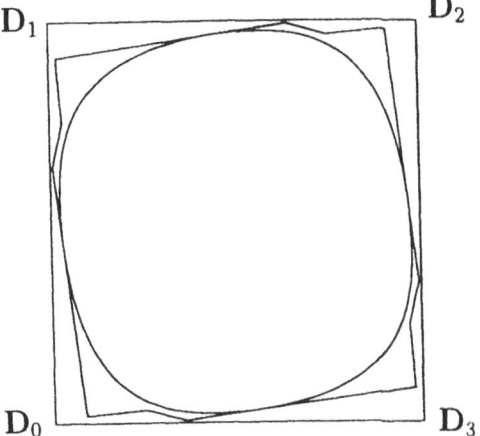

Figure 4. A closed 2HRS curve.

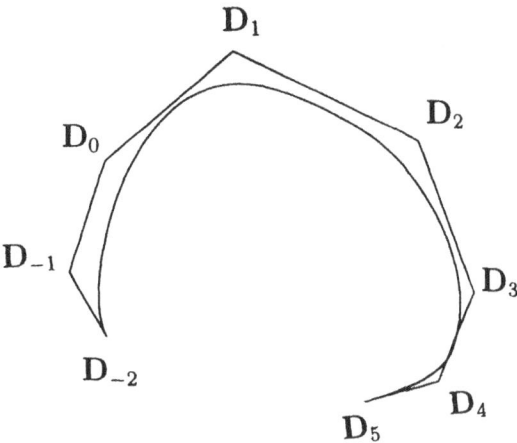

Figure 5. An open 2HRS curve.

5. Conclusions

In this paper, we have presented special classes of rational Bézier curves that correspond to 2-harmonic trajectory patterns. By piecing together 2-harmonic rational Bézier curves with C^3 continuity, we have developed a geometric algorithm for constructing a new class of spline curves called 2-harmonic rational spline (2HRS) curves.

Acknowledgment

This work was supported in part by US National Science Foundation grant MSS-9396265.

References

Fardanesh, B., and Rastegar, J., 1992, "A New Model Based Tracking Controller for Robot Manipulators Using The Trajectory Pattern Inverse Dynamics," *IEEE Trans. Robotics and Automation* **8**, *(2)*, pp. 279-285.

Farin, G., 1993. *Curves and Surfaces for Computer Aided Geometric Design.* 3rd ed., Academic Press.

Fu, K S., R.C. Gonzalez, and C.S.G. Lee, 1987, *ROBOTICS: Control, Sensing, Vision, and Intelligence*, McGraw-Hill, 580pp.

Ge, Q. J., and J. Rastegar, 1995a, Special classes of rational curves with low harmonic content for optimal robot manipulator trajectory synthesis with actuator dynamic response limitations, *Proc. 9th IFToMM World Congress on the Theory of Machines and Mechanisms*, Politecnico di Milano, Italy, August 1995.

Ge, Q.J., and J. Rastegar, 1995b, Low-harmonic rational Bézier and spline curves for joint trajectory synthesis with actuator dynamic response limitations, *Proc. 1995 IEEE Int'l Conf. on Robotics and Automation*, Nagoya, Japan, May 1995.

Hoschek, J. and D. Lasser, 1993, *Fundamentals of Computer Aided Geometric Design* (translated by L.L. Schumaker), A K Peters, 727pp.

Lin, C.S., Chang, P.R., and Luh, J.Y.S., 1983, "Formulation and Optimization of Cubic Polynomial Joint Trajectories for Industrial Robots," *IEEE Trans. Automatic Control*, AC-28(12):1066-1073.

Rastegar, J., Khorrami, F., and Retchkiman, Z, 1994, "Inversion of Nonlinear Systems Via Trajectory Pattern Method," *Proceedings of the 33rd Conf. on Decision and Control*, Orlando, Florida, to appear.

Tu, Q., and J. Rastegar, 1993, "Manipulator Trajectory Synthesis for Minimal Vibrational Excitation Due to Payload," *Trans. of the Canadian Society of Mechanical Engineers*, 17(4A):557-566.

Tu, Q., J. Rastegar, and R.J. Singh, 1994, "Trajectory Synthesis and Inverse Dynamics Model Formulation and Control of Tip Motion of a High Performance Flexible Positioning System," *Mechanism and Machine Theory*, 29(7):959-968.

COMPUTING THE IMMOBILIZING THREE-FINGER GRASPS OF PLANAR OBJECTS

JEAN PONCE
Dept. of Computer Science and Beckman Institute, University of Illinois, Urbana, IL 61801, USA

JOEL BURDICK
Dept. of Mechanical Engineering, California Institute of Technology, Pasadena, CA 91125, USA

AND

ELON RIMON
Dept. of Mechanical Engineering, Technion, Haifa, 32000 Israel

Abstract. This paper addresses the problem of computing frictionless three-finger immobilizing grasps of two-dimensional objects whose boundaries are described by polynomial splines. Using the mobility theory of Rimon and Burdick, we first develop a set of equations that describe the immobilization constraints. We then present a grasp planning algorithm which uses exact cell decomposition and homotopy continuation techniques to construct an explicit description of the immobilization regions (including sample points) in the contact configuration space. The problem of finding optimal immobilizing grasps reduces to hill climbing in each of these regions. We have implemented the proposed approach and present some preliminary results.

1. Introduction

This paper shows how to compute the entire set of frictionless grasps that immobilize a planar object whose boundary is specified by a polynomial spline curve. Force closure is one means to immobilize an object, and many investigators have developed methods for force-closure grasp planning. For example, Mishra, Shwartz, and Sharir have developed a linear time algorithm to determine force-closure grasps of frictionless polygonal objects (Mishra *et al.*, 1987). Nguyen (1988) and Markenscoff and Papadimitriou (1989) have also developed algorithms for finding force-closure grasps of polygonal objects. More recently, Ponce, Stam, and Faverjon (1993), Chen and Burdick (1992) and Blake (1992) have developed schemes for determining force-closure grasps of curved objects. A closely related problem is *fixture*

291

J.-P. Merlet and B. Ravani (eds.), Computational Kinematics, 291–300.

planning. See, for example the work, based on force-closure concepts, of Mishra (1991), Brost and Golberg (1994), and Wallack and Canny (1994).

Our work is a significant departure from previous work on fixture and grasp planning algorithms for two reasons. First, our calculations are not based on the concept of force closure, but on the recent mobility theory of Rimon and Burdick (1994a, 1994b). Conventional force-closure or form-closure theories imply that four frictionless fixtures are required to immobilize a planar object. However, by taking 2nd order (or curvature) effects into account, Rimon and Burdick have recently shown that *any* planar object with a smooth or polygonal boundary can in fact be immobilized with three fixtures which have convex (possibly flat in some cases) surfaces. As shown in Section 3, the 2nd order immobilization constraints define a subset of all equilibrium grasps through a set of non-linear constraints. When the boundary of the workpiece is described by a polynomial spline and the fingers are discs, these constraints are polynomial in the contact parameters, which naturally suggests an algebraic approach to grasp planning. We propose an exact cell decomposition scheme for the three-finger case, which allows us to find all regions of the three-dimensional contact configuration space that yield immobilizing grasps. This approach is reminiscent of Collins' cylindrical algebraic decomposition method for quantifier elimination (Collins, 1975), and relies on homotopy continuation (Morgan, 1987) for solving systems of polynomial equations.

2. Rigid Body Mobility Theory

We now review the essential components of the 2nd order mobility theory of (Rimon and Burdick, 1993; Rimon and Burdick, 1994a). While the theory is general for two or three-dimensional bodies, we restrict our attention to planar bodies here. Specifically, we consider a planar object $\mathcal{B}$, to be grasped by k frictionless fingers, $\mathcal{A}_1, \cdots, \mathcal{A}_k$. The analysis of $\mathcal{B}$'s mobility is formulated in its configuration space (*c-space*). The c-space is parameterized in terms of *hybrid coordinates* $q \in \mathbb{R}^3$ encoding both the position and orientation of $\mathcal{B}$.

2.1. 1ST AND 2ND ORDER MOBILITY

The fixtures are represented as c-space obstacles (or *c-obstacles*). For example, consider Figure 1, where $\mathcal{B}$ is an ellipse in contact with a single finger $\mathcal{A}_i$. The hybrid coordinates of $\mathcal{B}$ in c-space are $q = (d_x, d_y, \theta)$, and the c-obstacle due to $\mathcal{A}_i$ is the set of all configurations where $\mathcal{B}$ intersects the stationary "obstacle" $\mathcal{A}_i$. Thus, if q_0 is $\mathcal{B}$'s contact configuration, q_0 lies on the c-obstacle boundary, which is denoted $\mathcal{S}_i$. When $\mathcal{B}$ is in contact with k fingers, q_0 lies on the intersection of the surfaces $\mathcal{S}_i$ for $i = 1, ..., k$. The *free motions* of $\mathcal{B}$ are those local motions of $\mathcal{B}$ for which it either breaks away from or roll-slides on the surface of the finger bodies. In c-space, the free motions of $\mathcal{B}$ at q_0 are the c-space paths that emanate from q_0 and locally lie in the *free space*, defined as the complement of the c-obstacle interiors. In the following, $\hat{n}_i(q_0)$ denotes the outward pointing unit normal to $\mathcal{S}_i$ at q_0 (Figure 1(b)), and $T_{q_0}\mathcal{S}_i$ denotes the corresponding tangent plane.

The set $M_i^1(q_0)$ of 1st *order free motions* of $\mathcal{B}$ relative to $\mathcal{A}_i$ at q_0 is the set of tangent vectors $\dot{q}$ satisfying $\hat{n}_i(q_0) \cdot \dot{q} \geq 0$. This halfspace's boundary, which can be identified to $T_{q_0}\mathcal{S}_i$, is called the set of 1st *order roll-slide motions.* The

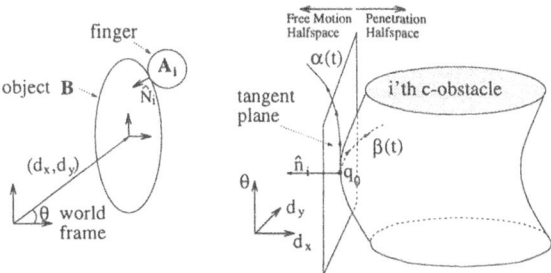

Figure 1. The 1st order approximation to the free motions of $\mathcal{B}$ at q_0.

halfspace's interior is termed the set of 1st *order escape motions*. For k fingers, the set of 1st order free motions is: $M^1_{1,\ldots,k}(q_0) = \cap^k_{i=1} M^1_i(q_0)$.

The distance between $\mathcal{B}$ and $\mathcal{A}_i$ increases to first order along escape motions, which implies that $\mathcal{B}$ locally breaks away from $\mathcal{A}_i$. To first order, $\mathcal{B}$ maintains zero distance from $\mathcal{A}_i$ along 1st order roll-slide motions, and *it is not possible* to determine from 1st order considerations whether $\mathcal{B}$ locally breaks away from or penetrates $\mathcal{A}_i$.

For example, the c-space curves $\alpha(t)$ and $\beta(t)$ in Figure 1 have the same tangent vector, and thus are equivalent to first order. Yet $\alpha(t)$ locally lies in the free space, while $\beta(t)$ does not. As we shall see, *all the free motions of $\mathcal{B}$ at an equilibrium grasp are necessarily roll-slide to 1st order*. This fact has the important consequence that the mobility of $\mathcal{B}$ at an equilibrium grasp depends on the second order properties of its local motions.

The second-order geometry of the free-motion curves and c-obstacle boundaries is determined by their curvature and curvature form, respectively. The *curvature form* of $\mathcal{S}_i$ at $q_0 \in \mathcal{S}_i$ is by definition $\kappa_i(q_0, \dot{q}) = \dot{q}^T[D\hat{n}_i(q_0)]\dot{q}$ for $\dot{q} \in T_{q_0}\mathcal{S}_i$. The matrix $D\hat{n}_i(q_0)$ encodes the curvature of the c-obstacle surface at q_0.

The set of 2nd *order free motions* of $\mathcal{B}$ relative to $\mathcal{A}_i$ at q_0, denoted $M^2_i(\dot{q}_0)$, is the set of pairs $(\dot{q}, \ddot{q})$ such that $\dot{q} \in T_{q_0}\mathcal{S}_i$ and $\kappa_i(q_0, \dot{q}) + \hat{n}_i(q_0) \cdot \ddot{q} \geq 0$. Among those, the pairs $(\dot{q}, \ddot{q})$ that satisfy $\kappa_i(q_0, \dot{q}) + \hat{n}_i(q_0) \cdot \ddot{q} = 0$ are called 2nd *order roll-slide motions*, and the remaining ones are termed 2nd *order escape motions*. For k fingers, the set of 2nd order free motions is: $M^2_{1,\ldots,k}(q_0) = \cap^k_{i=1} M^2_i(q_0)$.

We show in (Rimon and Burdick, 1993) that if $(\dot{q}, \ddot{q}) \in M^2_i(q_0)$ is a 2nd order escape motion, any c-space path $\alpha(t)$ with $\alpha(0) = q_0$, $\dot{\alpha}(0) = \dot{q}$, $\ddot{\alpha}(0) = \ddot{q}$, *locally lies in the free space* for $t \geq 0$. If $(\dot{q}, \ddot{q}) \in M^2_i(q_0)$ is a 2nd order roll-slide motion, *it is not possible* to determine from 2nd order considerations if $\alpha(t)$ locally lies in the freespace.

For planar bodies, the 1st order roll-slide motions can be characterized as follows. Let l_i denote the line collinear with the ith contact normal, and let ρ_i be the distance along l_i from the ith contact point, such that ρ_i is positive on $\mathcal{B}$'s side of the contact and negative on $\mathcal{A}_i$'s side. The 1st order roll-slide motions correspond to the collection of instantaneous rotations of $\mathcal{B}$ about an axis perpendicular to the plane and passing through l_i at distance ρ_i. In hybrid coordinates, tangent vectors are denoted $\dot{q} = (v, \omega)$, where v is translational velocity and ω angular velocity, and the 1st order roll-slide motions are instantaneous rotations of $\mathcal{B}$, which have the form $\dot{q} = (0, \omega)$.

Lemma 2.1 (Rimon and Burdick, 1993) *In the planar case, the c-space curvature of $\mathcal{S}_i$ at q_0 along the instantaneous rotation $\dot{q} = (0, \omega)$ of $\mathcal{B}$ about an axis located at distance ρ_i along the line l_i is:*

$$\kappa_i(q_0, \dot{q}) = \frac{1}{\kappa_{\mathcal{A}_i} + \kappa_{\mathcal{B}_i}}(\rho_i \kappa_{\mathcal{B}_i} - 1)(\rho_i \kappa_{\mathcal{A}_i} + 1)\omega^2. \tag{1}$$

2.2. A 2ND ORDER MOBILITY INDEX

At an equilibrium grasp the net wrench (i.e. force and torque) on $\mathcal{B}$ must be zero. The wrench due to a normal contact force applied by $\mathcal{A}_i$ on $\mathcal{B}$ is a positive multiple of the outward pointing finger c-obstacle normal $\hat{n}_i(q_0)$. Thus, the equilibrium condition in c-space is that there exist scalars $\lambda_1, \ldots, \lambda_k$ such that

$$\lambda_1 \hat{n}_1(q_0) + \cdots + \lambda_k \hat{n}_k(q_0) = 0 \quad \text{and} \quad \lambda_i \geq 0. \tag{2}$$

We assume that each $\hat{n}_i(q_0)$ in (2) is *essential* for the grasp, i.e., that each λ_i is strictly positive and that any $k - 1$ wrenches are linearly independent (Rimon and Burdick, 1994a). In this case, the coefficients λ_i are uniquely determined up to a scale factor.

While the individual c-obstacle curvature forms are in general not coordinate-invariant, it is shown in (Rimon and Burdick, 1994a) that their weighted sum defined below has a coordinate invariant structure which characterizes the mobility of $\mathcal{B}$ to 2nd order at the equilibrium grasp. The *c-space relative curvature form* associated with an equilibrium grasp is the quadratic form

$$\kappa_{\text{rel}}(q_0, \dot{q}) = \sum_{i=1}^{k} \lambda_i \kappa_i(q_0, \dot{q}) = \dot{q}^T [\sum_{i=1}^{k} \lambda_i D\hat{n}_i]\dot{q},$$

where $\dot{q} \in M^1_{1,\ldots,k}(q_0)$.

The 2nd *order mobility index* of an equilibrium grasp configuration, denoted $m^2_{q_0}$, is the number of non-negative eigenvalues of the c-space relative curvature matrix $\sum_{i=1}^{k} \lambda_i D\hat{n}_i$. A key interpretation of the 2nd order index is provided in (Rimon and Burdick, 1994a, Proposition 5.6): if $m^2_{q_0} = 0$, any local motion of $\mathcal{B}$ is either 1st order penetration, or it is 1st order roll-slide which is necessarily a 2nd order penetration motion. Thus $m^2_{q_0} = 0$ *implies that $\mathcal{B}$ is completely immobilized.*

3. Immobilizing Three-Finger Planar Grasps

In this section we use the mobility theory reviewed in the last section to characterize the set of immobilizing three-finger grasps of a planar object bounded by a parametric curve. We first develop a set of equations which describe the conditions under which $\mathcal{B}$ is held in equilibrium by the fingers $\mathcal{A}_1, \cdots, \mathcal{A}_k$. We then use the mobility index to place constraints on this set of grasps, thereby determining the subset of equilibrium grasps which are immobilizing. For simplicity's sake, we further assume that the fingers are discs with given radii. Our derivations are formulated in the *contact configuration space*: That is, let $x : I \subset \mathbb{R} \to \mathbb{R}^2$ denote the (piecewise) polynomial curve which bounds $\mathcal{B}$, and let $x_i = x(u_i)$ $(i = 1, 2, 3)$ denote the positions of the finger contact points along the boundary curve, we call the space of all triplets (u_1, u_2, u_3) the contact configuration space.

3.1. EQUILIBRIUM GRASPS

Here we denote by t_i (resp. n_i) the *unit* tangent (resp. normal) to the boundary of $\mathcal{B}$ at the contact point x_i, and by $w_i = (n_i, x_i \times n_i)^T$ the corresponding wrench due to a unit force applied along the contact normal at x_i. Recall that a necessary condition for an essential equilibrium grasp is that there exist three scalar $\lambda_1, \lambda_2, \lambda_3$ such that

$$\lambda_1 w_1 + \lambda_2 w_2 + \lambda_3 w_3 = 0, \quad \text{and} \quad \lambda_1, \lambda_2, \lambda_3 > 0, \tag{3}$$

and the wrenches w_i are pairwise linearly independent.

Equation (3) describes the set of all equilibrium grasps in the space of the parameters $u_1, u_2, u_3, \lambda_1, \lambda_2, \lambda_3$. However, we wish to characterize the equilibrium conditions directly in the contact configuration space. The following discussion shows how to eliminate the parameters $\lambda_1, \lambda_2, \lambda_3$. A necessary condition for the existence of coefficients λ_i, not all of them being equal to zero, such that (3) is satisfied is of course

$$\text{Det}(w_1, w_2, w_3) = 0, \tag{4}$$

which defines a two-dimensional surface S in the three-dimensional contact configuration space u_1, u_2, u_3 of the grasp. To ensure that the coefficients $\lambda_1, \lambda_2, \lambda_3$ are also positive, we form the 2D cross-products of the vectors in (3) with n_2 and n_1. This yields

$$\lambda_1 n_1 \times n_2 + \lambda_3 n_3 \times n_2 = 0 \quad \text{and} \quad \lambda_2 n_2 \times n_1 + \lambda_3 n_3 \times n_1 = 0. \tag{5}$$

From (5) it easily follows that the coefficients λ_i are defined (up to a scale factor) by

$$\lambda_i = n_{i+1} \times n_{i+2} = t_{i+1} \times t_{i+2}, \tag{6}$$

where index addition is performed modulo 3. In particular, (3) admits a solution if and only if (4) is satisfied, and

$$(t_1 \times t_2)(t_2 \times t_3) > 0 \quad \text{and} \quad (t_1 \times t_2)(t_3 \times t_1) > 0. \tag{7}$$

Hence, the set of equilibrium grasps in the contact configuration space is defined by (3) and the inequality constraints (7). These constraints only change sign when one traverses one of the cylindrical surfaces $t_i \times t_{i+1} = 0$.

To enforce the essential-grasp condition in the three-finger case, we further require that any two wrenches be linearly independent. It can be shown that this happens when the corresponding two fingers coincide or they form an antipodal point grasp. Numerous techniques are available for finding antipodal grasps (Chen and Burdick, 1992). Hence, in practice it may be easier to remove these sets of points from the final set of immobilizing grasps, rather than explicitly enforce the linear independence condition during the construction of the immobilizing regions. Alternatively, a sufficient condition for linear independence is:

$$t_i \times t_{i+1} \neq 0 \quad \text{for} \quad i = 1, 2, 3. \tag{8}$$

3.2. IMMOBILIZING GRASPS

The immobilizing grasps are simply the subset of the equilibrium grasps which are 2nd order immobile –i.e., the relative curvature matrix has all negative eigenvalues. In the three-finger case, the relative curvature form is:

$$\kappa_{\text{rel}} = \lambda_1 \kappa_1(q_0, \dot{q}) + \lambda_2 \kappa_2(q_0, \dot{q}) + \lambda_3 \kappa_3(q_0, \dot{q}), \qquad (9)$$

where $\dot{q} \in M^1(q_0)$. According to Lemma 2.1, this can be rewritten as:

$$\kappa_{\text{rel}} = \sum_{i=1}^{3} \frac{\lambda_i}{\kappa_{\mathcal{A}_i} + \kappa_{\mathcal{B}_i}} \big(\rho_i \kappa_{\mathcal{B}_i} - 1\big)\big(\rho_i \kappa_{\mathcal{A}_i} + 1\big)\omega^2 < 0, \qquad (10)$$

where ρ_i is the signed distance between the contact point x_i and the point x_0 where the lines of action of the three contact forces intersect, i.e., ρ_i is defined by $x_i - x_0 = \rho_i n_i$.

We already know from (6) that $\lambda_i = t_{i+1} \times t_{i+2}$, and a simple calculation shows that

$$\rho_i = \frac{(x_{i+1} - x_i) \cdot t_{i+1}}{t_i \times t_{i+1}}. \qquad (11)$$

Furthermore, if we denote the first and second derivative of the boundary curve position vector by x_i' and x_i'' respectively, we have

$$|x_i'|t_i = x_i' \quad \text{and} \quad |x_i'|^3 \kappa_{\mathcal{B}_i} = x_i' \times x_i''. \qquad (12)$$

In particular, it can be seen by substitution of (6), (11), and (12) into (10) that for (piecewise) polynomial boundary curves, the condition $\kappa_{\text{rel}} < 0$ can be expressed (after eliminating the radicals through appropriate squaring) by a polynomial equation in the curve parameters u_1, u_2, u_3.

4. Computing the Immobilizing Grasps

The set of equilibrium grasps maps onto a 2D surface S defined by (3) in the contact configuration space. The immobilizing grasps map to the regions of this surface which satisfy the constraints (7), (8), and (10). Constructing these regions is an instance of the more general problem of constructing the arrangement of two-dimensional regions defined by an equality constraint $f(u_1, u_2, u_3) = 0$ and n inequality constraints $g_i(u_1, u_2, u_3) \leq 0$, $i = 1, .., n$.

We propose an approach reminiscent of Collins' cylindrical algebraic decomposition method for quantifier elimination (Collins, 1975). We first compute the *occluding contour* of the surface S, i.e., the points of S such that $\partial f / \partial u_3 (u_1, u_2, u_3) = 0$ (Figure 2). The projection of the occluding contour onto the plane (u_1, u_2) is the *silhouette*. The silhouette has the following property. Let $n(u_1, u_2)$ denote the number of points of S above the point (u_1, u_2). $n(u_1, u_2)$ can only change when one crosses the silhouette.

We then construct the intersection curves of S with each one of the constraints $g_i(u_1, u_2, u_3) = 0$, project them onto the (u_1, u_2) plane, and construct the planar arrangement cut by these curves and the silhouette in the plane (u_1, u_2). Each region, R_i, in this planar arrangement has two important properties. First, since

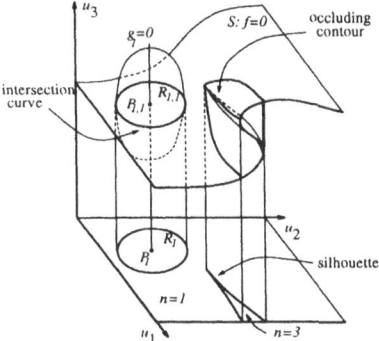

Figure 2. A surface S intersecting another surface, and the corresponding occluding contour, silhouette, and intersection curve. Here n is short for $n(u_1, u_2)$.

the silhouette does not cross R_i, $n(u_1, u_2)$ is constant over the region R_i. Let n_i denote the value of this constant. This allows us to associate with R_i a set of n_i regions $R_{i,j}$ ($j = 1, .., n_i$) of S whose boundaries can be constructed by back-projecting the boundary of R_i onto S. The second fundamental property of the arrangement is that, since the projected intersection curves do not cross the boundary of R_i, the sign of the original constraints is constant over each region $R_{i,j}$. Thus we can decide if $R_{i,j}$ is an immobilizing region by evaluating the constraints at a single sample point $P_{i,j}$ of $R_{i,j}$. The sample points $P_{i,j}$ can themselves be constructed by first constructing a sample point P_i for each region R_i, then intersecting a vertical ray through P_i with the surface S.

The main computational step of our approach is the construction of the regions R_i. This is a plane-sweep algorithm modified so it can handle algebraic curves. This requires computing the extrema of the input curves along some direction, as well as the curve singularities and pairwise projection intersections. In our case, all constraints are polynomial, and these points are found by solving square systems of multivariate polynomial equations. We use a distributed implementation of homotopy continuation (Morgan, 1987) for that purpose.

5. Optimal Grasps

The procedure outlined in the last section provides a starting point for optimization using any optimality criterion: since each region found by our algorithm satisfies all of the constraints necessary for immobilization, any optimization method can be used to find locally optimal immobilizing grasps in each of these regions.

To illustrate this idea, let us assume that B is an elastic object. We compute the moment component of the reaction wrench that arises due to the elastic deformation at the contacts when B undergoes a small rotation about the point of intersection of the contact normals. Our optimality criterion will be to maximize this moment.

We denote by ρ the signed distance between the center of rotation and the contact point, by $r_B = 1/\kappa_B$ the radius of curvature of the object boundary at the contact point, and by $r_A = 1/\kappa_A$ the radius of the finger (Figure 3). After a small rotation of angle α, the finger penetrates the object at a depth d. We assume that the reaction force is in the direction of the line joining the finger to the closest

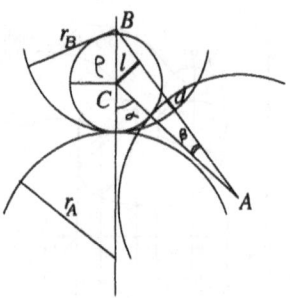

Figure 3. Geometry for moment computation.

contour point, with magnitude proportional to d. Approximating the boundary by its osculating circle, it follows that the force's direction is the radius of this circle going through the finger.

We first calculate d. Consider the triangle ABC in Figure 3. The angle between the edges CA and CB is $\pi - \alpha$. From trigonometry, we have:

$$\cos(\pi - \alpha) = \frac{CA^2 + CB^2 - AB^2}{2CA.CB}.$$

Clearly, $d = r_A + r_B - AB$. Using the above equation, we obtain that, for small values of α,

$$d = \frac{(r_A + \rho)(r_B - \rho)}{2(r_A + r_B)}\alpha^2 = -\frac{1}{2}\kappa\alpha^2.$$

We now compute the distance l between the line of action of the reaction force and the center of rotation C. Clearly, we have $l = (\rho + r_A)\sin\beta$, and using again trigonometry within the triangle ABC we obtain:

$$\frac{\sin\beta}{BC} = \frac{\sin(\pi - \alpha)}{AB}.$$

For small values of α, it follows that:

$$l = \frac{(r_A + \rho)(r_B - \rho)}{r_A + r_B}\alpha = -\kappa\alpha,$$

and we finally obtain the moment of the reaction force: $m = \frac{1}{2}\kappa^2\alpha^3$.

In particular, given three contacts, we can use the quantity $\sum_{i=1}^{3}\kappa_i^2$ as a cost function for optimization. This would result in grasps with the greatest rotational "stiffness." This criterion has the disadvantage of allowing some of the contacts to violate the immobilization constraints, and we have found it advantageous to use instead the cost function $\min_i \kappa_i^2$.

6. Implementation and Results

In our implementation, we have focussed on a simplified set of immobilizing grasps: since the coefficients λ_i are all positive for equilibrium grasps, immobilization will be guaranteed if *each* of the c-space curvatures κ_i is negative. To further simplify

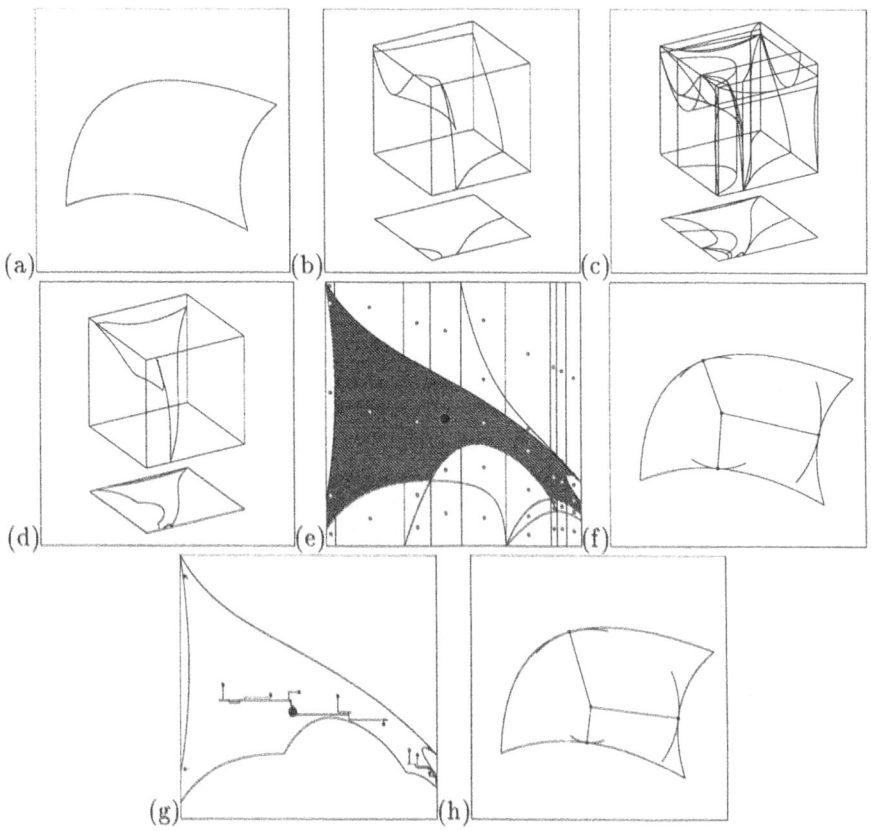

Figure 4. Implementation results. See text for details.

the equations, we also restrict our attention to point fingers ($\kappa_{\mathcal{A}_i} \to +\infty$). With these two simplifications, the degree of the equations is reduced, and the radicals disappear (see (Ponce *et al.*, 1995) for details).

We have constructed a preliminary implementation of the proposed approach. Figure 4(a) shows a piecewise-parabolic object, and Figure 4(b) shows the equilibrium surface S defined by (4). Figure 4(c) shows the intersections of S with the immobilization constraint surfaces. Figure 4(d) shows how S is trimmed by these constraints, and Figure 4(e) show the corresponding cell decomposition of the (u_1, u_2) plane (the shaded cells represent the immobilizing grasps). The immobilizing grasp corresponding to the sample shown as a large black disc in Figure 4(e) is shown in Figure 4(f). Note that the arcs in Figure 4(f) correspond to a circle whose radius is the distance from the concurrency point to the finger contact point.

We have also implemented a simple hill-climbing optimization scheme that uses the sample points in the immobilization regions as starting points for the optimization. To illustrate the method, we use a cost function which is a weighted combination of the moment criterion proposed in the previous section and a term which measures deviation from the optimal equilateral triangle criterion of Mirtich and Canny (1993). The result is shown in Figure 4(g), which shows the paths traced by the optimization procedure from the starting points (which are sample points generated during the cell decomposition). Five of the paths converge toward the same optimal grasp, indicated by a large black disc in Figure 4(g). The corresponding finger positions are shown in Figure 4(h).

Acknowledgments: Part of this work was conducted while J. Ponce was visiting the California Institute of Technology as a Beckman associate with the Center for Advanced Study of the University of Illinois at Urbana-Champaign. We gratefully acknowledge support from both institutions and from the Beckman Institute for Advanced Science and Technology. This work was also partially supported by ONR grant N00014-92-J-1920.

References

A. Blake. Computational modelling of hand-eye coordination. *Phil. Trans. R. Soc. Lond. B*, 337:351–360, 1992.

R.C. Brost and K.Y. Goldberg. A complete algorithm for synthesizing modular fixtures for polygonal parts. In *IEEE Int. Conf. Robot. and Automat.*, pp. 535–542, 1994.

I.M. Chen and J.W. Burdick. Finding antipodal point grasps on irregularly shaped objects. In *IEEE Int. Conf. Robot. and Automat.*, pp. 2278–2283, 1992.

G.E. Collins. *Quantifier Elimination for Real Closed Fields by Cylindrical Algebraic Decomposition*, volume 33 of *LNCS*. Springer-Verlag, 1975.

X. Markenscoff and C.H. Papadimitriou. Optimum grip of a polygon. *Int. J. Robot. Res.*, 8(2):17–29, 1989.

B. Mirtich and J.F. Canny. Optimum force-closure grasps. Technical Report ESRC 93-11/RAMP 93-5, University of California at Berkeley, 1993.

B. Mishra, J. T. Schwartz, and M. Sharir. On the existence and synthesis of multifinger positive grips. *Algorithmica*, 2:541–558, 1987.

B. Mishra. Workholding. In *Proceedings of IROS91*, pp. 53–57, 1991.

A.P. Morgan. *Solving Polynomial Systems using Continuation for Engineering and Scientific Problems*. Prentice Hall, 1987.

V-D. Nguyen. Constructing force-closure grasps. *Int. J. Robot. Res.*, 7(3):3–16, 1988.

J. Ponce, D. Stam, and B. Faverjon. On computing force-closure grasps of curved two-dimensional objects. *Int. J. Robot. Res.*, 12(3):263–273, 1993.

J. Ponce, J.W. Burdick, and E. Rimon. Computing the immobilizing three-finger grasps of planar objects. Technical Report UIUC-BI-AI-RCV-95-01, Beckman Institute, University of Illinois, 1995.

E. Rimon and J. W. Burdick. Towards planning with force constraints: On the mobility of bodies in contact. In *IEEE Int. Conf. Robot. and Automat.*, pp. 994–1000, 1993.

E. Rimon and J. W. Burdick. Mobility of bodies in contact: A new 2nd order mobility index for multiple-finger grasps. In *IEEE Int. Conf. Robot. and Automat.*, pp. 2329–2335, 1994.

E. Rimon and J. W. Burdick. Mobility of bodies in contact: How forces are generated by curvature effects. In *IEEE Int. Conf. Robot. and Automat.*, pp. 2336–2341, 1994.

A. Wallack and J.F. Canny. Planning for modular and hybrid fixtures. In *IEEE Int. Conf. Robot. and Automat.*, pp. 520–527, 1994.

A DYNAMIC FORMULATION FOR PLANAR MOTIONS USING GEOMETRIC KINEMATICS AND CAGD

KENNETH SPROTT

Research Assistant

University of California, Davis

ken@venus.engr.ucdavis.edu

AND

BAHRAM RAVANI

Professor

University of California, Davis

bravani@ucdavis.edu

Abstract.

This paper develops a method for dynamic analysis of planar motions that incorporates the kinematic geometry of motion as well as methods from the field of Computer Aided Geometric Design (CAGD). In this way, dynamic forces and the forces of constraint can be directly related to the kinematic geometry of the motion represented in terms of rolling of the two centrodes. The results are useful in dealing with problems associated with mechanics of manipulation as well as classical dynamics of planar bodies.

1. Introduction

One of the classical techniques of planar kinematics of instantaneous motions is based on incorporation of kinematic geometry of the motion in terms of rolling of the two centrodes. In this manner, motion properties can be derived from the geometric properties of the centrodes and their relative rolling motion. This technique is described in graphical form in classical kinematics books such as Rosenauer and Willis [6] and in more modern setting using instantaneous invariants in Bottema and Roth [1]. Although such formulations incorporating kinematic geometry of motion

J.-P. Merlet and B. Ravani (eds.), Computational Kinematics, 301–310.

have provided a powerful tool for kinematics analysis, they have not been directly incorporated into analysis of forces produced by the relative motion of the two bodies and the constraints. Analysis of forces is usually performed by dynamic (or static) analysis based on different formulations of equations of motion which, in general, are not in terms of the kinematic geometry of the relative motion.

In this paper, we present a method for formulation of dynamics of planar motions that would take advantage of the description of the kinematics in terms of rolling of the two centrodes of the motion. In this manner, the constraint forces and those resulting from the motion can be related to the geometry of the planar motion described in terms of rolling of the two centrodes. We show that the classical formulation of instantaneous planar kinematics in terms of rolling of the two centrodes can be re-formulated in a modern setting based on a Lie algebraic framework when one uses a metric based on kinetic energy and a natural parameter for the planar motion, in terms of the rotation angle. The result is a dynamic formulation that distinguishes the path dependent components of a force system and incorporates kinematic geometry of the relative motion directly into the formulation. This type of formulation is useful in applications where the moving body is constrained to follow a particular path such as in the rigid body guidance problem. It is also useful in problems associated with mechanics of manipulation where one is interested in determining forces that can produce a desired motion in the presence of forces of constraints. We use this compact kinematic description in conjunction with methods from the field of Computer Aided Geometric Design (CAGD) to develop a procedure for approximating a point path of the moving body. This approximated path is useful in both a general description of the motion of a moving body and in a practical implementation of the dynamic formulation.

We begin this paper by providing some background information necessary for the understanding of the dynamic formulation. We then present the formulation and show the point-path approximation technique. Finally, we use these techniques in a simple example as an illustration of the practical application of the formulation.

2. Dynamic Formulation

2.1. BACKGROUND

In this section, we briefly present some of the background material used in the development of the dynamic formulation.

In geometric kinematics, the natural parameter used to describe a motion is chosen as the rotation angle of the moving body. It can be shown [1] that this parameterization greatly simplifies the analysis of the geometric

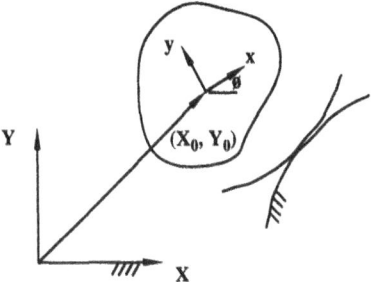

Figure 1.   Reference Frames for Geometric Kinematics

properties of the motion. A motion parameterized in this manner can be represented as:

$$\begin{bmatrix} X \\ Y \\ 1 \end{bmatrix} = \begin{bmatrix} \cos\phi & -\sin\phi & X_0(\phi) \\ \sin\phi & \cos\phi & Y_0(\phi) \\ 0 & 0 & 1 \end{bmatrix} \begin{bmatrix} x \\ y \\ 1 \end{bmatrix} \tag{1}$$

where $[X\ Y\ 1]^T$ represents the fixed coordinate frame and $[x\ y\ 1]^T$ represents the moving frame. This representation is an element of the special Euclidean group denoted by SE(2). It can be shown [5] that since this group has both the properties of a group and the properties of a smooth manifold that it is a Lie group. The Lie group structure of SE(2) will be utilized to develop the dynamic formulation.

The Lie algebra g associated with a Lie group G is a vector space together with a bilinear map $[*, *]: g \times g \to g$, called the Lie bracket, such that (i) $[x, y] = -[y, x]$ and (ii) $[[x, y], z] + [[y, z], x] + [[z, x], y] = 0$. An element of the Lie algebra is the representation of velocity vectors at the identity element for paths on the Lie group. The Lie algebra for SE(2), denoted *se(2)*, is of the form:

$$\xi = \begin{bmatrix} 0 & -\omega & v_x \\ \omega & 0 & v_y \\ 0 & 0 & 0 \end{bmatrix}$$

The Lie bracket is defined as the matrix commutator $[A, B] = AB - BA$.

The representation given by (1) is a one-parameter subgroup of the group of displacements. These sub-groups are continuous paths on the smooth manifold defined by the Lie group structure. Velocity vectors for these paths at a point can be found by differentiating. If we let g be the element of the group, the velocity vector of a point is given by:

$$g' = \begin{bmatrix} -\sin\phi & -\cos\phi & X_0' \\ \cos\phi & -\sin\phi & Y_0' \\ 0 & 0 & 0 \end{bmatrix}$$

where the $'$ denotes differentiation with respect to the angle ϕ. The velocity vectors at a point can be associated with an element of the Lie algebra by using either left or right translations. For the velocity vector g', the associated element of the Lie algebra can be found by using left translations as $V_\phi^b = g^{-1}g'$. This results in:

$$V_\phi^b = \begin{bmatrix} 0 & -1 & X_0' \cos\phi + Y_0' \sin\phi \\ 1 & 0 & -X_0' \sin\phi + Y_0' \cos\phi \\ 0 & 0 & 0 \end{bmatrix}$$

This represents the instantaneous velocity of the moving body as expressed in the body fixed coordinate frame. This velocity is left invariant which means that it is invariant to changes in the fixed (inertial) frame. For a general motion, an element of $se(2)$ can be shown [5] to be isomorphic to a vector in $\Re^3$ that is,

$$V_\phi^b \mapsto v_\phi = ((X_0' \cos\phi + Y_0' \sin\phi), (-X_0' \sin\phi + Y_0' \cos\phi), 1) \qquad (2)$$

A complete description of Lie groups and Lie algebras can be found in Chevallier [3].

2.2. DYNAMICS

In this section, we use the Lie group structure described in the previous section with a kinetic energy expression to write the dynamic equations. The kinetic energy for a motion parameterized with the rotation angle is written as $T(\phi) = \frac{1}{2}v_\phi^T M v_\phi$ where M is the generalized mass matrix and contains both mass and rotational inertia terms. It should be noted that this expression is not the true kinetic energy for the system as the velocity is the derivative with respect to the rotation angle not time.

Since SE(2) is a Lie group it is both a group and a smooth manifold. A set of basis vectors $\{e_1, e_2, \ldots, e_n\}$ can be defined for the tangent space of the manifold. With these basis vectors, there exist numbers C_{jk}^i such that

$$[e_j, e_k] = C_{jk}^i e_i$$

where these numbers are the *structure constants* of the Lie group. The kinetic energy expression defines a metric on the tangent space as

$$g(v_\phi, v_\phi) = 2T = v_\phi^T M v_\phi$$

and this metric endows the manifold with an inner product, $g_{ij} = <e_i, e_j>$. With the inner product defined, the force relationships can be found by using the covariant derivative, or connection, and the metric. The connection in the presence of this metric will be:

$$g_{il}\frac{dv_\phi^l}{d\phi} + \gamma_{ijk}v_\phi^j v_\phi^k = \widehat{Q}_i \tag{3}$$

where γ_{ijk} are the connection coefficients. These are defined as:

$$\gamma_{ijk} = \frac{1}{2}(\frac{\partial g_{ij}}{\partial \pi^k} + \frac{\partial g_{ik}}{\partial \pi^j} - \frac{\partial g_{jk}}{\partial \pi^i}) - \frac{1}{2}(g_{kl}C_{ji}^l + g_{jl}C_{ki}^l + g_{il}C_{jk}^l)$$

where the π^i are the coordinates on the manifold. The $\widehat{Q}_i$ are the *geometric forces*. It can be shown [2] that this same derivation can be used with a time-dependent velocity to write the generalized force relationships as:

$$g_{il}\frac{dv^l}{dt} + \gamma_{ijk}v^j v^k = Q_i \tag{4}$$

where the metric and connection coefficients are the same as in (3) and the Q_i are the generalized forces.

To complete the development of the dynamic formulation, the geometric forces must be related to the generalized forces for the system. This is accomplished by using a change of variables and the chain rule. If we let ω be the angular velocity and α be the angular acceleration then:

$$\frac{dv_\phi^j}{d\phi} = (\ddot{x}^j - v_\phi^j\alpha)\frac{1}{\omega^2} \tag{5}$$

and

$$v_\phi^j = \frac{\dot{x}^j}{\omega} \tag{6}$$

These can be substituted into (3) and, after rearranging terms, it becomes:

$$\widehat{Q}_i\omega^2 + g_{il}[v_\phi]^l\alpha = g_{il}[\ddot{x}]^l + \gamma_{ijk}\dot{x}^j \dot{x}^k = Q_i \tag{7}$$

which establishes the required relationship.

The coordinate frames can now be chosen as a body-fixed frame with origin at the center of mass and axes aligned with the principle axes of the body. With this choice, the metric becomes:

$$g_{ij} = \begin{bmatrix} m & 0 & 0 \\ 0 & m & 0 \\ 0 & 0 & I_{cg} \end{bmatrix}$$

where m is the mass of the body and I_{cg} is the rotational inertia. Using this metric and substituting for the velocity terms, the geometric force becomes:

$$\begin{aligned}
\hat{Q}_1 &= m(X_0'' \cos\phi + Y_0'' \sin\phi) \\
\hat{Q}_2 &= m(-X_0'' \sin\phi + Y_0'' \cos\phi) \\
\hat{Q}_3 &= 0
\end{aligned}$$

$$(8)$$

These expressions can be substituted into (7) to obtain:

$$\begin{aligned}
m(X_0'' \cos\phi + Y_0'' \sin\phi)\omega^2 + m(X_0' \cos\phi + Y_0' \sin\phi)\alpha &= Q_1 \\
m(-X_0'' \sin\phi + Y_0'' \cos\phi)\omega^2 + m(-X_0' \sin\phi + Y_0' \cos\phi)\alpha &= Q_2 \\
I_{cg}\alpha &= Q_3
\end{aligned}$$

$$(9)$$

This is the desired dynamic formulation of the equations of motion using the geometric parameterization. It should be emphasized that the generalized forces, Q_i, are written in terms of the body fixed frame. In the inertial frame these forces can be written in vector/matrix form as:

$$m\omega^2 \begin{bmatrix} X_0'' \\ Y_0'' \end{bmatrix} + m\alpha \begin{bmatrix} X_0' \\ Y_0' \end{bmatrix} = \begin{bmatrix} \cos\phi Q_1 - \sin\phi Q_2 \\ \sin\phi Q_1 + \cos\phi Q_2 \end{bmatrix} = \begin{bmatrix} \bar{Q}_1 \\ \bar{Q}_2 \end{bmatrix}$$

where $\bar{Q}_1$ and $\bar{Q}_2$ are the generalized forces written in terms of the inertial frame coordinates. One of the benefits of these expressions is that the path information, X_0 and Y_0, can be obtained from the geometric properties of the centrodes [1]. Information obtained from the kinematic analysis of the motion can now be used in a dynamic analysis.

Remarks:

1. The geometric parameterization is not valid for a pure translation. This is obvious since the derivative with respect to the rotation angle is not defined in this case. However, pure translation is a special case in geometric kinematics as well.

2. In both (3) and (4) if the Q_i or $\hat{Q}_i$ are zero then the equations define the requirements for a geodesic. From (8) it is obvious that a sufficient condition for a geodesic would be that $X_0'' = Y_0'' = 0$.

This last remark can be used to develop a method of approximating a point path of a motion.

2.3. PATH APPROXIMATION

It has already been shown that a sufficient condition for a path to be a geodesic is that $X_0'' = Y_0'' = 0$. This condition can be used to show that a geodesic path can be described as:

$$X_0 = k_1\phi + c_1$$
$$Y_0 = k_2\phi + c_2$$

where k_i and c_i are real constants. These geodesics can be used to generate a smooth motion between several different displacements. A computationally efficient method is to generate a cubic Bezier curve between two points. A cubic Bezier curve is an affine combination of four control points in some space. The four control points form a control polygon for the curve with the line between the first two points tangent to the curve at the zero point and the line between the last two points tangent to the curve at the other end. In the case of the motion, we wish to generate a trajectory $\mathbf{P}(\phi) = (X_0(\phi), Y_0(\phi))$ with the geometric parameter ϕ being the global parameter for an affine interpolation. The global parameter must be mapped into the interval $[0, 1]$ then the trajectory can be generated using Bezier curves. The geometric parameter is mapped by using:

$$u = \frac{\phi - \phi_i}{\phi_f - \phi_i} = \frac{\phi - \phi_i}{\Delta\phi} \tag{10}$$

For a first order Bezier curve, we have only two control points (x_0, y_0) and (x_1, y_1) and the interpolation is given by:

$$\mathbf{p}(u) = \begin{bmatrix} ux_1 + (1-u)x_0 \\ uy_1 + (1-u)y_0 \end{bmatrix}$$

When equation (10) is substituted into this we get:

$$\mathbf{p}(\phi) = \begin{bmatrix} \frac{(\phi-\phi_i)}{\Delta\phi}x_1 + (1 - \frac{\phi-\phi_i}{\Delta\phi})x_0 \\ \frac{(\phi-\phi_i)}{\Delta\phi}y_1 + (1 - \frac{\phi-\phi_i}{\Delta\phi})y_0 \end{bmatrix}$$

When this is simplified we get:

$$\mathbf{p}(\phi) = \begin{bmatrix} \frac{(x_1-x_0)}{\Delta\phi}(\phi - \phi_i) + x_0 \\ \frac{(y_1-y_0)}{\Delta\phi}(\phi - \phi_i) + y_0 \end{bmatrix}$$

Which is what we expect for the straight lines between two points.

Now given four control points (x_i, y_i), $i = 0, 1, 2, 3$, we can generate a cubic Bezier curve with the affine parameter u:

$$\mathbf{p}(u) = \begin{bmatrix} u^3x_3 + 3u^2(1-u)x_2 + 3u(1-u)^2x_1 + (1-u)^3x_0 \\ u^3y_3 + 3u^2(1-u)y_2 + 3u(1-u)^2y_1 + (1-u)^3y_0 \end{bmatrix}$$

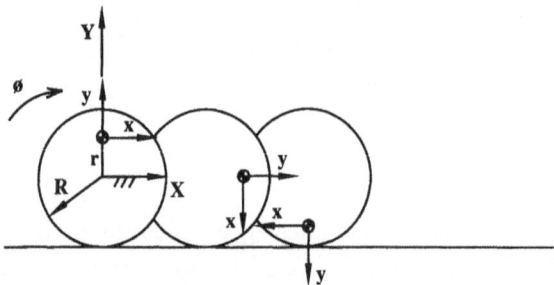

Figure 2.   Disk with Offset CG

If we now substitute (10) into this expression we get a cubic polynomial using the geometric parameter ϕ.

$$\mathbf{p}(\phi) = \left[\begin{array}{l} (\frac{(\phi-\phi_1)}{\Delta\phi})^3 x_3 + 3(\frac{(\phi-\phi_1)}{\Delta\phi})^2(1 - \frac{\phi-\phi_1}{\Delta\phi})x_2 + 3\frac{(\phi-\phi_1)}{\Delta\phi}(1 - \frac{\phi-\phi_1}{\Delta\phi})^2 x_1 + (1 - \frac{\phi-\phi_1}{\Delta\phi})^3 x_0 \\ (\frac{(\phi-\phi_1)}{\Delta\phi})^3 y_3 + 3(\frac{(\phi-\phi_1)}{\Delta\phi})^2(1 - \frac{\phi-\phi_1}{\Delta\phi})y_2 + 3\frac{(\phi-\phi_1)}{\Delta\phi}(1 - \frac{\phi-\phi_1}{\Delta\phi})^2 y_1 + (1 - \frac{\phi-\phi_1}{\Delta\phi})^3 y_0 \end{array} \right]$$

$$(11)$$

This expression gives us a cubic polynomial in the geometric parameter for the interval (x_0, y_0) (x_3, y_3) which can be substituted directly into the original displacement expression. Note that the control points for the Bezier interpolation are nothing more than the location of the origin of the moving frame in the displacement expression. It should also be noted that the interpolation is valid for a specific interval only which means that this is strictly a local parameterization. However, for a desired motion it is possible to create a spline through some set of desired positions then examine the properties of the motion for each of the intervals of the spline. Both the spacing of the precision positions and the location of the Bezier control points must be used in order to maintain continuity along the motion. An example will serve to illustrate these concepts.

3. Example

The dynamic formulation is particularly useful in applications where the moving body is constrained to follow a particular path. In this example we will examine the forces on a disk with an offset center of gravity (cg) as it rolls without slipping on a flat surface. The path of the cg will be a prolate cycloid. It is only necessary to approximate the path for a half rotation since the path repeats in this period.

To begin, we fix the origin of the moving coordinate frame to the cg as shown in figure 2. Next we approximate the path of the cg using two equally spaced cubic Bezier curve segments. The first segment will be for the interval $0 \leq \phi \leq \frac{\pi}{2}$ and the second segment will be for the interval

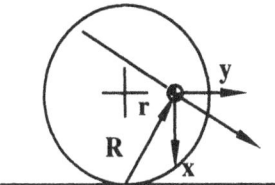

Figure 3. Path tangent for intermediate position

$\frac{\pi}{2} \leq \phi \leq \pi$. The control points for the Bezier curves are found from the path constraints and the continuity conditions. The end points of the segments are on the path and the interior points will lie along the path tangents. The coordinates of the points are written in terms of the fixed coordinate frame.

For the first segment, the endpoints are found from the path requirements. The first point, (x_0, y_0) at $\phi = 0$, is specified as $x_0 = 0$ and $y_0 = r$. The final point, (x_3, y_3) at $\phi = \frac{\pi}{2}$, is found by using the requirement that the disk roll without slipping. With this condition, the center of the disk will be at $(\frac{R\pi}{2}, 0)$ after rolling $\frac{\pi}{2}$ radians. The point coordinates are then located at $((\frac{R\pi}{2} + r), 0)$. The remaining two points are found from the tangent requirements at the endpoints. The path tangent is found from the requirements of the motion, that is the tangent will be perpendicular to the ray through the point on the path and the velocity pole (see figure 3). The magnitude of the tangent will be the length of this ray. This value can be compared to the expression found from taking the the derivative of (11) and evaluating it at $\phi = 0$ and $\phi = \frac{\pi}{2}$. For the point (x_1, y_1), the tangent will be $[(R+r)\ \ 0]^T$ at $\phi = 0$. The comparison with the required derivative yields:

$$\frac{d\mathbf{p}}{\phi} = \frac{1}{\Delta\phi}\begin{bmatrix} R + r \\ 0 \end{bmatrix} = \begin{bmatrix} 3x_1 - 3x_0 \\ 3y_1 - 3y_0 \end{bmatrix}$$

which after substituting for (x_0, y_0) yields:

$$\begin{bmatrix} x_1 \\ y_1 \end{bmatrix} = \begin{bmatrix} \frac{(R+r)\pi}{6} \\ R + r \end{bmatrix}$$

Similarly, the point (x_2, y_2) can be found by noting that the tangent at $\phi = \frac{\pi}{2}$ is $[R\ \ (-r)]^T$. This yields the coordinates $((\frac{R\pi}{3} + r), \frac{r\pi}{6})$. The same technique is used to find the four control points for the second segment. It can be shown [4] that since the interval, $\Delta\phi$, is the same for the two segments and the tangents match at $\phi = \frac{\pi}{2}$ then this spline will have C^2 continuity. This will be sufficient for the force relationships.

The geometric force can be found by using (8) and by taking derivatives of the paths given by the four control points and (11). The generalized force can be found by using (9) and the derivatives of the paths. The forces on the moving body can now be evaluated as a function of the angular velocity and angular acceleration. An example might be if the disk were an unbalanced wheel on a vehicle undergoing a constant torque as the vehicle accelerates. In this case, the angular velocity and acceleration are found from the Q_3 expression in (9). These are then used to evaluate the forces Q_1 and Q_2. It should be noted that the path information is an approximation of the actual path. This will generate errors in the dynamic formulation. However, these errors can be mitigated by using more curve segments in the approximation.

4. Conclusion

In this paper, we have combined concepts from the field of CAGD and kinematics to provide a formulation of dynamical equations for planar motion that would relate forces to the kinematic geometry motion. The method developed has applications in motion planning problems as well as in classical dynamics of rigid bodies. Generalization of this work to three dimensional problems is presently under development [7].

References

[1] Bottema, O., and Roth, B. *Theoretical Kinematics*, Dover Publications, Inc., New York, 1979

[2] Brockett, R. W., Stokes, A., and Park, F. A geometrical formulation of the dynamical equations describing kinematic chains, In *Proceedings, IEEE Conference on Robotics and Automation*, pages 671-677, 1993

[3] Chevallier, D. P. Lie Algebras, Modules, Dual Quaternions, and Algebraic Methods in Kinematics, *Mechanisms and Machine Theory*, pages 613-627, vol. 26, 1991

[4] Farin, G. *Curves and Surfaces for Computer Aided Geometric Design*, Academic Press, Inc., San Diego, Ca., 1993

[5] Murray, R. M., Li, Z., and Sastry, S. S. *A Mathematical Introduction to Robotic Manipulation*, CRC Press, Boca Raton, Florida, 1994

[6] Rosenauer, N. and Willis, A. H. *Kinematics of Mechanisms*, Dover Publications, Inc., New York, 1967

[7] Sprott, K. Ph.D dissertation *to be published*, University of California, Davis, 1995